Legal Liabilities in Safety and Loss Prevention

Occupational Safety and Health Guide Series

Series Editor: Thomas D. Schneid

Eastern Kentucky University, Richmond, USA

The aim of this series is to develop books to address the myriad of unique topics which are encompassed within the areas of responsibility of a safety professional. The scope of the series is broad, addressing traditional occupational safety and health topics as well as emerging and correlating topic areas such as ergonomics, security, labor, school safety, employment, workplace violence, and other topics. Safety professionals often assume a broad spectrum of responsibilities, depending on the organization, which go beyond the traditional safety professional's role. This series should become the "one stop" source for safety professionals to locate texts on traditional as well as emerging and correlating subjects related to the safety and health function.

Safety Law
Legal Aspects in Occupational Safety and Health
Thomas D. Schneid

Physical Hazards of the Workplace
Barry Spurlock

Loss Prevention and Safety Control
Terms and Definitions
Dennis P. Nolan

Discrimination Law Issues for the Safety Professional
Thomas D. Schneid

Labor and Employment Issues for the Safety Professional
Thomas D. Schneid

The Comprehensive Handbook of School Safety
Edited by E. Scott Dunlap

Creative Safety Solutions
Thomas D. Schneid

Physical Security and Safety
A Field Guide for the Practitioner
Edited by Truett A. Ricks, Bobby E. Ricks and Jeffrey Dingle

Workplace Safety and Health
Assessing Current Practices and Promoting Change in the Profession
Thomas D. Schneid

For more information about this series, please visit: https://www.crcpress.com/Occupational-Safety–Health-Guide-Series/book-series/CRCOCCSHGS

Legal Liabilities in Safety and Loss Prevention

A Practical Guide, Third Edition

Thomas D. Schneid

CRC Press
Taylor & Francis Group
Boca Raton London New York

CRC Press is an imprint of the
Taylor & Francis Group, an **informa** business

CRC Press
Taylor & Francis Group
6000 Broken Sound Parkway NW, Suite 300
Boca Raton, FL 33487-2742

© 2019 by Taylor & Francis Group, LLC
CRC Press is an imprint of Taylor & Francis Group, an Informa business

No claim to original U.S. Government works

Printed on acid-free paper

International Standard Book Number-13 978-1-138-50165-2 (Paperback)

Library of Congress Cataloging-in-Publication Data

Names: Schneid, Thomas D., author.
Title: Legal liabilities in safety and loss prevention : a practical guide / authored by
   Thomas D. Schneid.
Description: Third edition. | Boca Raton, FL : CRC Press, an imprint of Taylor & Francis
   Group, 2019. | Series: Occupational safety and health guide series | Includes
   bibliographical references and index.
Identifiers: LCCN 2018057912| ISBN 9781138501652 (pbk. : alk. paper) |
   ISBN 9781138501690 (hardback : alk. paper) | ISBN 9781315144504 (e-book)
Subjects: LCSH: Industrial safety—Law and legislation—United States. | Industrial
   hygiene—Law and legislation—United States. | Employers' liability—United States. |
   Risk management—Law and legislation—United States.
Classification: LCC KF3570 .S36 2019 | DDC 344.7304/65—dc23
LC record available at https://lccn.loc.gov/2018057912

Visit the Taylor & Francis Web site at
http://www.taylorandfrancis.com

and the CRC Press Web site at
http://www.crcpress.com

Contents

Preface

Reading, Analyzing, and Briefing a Court Decision:

For those safety and loss prevention professionals who are unfamiliar with reviewing, analyzing, and briefing a court decision, the following synopsis is provided to assist you through this process.

READING A COURT DECISION

- Read the case in total to acquire a "flavor" for the type of case.
- Determine what type of case you are reading, e.g., civil, criminal, administrative, etc.
- Examine the title of the case and identify the court making the decision, e.g., US Supreme Court, 4th Circuit Court of Appeals, etc.
- Identify the date of the decision.
- Review the brief summary or headnote, usually provided at the beginning of each decision.
- If the case is an appeal, note whether the decision was unanimous or handed down by a divided court. Note if there is a dissenting opinion provided by the judges who did not agree with the majority decision.
- Read the case again with specificity. Identify the parties, issues, arguments, and decision.
- Always look up legal terms you do not fully understand in a legal dictionary or a reputable website.
- Brief your case in order to fully understand the issues, facts, and decision in the case.

ANALYZING THE CASE

Case law is an accumulation of court decisions that provide guidance and direction on current and future cases and decisions. Cases usually start at the district or lowest level of the courts and are often appealed to a higher appellate court, up to and including the U.S. Supreme Court. There are federal court decisions as well as individual state court decisions and administrative decisions, such as the OSHRC. *Review - Commission*

As a general rule, when analyzing a case, you should read the case in detail one time to acquire a "flavor" for the case. On the second reading, the issues in the case, as well as the facts of the case and the decision, should be identified. On the third reading of the case, you should take notes for future reference, identifying the issues of the case, the important or pertinent facts of the case, the decision of the court, and any dissent.

It is important to carefully analyze each issue in the case and determine why the court decided the case in this manner. Do you agree or disagree with the court's decision and why? Was the court's decision overturned at any level of appeal?

BRIEFING THE CASE

The primary reason for "briefing" a case is to assist you in understanding the particular legal issues and their significance. There are various methods of briefing a case, and the following format is only offered as an example of one of these methods. No matter what method you adopt, it is vital that you read the case thoroughly at least one time to acquire a general idea of what it is about before beginning to take notes and develop your brief.

1. List the case name, the court, and the date of the decision.
   Joe Anyone v. Tom's Widget Company, Inc.
   Madison County Circuit Court—4th Division
   September 1, 2010
2. Issues: In 1–2 sentences, identify the key issue or issues in the case.
3. Facts: Usually in 1–2 paragraphs, summarize the key facts involved.
4. Holding (or Decision): Identify the decision of the court.
5. Dissent (or Dissenting opinion): Identify any dissenting opinions provided by the minority in the decision. Note: Usually, dissenting opinions are provided only on appellate cases.
6. My Opinion: Do you agree or disagree with the decision and why?

A brief should not be more than one (1) page in length; thus, it should be very concise and "to the point." Your brief can be used to assist you in your course work as well as to refresh your memory regarding the case.

EXAMPLE CASE BRIEF

Case Name: Marshall v. Barlow's Inc., 436 U.S. 307 (1978)

Issue: Is Section 8(a) of the OSH Act unconstitutional in that it violates the Fourth Amendment?   *respondent*

Facts: Appellee (Barlow's) initially brought this action to obtain injunctive relief against a warrantless inspection of its business by OSHA. The inspection was permitted under Section 8(a) of the OSH Act, which authorized agents of the Secretary of Labor to search the work area of any facility within OSHA's jurisdiction for safety hazards and violations without obtaining a search warrant or other processes. A three-judge Idaho District Court rules in favor of Barlow's and concluded that the Fourth Amendment required a warrant for this type of search and that the statutory authorization for warrantless inspections was unconstitutional. This appeal resulted.

Holding: Yes, Section 8(a) of the OSH Act was unconstitutional in that it violated the Fourth Amendment. The U.S. Supreme Court affirmed the decision of the Idaho District court and granted Barlow's an injunction enjoining the enforcement of the Act to that extent. The Court stated that the rule against warrantless searches applies to commercial premises as well as private homes. Although an exception to this rule is applied to certain "carefully defined classes of cases" including closely regulated businesses such as firearms and liquor industries, this exception does not automatically apply to all businesses engaged in interstate commerce.

My Opinion: I agree with the Court in this case. OSHA should be required to provide a reason for the inspection and show proof supporting the reason for an inspection if the employer or business does not waive his/her rights. The requirement for OSHA to acquire an administrative search warrant provides the employer or business the opportunity to argue his/her case before permitting the inspection.

Acknowledgments

My thanks to my parents, Bob and Rosella, for their sacrifice to ensure that all of their children achieved the education necessary for each to achieve success in their chosen professions.

To my wife, Jani, who endured my work schedule in completing my writing activities.

To my daughter Shelby who published her first book this year, keep striving and working.

To my daughter Madison who has completed her graduate work and is working in the safety field, you will do well. Safety is a learning experience each and every day.

To my daughter Kasi who is completing her undergraduate work, continue to explore and find the work which will make you happy over the long run.

And to my students, I hope this text expands your horizons and prepares you to enter the best job, in my opinion, in the world. Few jobs permit you to have a daily impact on the safety and health of others as well as an impact on your company or organization and the world. I hope you enjoy the text and have a great learning experience.

Author

Thomas D. Schneid is a tenured professor in the Department of Safety, Security, and Emergency Management (formerly Loss Prevention and Safety) at Eastern Kentucky University and serves as the department chair and graduate program director for the online and on-campus Master of Science degree in Safety, Security, and Emergency Management.

Tom has earned a BS in education, an MS and CAS in safety, an MS in international business, and a PhD in environmental engineering as well as his Juris Doctor (JD in law) from West Virginia University and LLM (Graduate Law) from the University of San Diego. Tom is a member of the bar for the US Supreme Court, 6th Circuit Court of Appeals, and a number of federal districts as well as the Kentucky and West Virginia Bar.

Tom has authored or co-authored twenty one texts, including the Americans with Disabilities Act casebook, Fire Law, Fire and Emergency Law Casebook, Food Safety Law, Human Resources for Safety Professionals, and Corporate Safety Compliance. In addition, he has authored more than 100 articles on legal issues and safety-related topics and has presented at numerous conferences worldwide.

Morgantown

1 Overview and History

CHAPTER OBJECTIVES

1. Acquire an understanding of the history of the OSH Act.
2. Acquire an understanding of the OSH Act itself.
3. Understand the boundaries of the OSH Act.
4. Understand the General Duty Clause.

THE FEDERAL OCCUPATIONAL SAFETY AND HEALTH ACT OF 1970

Before the Federal Occupational Safety and Health Act (OSH Act) of 1970 was enacted, safety and health issues were limited to safety and health laws for specific industries and laws that governed federal contractors. It was during this period, prior to the enactment of the OSH Act in 1970, that Congress gradually began to regulate specific areas of safety and health in the American workplace through such laws as the Walsh–Healey Public Contracts Act of 1936, the Labor Management Relations Act (Taft–Hartley Act) of 1947, the Coal Mine Safety Act of 1952, and the McNamara–O'Hara Public Service Contract Act of 1965.

With the passage of the then-controversial OSH Act in 1970, federal and state government agencies became actively involved in managing health and safety in the private sector workplace. Employers were placed on notice that unsafe and unhealthful conditions and acts would no longer be permitted to endanger the health, and often the lives, of American workers. In many circles, the Occupational Safety and Health Administration (OSHA) became synonymous with the "safety police," and employers were often forced, under penalty of law, to address safety and health issues in their workplaces.

Today, the OSH Act itself is virtually unchanged since its 1970 roots. The basic methods for enforcement, standards development and promulgation, as well as adjudication, remain intact. In approximately the past 40 years, OSHA has, however, added many new standards that are based primarily on the research conducted by the National Institute for Occupational Safety and Health (NIOSH) and recommendations from labor and industry. In addition, OSHA has revisited several of the original standards in order to update or modify the particular standard. The Occupational Safety and Health Review Commission (OSHRC) and the courts have been very active in resolving many disputed issues and clarifying the law as it stands.

There is a trend within Congress, industry, and labor to believe that, in order to achieve the ultimate goal of reducing workplace injuries, illnesses, and fatalities, additional changes to the OSH Act and the structure of OSHA are needed. OSHA has taken up the challenge and has moved toward performance-based standards. OSHA has also attempted to address many of the new hazards created by our technological advances and the changing workplace. Professionals working in safety and loss prevention should study the past in order to plan and set their course for the future. Change is inevitable; however, we can anticipate that new standards will be based on the knowledge obtained from past victories and mistakes. Change is necessary in order to achieve our ultimate goal—a safe and healthful workplace for all.

LEGISLATIVE HISTORY

Throughout the history of the United States, the potential for the American worker to be injured or killed on the job has been a brutal reality. Many disasters, such as that at Gauley Bridge, West Virginia, fueled the call for laws and regulations to protect the American worker. As early as the 1920s, many states recognized the safety and health needs of the industrial worker and began to enact workers' compensation and industrial safety laws. The first significant federal legislation was the Walsh–Healey Public Contracts Act of 1936, which limited working hours and the use of child and convict labor. This law also required that contracts entered into by any federal agency for more than $10,000 contain the stipulation that the contractor would not permit conditions that were unsanitary, hazardous, or dangerous to employees' health or safety.

In the 1940s, the federally enacted Labor Management Relations Act (Taft–Hartley Act) provided workers with the right to walk off a job if it was "abnormally dangerous." Additionally, in 1947, President Harry S. Truman created the first Presidential Conference on Industrial Safety.

In the 1950s and 1960s, the federal government continued to enact specialized safety and health laws to address particular circumstances. The Coal Mine Safety Act of 1952, the Maritime Safety Act, the McNamara–O'Hara Public Service Contract Act (protecting employees of contractors performing maintenance work for federal agencies), and the National Foundation on the Arts and Humanities Act (requiring recipients of federal grants to maintain safe and healthful working conditions) were passed during this time.

The federal government's first significant step in developing coverage for workplace safety and health was the passage of the Metal and Nonmetallic Mine Safety Act of 1966. Following the passage of this Act, President Lyndon B. Johnson, in 1968, called for the first comprehensive occupational safety and health program as part of his Great Society program. Although this proposed plan never made it to a vote in Congress, the seed was planted for future legislation.

One particular incident shocked the American public and federal government into action. In 1968, a coal mine fire and explosion in Farmington, West Virginia, killed 78 miners. Congress reacted swiftly by passing a number of safety and health laws, including the Coal Mine Health and Safety Act of 1969, the Contract Work Hours and Safety Standards Act of 1969 (also known as the Construction Safety Act), and the Federal Railway Safety Act.

In 1970, fueled by the new interest in workplace health and safety, Congress pushed for more comprehensive laws to regulate the conditions of the American workplace. To this end, Congress passed the OSH Act of 1970 and it became effective April 28, 1971. The overriding purpose and intent of the OSH Act was "to assure so far as possible every working man and woman in the nation safe and healthful working conditions and to preserve our human resources."

COVERAGE AND JURISDICTION OF THE OSH ACT

The OSH Act covers virtually every American workplace that employs one or more employees and engages in a business that affects interstate commerce in any way. The OSH Act covers employment in every state, the District of Columbia, Puerto Rico, Guam, the Virgin Islands, American Samoa, and the Trust Territory of the Pacific Islands. The OSH Act does not, however, cover employees in situations in which other state or federal agencies have jurisdiction that requires the agencies to prescribe or enforce their own safety and health regulations. Additionally, the OSH Act exempts residential owners who employ people for ordinary domestic tasks, such as cooking, cleaning, and child care. It also does not cover federal, state, and local governments or Native American reservations.

The OSH Act requires that every employer engaged in interstate commerce furnish employees "a place of employment ... free from recognized hazards that are causing, or are likely to cause, death or serious harm." To help employers create and maintain safe working environments and to enforce laws and regulations that ensure safe and healthful work environments, Congress created OSHA, a new agency under the direction of the Department of Labor.

Today, OSHA is one of the most widely known and powerful enforcement agencies within the federal government structure. OSHA has been granted broad regulatory powers to promulgate regulations and standards, investigate and inspect workplaces, issue citations, and propose penalties for safety violations in the workplace.

The OSH Act also established an independent agency, the Occupational Safety and Health Review Commission (OSHRC), to review OSHA citations and decisions. The OSHRC is a quasi-judicial and independent administrative agency composed of three commissioners, appointed by the president, who serve staggered six-year terms. The OSHRC has the power to issue orders; uphold, vacate, or modify OSHA citations and penalties; and direct other appropriate relief and penalties.

The education arm of the OSH Act is the National Institute for Occupational Safety and Health (NIOSH), which was created as a specialized education agency of the existing National Institutes of Health. NIOSH conducts occupational safety and health research and develops criteria for new OSHA standards. NIOSH may conduct workplace inspections, issue subpoenas, and question employees and employers, but it does not have the power to issue citations or penalties.

STATE SAFETY PLANS

Notwithstanding OSH Act enforcement through the previously noted federal agencies, OSHA encourages individual states to take responsibility for OSHA administration and enforcement within their respective boundaries. Each state possesses the ability to request and be granted the right to adopt state safety and health regulations

[handwritten margin notes top:]
1- Fed State Plans
2- State Plans
3 - Fed Plans

[handwritten margin notes left side:]
KY OSH private sector workers and all state + local govt workers

and enforcement mechanisms. In section 18(b), the OSH Act provides that any state "which, at any time, desires to assume responsibility for development and the enforcement therein of occupational safety and health standards relating to any ... issue with respect to which a federal standard has been promulgated ... shall submit a state plan for the development of such standards and their enforcement." Before a state plan can be placed into effect, the state must develop and submit its proposed program to the secretary of labor for review and approval. The secretary must certify that the state plan's standards are "at least as effective" as the federal standards and that the state will devote adequate resources to administering and enforcing its standards.

In most state plans, the state agency has developed more stringent safety and health standards than OSHA and has usually developed more stringent enforcement schemes. The secretary of labor has no statutory authority to reject a state plan if the proposed standards or enforcement scheme are more strict than the OSHA standards, but can reject the state plan if the standards are below the minimum limits set under OSHA standards. These states are known as state plan states and territories. By 2009, 24 states and two territories have approved functional state plan programs. Employers in state plan states and territories must comply with their state's regulations; federal OSHA plays virtually no role in direct enforcement.

OSHA does, however, possess an approval and oversight role regarding state plan programs. OSHA must approve all state plan proposals prior to their enactment. It also maintains oversight authority to "pull the ticket" of any state plan program at any time if the program is not achieving the identified prerequisites. Enforcement of this oversight authority was observed after a fire in 1991 that resulted in several workplace fatalities at the Imperial Foods facility in Hamlet, North Carolina. Following this incident, federal OSHA assumed jurisdiction and control over the state plan program in North Carolina and made significant modifications to this program before returning the program to state control.

Safety and loss prevention professionals need to ask the following questions when determining jurisdiction under the OSH Act:

1. Am I a covered employer under the OSH Act?
2. If I am a covered employer, what regulations must I follow to ensure compliance?

The answer to the first question is "yes" for virtually every class of private sector employers. Any employer in the United States that employs one or more persons and is engaged in a business that, in any way, affects interstate commerce is within the scope of the federal OSH Act. The phrase "interstate commerce" has been broadly interpreted by the U.S. Supreme Court, stating that interstate commerce "goes well beyond persons who are themselves engaged in interstate or foreign commerce." In essence, anything that crosses state lines, whether it is a person, a material good, or a service, places the employer in interstate commerce. Although there are exceptions to this general statement, interstate commerce has been "liberally construed to effectuate the congressional purpose" of the OSH Act.

When identifying coverage under the OSH Act, an employer must distinguish between a state plan jurisdiction and federal OSH Act jurisdiction. If its facilities or operations are located within a state plan state, an employer must comply with the regulations of its state. Safety and loss prevention professionals should contact each state's Department of Labor to acquire the pertinent regulations and standards. If facilities or operations are located in a federal OSHA state, the applicable standards and regulations can be acquired from any area OSHA office or the Code of Federal Regulations.

A common jurisdictional mistake occurs when an employer operates multiple facilities in different locations. Safety and loss prevention professionals should ascertain which state or federal agency has jurisdiction over each facility or operation and which regulations and standards apply.

OSHA STANDARDS AND THE GENERAL DUTY CLAUSE

PROMULGATION OF STANDARDS

The OSH Act requires that a covered employer must comply with specific occupational safety and health standards, as well as all rules, regulations, and orders pursuant to the OSH Act that apply to the workplace. The OSH Act also requires that all standards be based on research, demonstration, experimentation, or other appropriate information. The secretary of labor is authorized under the Act to "promulgate, modify, or revoke any occupational safety and health standard." The OSH Act also describes the procedures that the secretary must follow when establishing new occupational safety and health standards.

The OSH Act authorizes three ways to promulgate new standards: (1) national consensus standards, (2) informal (standard) rulemaking, and (3) emergency temporary standards. From 1970 to 1973, the secretary of labor was authorized in section 6(a) of the Act to adopt national consensus standards and establish federal safety and health standards without following lengthy rulemaking procedures. Many of the early OSHA standards were adapted from other areas of regulation, such as the National Electric Code and American National Standards Institute (ANSI) guidelines. However, this promulgation method is no longer in effect.

The usual method of issuing, modifying, or revoking a new or existing OSHA standard is described in section 6(b) of the OSH Act and is known as informal rulemaking. This method requires providing notice to interested parties, through subscription in the *Federal Register*, of the proposed regulation and standard, and allows parties the opportunity for comment in a nonadversarial, administrative hearing. The proposed standard can also be advertised through magazine articles and other publications, thus informing interested parties of the proposed standard and regulation. This method differs from the requirements of most other administrative agencies that follow the Administrative Procedure Act because the OSH Act provides interested persons the opportunity to request a public hearing with oral testimony. It also requires the secretary of labor to publish a notice of the time and place of such hearings in the *Federal Register*.

Although not required under the OSH Act, the secretary of labor has directed, by regulation, that OSHA follow a more rigorous procedure for comment and hearing than other administrative agencies. When notice and request for a hearing are received, OSHA must provide a hearing examiner to listen to any oral testimony offered. All oral testimony is preserved in a verbatim transcript. Interested persons are provided an opportunity to cross-examine OSHA representatives or others on critical issues. The secretary must state the reasons for the action to be taken on the proposed standard, and the statement must be supported by substantial evidence in the record as a whole.

The secretary of labor has the authority to disallow oral hearings and to call for written comment only. Within 60 days after the period for written comment or oral hearings has expired, the secretary must decide whether to adopt, modify, or revoke the standard in question. The secretary may also decide not to adopt a new standard. The secretary must then publish a statement explaining the reasons for any decision in the *Federal Register*. OSHA regulations further mandate that the secretary provide a supplemental statement of significant issues in the decision. Safety and health professionals should be aware that the standard adopted and published in the *Federal Register* may be different from the proposed standard. The secretary is not required to reopen hearings when the adopted standard is a "logical outgrowth" of the proposed standard.

The final method for promulgating new standards, which is most infrequently used, is the emergency temporary standard permitted under section 6(c). The secretary of labor may immediately establish a standard if it is determined that employees are subject to grave danger from exposure to substances or agents known to be toxic or physically harmful and that an emergency standard would protect the employees from the danger. An emergency temporary standard becomes effective upon publication in the *Federal Register* and may remain in effect for six months. During this six-month period, the secretary must adopt a new, permanent standard or abandon the emergency standard.

Only the secretary of labor can establish new OSHA standards; however, recommendations or requests for an OSHA standard can come from any interested person or organization, including employees, employers, labor unions, environmental groups, and others. When the secretary receives a petition to adopt a new standard or to modify or revoke an existing standard, he or she usually forwards the request to NIOSH and the National Advisory Committee on Occupational Safety and Health (NACOSH). Alternatively, the secretary may use a private organization such as ANSI for advice and review.

THE GENERAL DUTY CLAUSE

As previously stated, the OSH Act requires that an employer maintain a place of employment free from recognized hazards that are causing, or are likely to cause, death or serious physical harm, even if there is no specific OSHA standard addressing the circumstances. Under section 5(a)(1), the general duty clause, an employer may be cited for a violation of the OSH Act if the condition causes harm or is likely to cause harm to employees, even if OSHA has not promulgated a standard specifically addressing the particular hazard. The general duty clause is a catchall standard encompassing all potential hazards that have not been specifically addressed in the OSHA standards. For example, if a company is cited for an ergonomic hazard and there is no ergonomic standard to apply, the hazard will be cited under the general duty clause.

Prudent safety and loss prevention professionals often take a proactive approach in maintaining their competency in this expanding area of OSHA regulations. As noted previously, notification of a new OSHA standard, modification of an existing standard, revocation of a standard, or establishment of an emergency standard must be published in the *Federal Register*. Safety and health professionals can use the *Federal Register*, or other professional publications that monitor this area, to track the progress of proposed standards. With this information, safety and health professionals can provide testimony to OSHA and, when necessary, prepare their organizations for acquiring resources and personnel necessary to achieve compliance and get a head start on developing compliance programs to meet requirements in a timely manner.

SELECTED CASE STUDY

Note: Cases may be modified for the purpose of this text.

SECRETARY OF LABOR V. DIERZEN-KEWANEE HEAVY INDUSTRIES, LTD.

OSHRC Docket No. 07-0675 and 07-0676 (2009)

(Case selected from the OSHRC website and edited for the purposes of this text)

Decision and Order

Dierzen-Kewanee Heavy Industries, LTD (Dierzen), manufactures light curved-bed dump truck bodies. It operates out of space in a former boiler factory in Kewanee, Illinois. Dierzen began its operations in 2003 with three employees. Four and one-half years later, it employs 36 individuals (Tr. 11-12, 134).

The Peoria Area Office of the Occupational Safety and Health Administration (OSHA) has a difficult history with Dierzen. OSHA sought assurance that violations listed in previous OSHA citations had been corrected. When Dierzen ignored OSHA's requests for it to send in abatement information, OSHA scheduled Dierzen for a follow-up inspection in October 2006. As a result of that inspection, OSHA issued the instant citations on April 10, 2007. Categorizing the standards as primarily related to "safety" (Docket No. 07-0675) or to "health" (Docket No. 07-0676), OSHA issued serious, repeat, willful, and "other than serious" citations and penalties to Dierzen. Dierzen did not contest the existence of any of the violations, which thus became a final order "not subject to review by any court or agency" [29 U.S.C. 659(a)]. Dierzen contested only the reasonableness of OSHA's proposed monetary penalties.

The parties participated in lengthy settlement judge proceedings, but were ultimately unable to reach a resolution on the appropriate penalties. The case was re-assigned to the undersigned judge to conduct the hearing and to issue a decision in the matter. A hearing was held on July 8, 2008, at which the parties presented evidence and argued their positions on the record. It was determined the proposed penalties should not be reduced based upon purported financial difficulties. For the reasons discussed below, the assessed penalties afforded some reduction from OSHA's proposed penalties.

Background

OSHA's first inspection of Dierzen began on April 12, 2005. OSHA sent both a safety specialist and an industrial hygienist to conduct a safety and health inspection of the manufacturing facility. As a result of that inspection, on June 7, 2005, OSHA cited numerous violations, i.e., ten safety violations (Exh. C-1) and 36 health violations (Exh. C-2, numbered 1 through 32). Dierzen contested the citations, and the case proceeded under Review Commission jurisdiction toward hearing. Shortly before the scheduled hearing, on April 28, 2006, the parties resolved the matter by stipulation and agreement (Exh. C-3). The settlement substantially reduced the penalties to $10,000, which the company was to pay under an extended installment agreement. Dierzen agreed the violations had been or would be abated within the specified time frame. Dierzen paid only the first installment of the reduced penalty and refused to make further payments. Significantly for this case, Dierzen also refused to provide information verifying the violations had been corrected.

OSHA attempted to secure the abatement information through repeated requests by telephone and by letter. Dierzen did not respond. OSHA sought to secure the information by issuing a separate July 2006 citation to Dierzen, which asserted the company failed to provide OSHA with abatement information. Dierzen did not respond to that citation. OSHA then determined to conduct a "follow-up" inspection to check the status of abatement of the 2005 citations (Tr. 51).

OSHA's safety inspector William Hancock and its industrial hygienist Sue Ellen DeManche began the follow-up inspection on October 11, 2006. A follow-up inspection is limited to review of the earlier-cited items and to other apparent violations "in plain sight." Following the inspection, on April 10, 2007, OSHA cited Dierzen with willfully violating four standards and with repeatedly violating one other in the "safety" case. In the "health" case, OSHA cited two serious, seven willful, fourteen repeat, and three "other" violations. Only the amount of the penalties was at issue.

Discussion

Under § 17(j) of the Act, penalties are calculated with "due consideration" given to (1) the size of the employer's business, (2) the gravity of the violation, (3) the good faith of the employer, and (4) the history of previous violations. OSHA seeks to standardize penalties throughout the nation by providing guidance to its personnel in a Field Inspection Reference Manual (FIRM). The evidence established the secretary considered these four statutory factors and followed OSHA's FIRM to arrive at its proposed penalties (Tr. 23, 52–54, 65, 129). The Commission, however, is the final arbiter of penalties in all contested cases. Secretary v. OSHRC and Interstate Glass Co., 487F.2d 438 (8th Cir. 1973). The Commission must determine a reasonable and appropriate penalty in light of § 17(j) of the Act and may arrive at a different formulation than the secretary in assessing the statutory factors.

1. Size, Good Faith, and Past History
   Size:
   The employer's size is the first of the mandated penalty considerations. Dierzen argues the statute's phrase "size of the business" requires the judge to weigh its status as a newly formed business operation. Dierzen asserts it was promised start-up financing from the state that never materialized. If it had the promised money, it posits, it could have taken care of the facility's safety and health dangers after OSHA pointed them out in the earlier citations. Dierzen suggests it did not have the resources to come into compliance with the safety and health requirements or to pay the OSHA penalties. It argues only a nominal penalty is appropriate.

   The secretary opposes such a formulation of "size" and contends it is generally inappropriate to consider an employer's financial condition in penalty calculations under the Act. She asserts "size" refers only to the number of employees employed.

   Even if the secretary's penalty formula does not result in penalties that are punitive in nature, a penalty may be unduly burdensome or excessive in a specific case. The Commission has not finally determined whether an employer's poor financial condition can properly weigh toward a penalty reduction. In rare occasions the Commission has stressed the impact of a total penalty on the viability of a business and reduced the penalty accordingly. See Colonial Craft Reproductions, 1 BNA OSHC 1063, 1065 (No. 881, 1972) (full adjustment of the penalty for size avoids "destructive penalties" where a safe and healthful workplace was secured); Specialists of the South, Inc., 14 BNA OSHC 1910 (No. 89-2241, 1990) (smaller combined penalty approved for impressive, co-operative employer).

   In such cases the employer has met prerequisites. First, the employer has actually proven its precarious financial condition. Dierzen provided no real evidence of a negative financial status. It offered no documentary evidence. Renee Goff, one of the earliest employees who is now "safety administrator," testified Louie Dierzen instructed there was insufficient money for abatement with expensive equipment. This is hardly sufficient proof to document Dierzen's financial data.

   Of equal importance, an employer must establish it deserves to have its poor finances affect the penalty. In Interstate Lead Company, 15 BNA OSHC 1989, 2000 (No. 89-2088P, 89-3296, 1992), Judge James D. Burroughs succinctly summarized this concept:

   As a practical matter, the financial condition in certain cases must be considered. OSHA and the Commission were not created to eliminate business activity, as some employers contend. OSHA was created to preserve the health and safety of working men and women of this nation. They constitute resources in which the nation has a vital interest in protecting. Where an employer approaches its responsibility under the Act in good faith, has no detrimental history, and seeks to abate violations, it is only practical that some considerations be given to an employer's negative financial condition and the effect of penalties assessed on the viability of the business.

However, if the employer has not acted in good faith but uses a precarious financial condition as an excuse to ignore the safety and health of its employees, the extraordinary relief is not warranted.

Dierzen consistently demonstrated a cavalier attitude and a lack of cooperation toward achieving safety and health in its facility. It was unresponsive to the Act's ordinary enforcement mechanism of citation and penalty, and it simply ignored or stalled OSHA and reneged on its agreement to come into compliance and pay earlier penalties. Following the 2005 inspection Dierzen exerted minimal to no effort to correct the violations. During the follow-up inspection owner Louis Dierzen advised Hancock he was trying to correct things he could, but that he could not correct many of the violations because he did not have the money (Tr. 28-29). The facts do not bear out this assertion. Russ Spencer, an engineer who contracts with Dierzen, was Dierzen's designated representative during the inspection (Tr. 17). Spencer told Hancock he developed an abatement plan, together with some time frames to correct the violations. Owner Louie Dierzen told Spencer to send the plan to Dierzen's attorney, which Spencer did. He never heard or did anything further. Spencer understood from Louie Dierzen he would get no money to abate violations. Dierzen did not seek to abate those violations which required minimal expenditure, and it did not seek to protect employees in any alternate way.

No one relishes imposing the type of penalty which may jeopardize a small business, especially a small manufacturing concern. Yet the Act never contemplated employees should risk their health and safety simply because their employer is a poorly funded concern.

Under the FIRM OSHA reduced its initial penalty calculations by 40 percent, because having 36 employees fits within a range corresponding to a "small" employer (Tr. 164). Looking at the specific number of employees, rather than the number in a range, convinces this judge to further reduce some penalties because Dierzen is a very small employer.

Good Faith:
OSHA is correct that Dierzen has not acted in good faith and is not entitled to a good faith credit.

Past History:
Dierzen's past history is negative. OSHA discussed the earlier violations in the 2005 closing conference. The violations were further explained in the

written citations, which were litigated until the eve of trial. When Dierzen settled the case on apparently favorable terms, it had a second chance to comply with the Act, which it ignored. No credit is warranted for past history.

Classification of Violations of Willful or Repeat:
Of most significance to the over-all penalty in this case is that many violations were willful or repeat. As the Act designates, these carry an enhanced penalty. The OSH Act sets a maximum of $7,000 for each serious violation, but a $70,000 maximum penalty for each willful or repeat violation [29 U.S.C. 666(a)]. Dierzen does not contest the characterizations of the individual violations. Even if it did, the facts establish the violations are properly classified. Dierzen acted with conscious disregard that the precise conditions found in 2005, and still existing in 2006, constituted violations of the safety and health standards. In numbers of employees, Dierzen grew by 1000 percent from 2003 to 2006. If Dierzen directed a portion of the effort necessary to achieve that remarkable feat toward the employees' safety and health, this case would not exist.

Following its procedures, OSHA properly proposed $33,000 for each willful violation in both the safety and health cases. After consideration of the degree of willfulness, the gravity of each violation, and some mitigating efforts by certain Dierzen personnel, this judge arrived at a different assessment than OSHA proposed. This does not signify Dierzen's actions and omissions are viewed with less disfavor.

Many violations were also classified as repeat. With a significant effect of lessening the over-all penalty, OSHA considered violations repeat rather than willful. OSHA asserted the violation was repeat if Dierzen sought to address the violations in some way, even if the effort was not effective (Tr. 129). Dierzen is a small employer and was in repeat violation for the first time, meaning OSHA doubled its initial penalty calculation. For the repeat health violations, DeManche concluded there was less likelihood the violations would result in accidents since few employees were exposed over a relatively short duration of exposures (Tr. 129). The assessed penalties reflect that view.

2. Gravity of Individual Violations and Assessed Penalties
Of the four statutory penalty factors, the gravity of the violation is usually the most significant. See e.g., Orion Constr., Inc., 18 BNA OSHC 1867, 1868 (No. 98-2014, 1999). Gravity addresses the setting and the circumstances of the violations, i.e., the degree to which the standard was violated and the harm anticipated. The specific considerations include such facts as the number of employees exposed to the conditions, the duration of exposure, the degree of probability that an accident would occur, or precautions taken against injury. Agra Erectors Inc., 19 BNA OSHC 1063, 1065 (Docket No. 98-0866, 2000).

OSHRC Docket No. 07-0675—

The Safety Case

The April 10, 2007, safety citation asserted four willful and one repeat violations. The gravity and the assessed penalty are discussed below.

Willful Citation No. 1, Items 1–4

Willful Item 1—§ 1910.23(c)(1)　Dierzen willfully violated the fall protection standard of § 1910.23(c)(1) (item 1). Employees welded down onto truck bodies from an open-sided skeletal frame fixture 7 feet, 9 inches, above a concrete floor. At any given time five employees weld from atop the frame fixture for several hours a week. Dierzen did not provide fall protection. Falls from the frame would likely result in broken bones, concussions, or other serious injury. Spencer admitted Dierzen did not attempt to correct the violation. Nor did it make any modifications to lessen the hazard or to provide alternative fall protection (Tr. 21-22, 45, 48). Considering these facts and the statutory elements discussed above, a penalty of $13,500.00 is assessed.

Willful Item 2—§ 1910.147(c)(4)(i)　Dierzen willfully violated the lockout/tagout (LOTO) requirements of § 1910.147(c)(4)(i) (item 2) by refusing to develop written LOTO procedures for hazardous energy sources, i.e., the "computer numerically controlled" (CNC) lathe (used to turn down metal parts and form and cut them into the desired sizes) and the CNC plasma cutter (used for cutting sheets of steel plate). Two maintenance workers serviced the equipment for several hours each day. Without specific written LOTO procedures to identify and control the energy sources, the machinery could unexpectedly start and crush, cut, or amputate the fingers or hands (Tr. 24-29). Considering these facts, the statutory elements, and the relationship between this and the following violation, a penalty of $8,000.00 is assessed.

Willful Item 3—§ 1910.147(c)(7)(i)　In item 3, a violation of § 1910.147(c)(7)(i), Dierzen willfully failed to provide LOTO training. Dierzen should have trained the two maintenance employees and others on the shop floor to follow procedures on how to lock and tag out equipment. Although the Cincinnati shear was initially locked out for service, another employee bypassed the lock in order to operate the shear. Dierzen did not itself train anyone, but a maintenance employee learned about the procedures from another employer (Tr. 29-34). The duration of the exposure and the potential injuries are described above. Considering these facts and the statutory elements, a penalty of $8,000.00 is assessed.

Willful Item 4—§ 1910.212(a)(3)(ii)　Dierzen willfully violated § 1910.212(a)(3)(ii) (item 4) when it failed to guard the points of operation of the Cincinnati 400-ton press brake and Cincinnati Model 1810 shear. The anticipated hazard for the operators, whose fingers could be as close as two inches from the die or knife blade of the shear, was amputation of the fingers or other parts of the hand. One operator used these pieces of equipment, and he was exposed to the hazard for approximately 4 hours a week (Tr. 34-37). Considering these facts and the statutory elements, a penalty of $13,500.00 is assessed.

Repeat Citation No. 2, Item 1

A repeated violation also carries an increased monetary penalty. As a first repeat citation for an employer with 250 or fewer employees, OSHA doubled Dierzen's recommended penalty.

Repeat Item 1—§ 5(a)(1) of the OSH Act

Dierzen violated the "general duty clause" of § 5(a)(1) of the Act when it failed to protect employees from the recognized hazard associated with failing to guard the foot treadle of the Cincinnati 400-ton press brake. If the foot treadle were accidentally hit or pressed, the machine could cycle while the operator's hands were in the equipment, perhaps amputating the operator's fingers or hands. The repeat classification is based on a previous § 5(a)(1) violation for the same condition, except that the equipment (a Cincinnati 225-ton press) differed. One employee was exposed to the hazard for 4 hours each week (Tr. 37-40). Considering these facts and the statutory elements, a penalty of $3,000.00 is assessed.

Docket No. 07-0676—

The Health Case

At issue in the April 10, 2007, health citations are penalties for three serious, seven willful, and 14 repeat violations. Although a "follow-up" inspection, OSHA's industrial hygienist Sue Ellen DeManche noted three violations that had not been previously cited but were in plain sight as she checked for the follow-up items. These violations are classified as serious.

Serious Citation No. 1, Items 1–3

Serious Item 1—§ 1910.107(g)(3)   Dierzen failed to dispose of rags and waste saturated with paint finishing materials from the spraying room in violation of § 1910.107(g)(3). It could have used a closed metal waste container or other approved waste disposal methods. Instead, Dierzen left the solvent and paint-soaked paper and rags in an open plastic trash can. The readily combustible materials, piled together, could burst into flames from a random spark or they could accelerate or intensify a fire. The likely injury is severe burns and smoke inhalation. Two employees were exposed to the hazard for approximately 3 hours each workday (Tr. 63-64, 115). Considering these facts and the statutory elements, a penalty of $900.00 is assessed.

Serious Item 2—§ 1910.303(b)(2)   Section 1910.303(b)(2) requires electrical equipment to be used and installed in accordance with the labeled instructions. In the welding area, DeManche stepped near an energized 240-volt metal "handy box" improperly used as an outlet device for the arc welder. Manufacturer's instructions specify handy boxes must be mounted in a wall. When on the floor, the outlet is subject to being stepped on, tripped over, kicked, or knocked around, pulling on the energized conductors. Also, the employee would hold the handy box to plug in the arc welder. Coming into contact with any bare wires of the improperly protected conductors could cause shocks or burns. Twenty-two welders worked in the area for

8 hours each day (Tr. 65-67, 116). Considering these facts and the statutory elements, a penalty of $1,200.00 is assessed.

Serious Item 3—§ 1910.305(b)(1)   Section 1910.305(b)(1) requires the unused openings in circuit breaker boxes to be effectively closed. A circuit breaker box contained open slots, exposing live wires, which an employee could inadvertently contact while accessing the box. The circuit breaker box was energized at 240 volts. Dierzen's 22 welders could be exposed to severe shock if they made contact with the parts during their 8-hour workdays (Tr. 67-68, 117). Considering these facts and the statutory elements, a penalty of $900.00 is assessed.

Willful Citation No 2, Items 1a–4b

Willful grouped Items 1a, 1b, and 1c—§ 1910.106(d)(4)(iii), § 1910.107(c)(5), and § 1910.107(c)(6)

For penalty purposes the secretary grouped three willful violations related to multiple unapproved electrical wiring, outlets, and cords, located in and adjacent to the paint spraying area. Dierzen violated § 1910.106(b)(4)(iii) (item 1a) when it stored flammable liquids in a room with unapproved electrical equipment that was adjacent to the spray area. Dierzen violated § 1910.107(c)(5) (item 1b) when using unapproved electrical equipment in the spray area, which could have deposits of readily ignitable residues, explosive vapors, and solvents. Dierzen violated § 1910.107(c)(6) (item 1c) by using unapproved electrical wiring, installations, and flexible cords in this hazardous location. The grouped violations share the common hazard in the areas where flammable vapors collect that a spark from unapproved electrical equipment could start a fire. Two employees wiped down the truck exteriors with solvents, mixed paints and solvents, and spray painted for 3 hours each workday. They used the electrical equipment frequently. A fire could expose employees to burns and smoke inhalation (Tr. 69-72, 118). Considering these facts and the statutory elements, a penalty of $20,000.00 for the three grouped violations is assessed.

Willful Item 2—§ 1910.134(i)(7)   To remove paint and rust from the trucks, Dierzen's abrasive blaster pressure-sprayed abrasive silica sand through the hose. The operator wears an air line respirator powered by an oil lubricated compressor. Dierzen had not utilized either a carbon monoxide or a high-temperature alarm to monitor the air coming into the respirator in violation of § 1910.134(i)(7). If the compressor heated excessively or malfunctioned in other ways, carbon monoxide could be sent directly into the respirator without anyone becoming aware. One employee was exposed while blasting approximately 1½ hours a day, three times a week. Dierzen made no effort to correct the violation which could lead to carbon monoxide poisoning, even though abatement could have been quickly, easily, and inexpensively achieved (Tr. 72-74). Considering these facts and the statutory elements, a penalty of $10,000.00 is assessed.

Willful Item 3—§ 1910.244(b)   In willful violation of § 1910.244(b) Dierzen failed to equip the operating valve of the abrasive blasting equipment with a "deadman's switch." The switch immediately deactivates the sprayer when the operator ceases to manually depress it. The potential hazard is that if the operator loses control of the hose, it could continue pressure spraying the operator or others with silica sand, leading to severe abrasions. The abrasive blaster was exposed to the hazard 1½ hours a day, 3 days a week. Placing the deadman's switch to the nozzle is neither difficult nor expensive (Tr. 75-76). Considering these facts and the statutory elements, a penalty of $10,000.00 is assessed.

Willful Items 4a and 4b—§§ 1910.1000(c) and § 1910.1000(e)   The two grouped violations of § 1910.1000(c) (item 4a) and § 1910.1000(e) (item 4b) concern over-exposure to silica dust. During the sampled period, the abrasive blaster was exposed to respirable silica over twice the permissible exposure limit (PEL). Despite the known silica hazard, Dierzen did not seek any feasible administrative or engineering controls to prevent the over-exposure. Exposure to respirable silica above the PEL leads to silicosis, decreased lung capacity, cancer, and potentially to death. The operator was exposed 1½ hours a day, 3 days a week. The only change in the cited conditions between 2005 and 2006 is that production and sand blasting increased. Dierzen considered splitting the work with another employee, but never did so. It considered no other controls (Tr. 76-79, 119, 127). Although not relevant to the existence of the violation, the fact that the blaster wore an airline respirator lessens the gravity of the exposure. Considering these facts and the statutory elements, a penalty of $20,500.00 is assessed.

Repeat Citation No. 3, Items 1—12

Repeat Item 1—§ 1910.23(a)(5)   In violation of § 1910.23(a)(5) Dierzen failed to guard a concrete pit in the spray paint room which existed at the time Dierzen purchased the facility. The pit was 21 inches wide by 110 inches long by 51 inches deep. Dierzen's painter told OSHA Dierzen covered the pit with boards at one point, but the boards gradually disappeared. The pit was open during the inspection. The painter and helper were exposed to the hazard 3 hours a day, 5 days week, while they prepped and painted the trailers or cleaned the equipment and mixed paints. They spent much time walking and spraying in the area, looking upward. A 4-foot fall into the pit could result in broken bones (Tr. 79, 82, 120). Considering these facts and the statutory elements, a penalty of $2,000.00 is assessed.

Repeat Item 2—§ 1910.106(d)(4)(i)   Dierzen used the room adjacent to and opening into the spray paint area to store flammable liquids in violation of § 1910.106(d)(4)(i). The storage room should have had a self-closing door between the two rooms, but the door was kept open. The purpose of the standard is to prevent flammable vapors from igniting. The spray painter and his helper were exposed to the hazard while they regularly secured materials from the storage room for the period stated above (Tr. 82-84). Considering these facts and the statutory elements, a penalty of $3,000.00 is assessed.

Repeat Items 3a—3b, § 1910.134(c)(1) and 1910.134(c)(3)   Repeat items 3a through 6 each relate to airborne respirable chemicals. Dierzen violated § 1910.134(c) (1) (item 3a) by failing to develop and implement a written respiratory protection

program. It neither hired a trained administrator nor trained any of its employees
to oversee the program and conduct evaluations, in violation of § 1910.134(c)(3)
(item 3b). Dierzen utilized two types of respirators, an airline respirator for the abrasive blasting and a half-mask respirator for painting. The purpose of a program is
to ensure the respiratory protection requirements are met for the range of hazardous
airborne contaminants in Dierzen's facility, e.g., to ensure the proper use of respirator filters and cartridges, and that training, medical evaluations, and fit testing are
completed. Without a program employees are less likely to control their exposure
to airborne contaminants during their 1 ½ to 3 hours of exposure (Tr. 84–87, 121).
Considering these facts and the statutory elements, a penalty of $2,400.00 is assessed
for the two grouped violations.

Repeat Item 4—§ 1910.134(e)(1)   In repeat violation of § 1910.134(e)(1) Dierzen
did not provide medical evaluations for the painter and the helper to assure they were
medically able to wear respirators. The anticipated hazard is that an employee with
a medical condition affected by the respirator could suffer pulmonary stress, short-
ness of breath, or dizziness. Although Dierzen did not provide the abrasive blaster
with a medical examination, he had the examination from a previous employer and
learned he could wear the respirator. The frequency and duration of the exposure is
described above (Tr. 87-89). Considering these facts and the statutory elements, a
penalty of $1,800.00 is assessed.

Repeat Item 5—§ 1910.134(f)(1)   In violation of § 1910.134(f)(1) Dierzen did
not provide the painter or his helper with the quantitative or qualitative fit test to
assure a correct size and a correct seal are optimal. Dierzen could have provided
the "quantitative fit test" (where computerized equipment counts the particles inside
and outside the employee's respirator) or the "qualitative fit test" (where employees
identify when they a smell an odor). Dierzen did neither. The resulting injury would
most likely be a temporary dizziness or headaches. The frequency and duration of
the exposure for the painter and helper is described above. Although Dierzen did
not perform a fit test for the abrasive blaster, he had been fit tested by a previous
employer (Tr. 90-93). Considering these facts and the statutory elements, a penalty
of $1,800.00 is assessed.

Repeat Item 6—§ 1910.134(k)(3)   Dierzen offered no respiratory protection
training for employees required to wear respirators in violation of § 1910.134(k)(3).
The three employees were exposed to a variety of airborne contaminants that could
lead to silicosis, lung scarring, carbon monoxide poisoning, dizziness, or short-
ness of breath. Without proper training on the use and care of the respirator it may
become useless to the employee. For example, one employee wore an ineffective
organic vapor cartridge for silica and left the respirator in the open environment
to collect dust and contaminants. An employee was wiping out the inside of his
respirator with alcohol, which can degrade the plastic and interfere with seal. The
blaster did not understand his potential for carbon monoxide poisoning. The frequency and duration of the exposure for the three employees is described above.
Although Dierzen did not provide the training, one employee was trained by another

employer (Tr. 93-96). Considering these facts and the statutory elements, a penalty of $1,800.00 is assessed.

Repeat Item 7—§ 1910.157(e)(3)   Dierzen did not annually check the maintenance of at least eight fire extinguishers in violation of § 1910.157(e)(3). Throughout the facility the fire extinguishers had not been tested since either 2001 and 2002. A couple of the fire extinguishers were discharged and had not been refilled. Two or three extinguishers had been purchased since the 2005 inspection, but they remained in their boxes in the office. Few knew of their existence, and they were not available to employees on the floor. The injury was smoke and burn related from a fire which was quickly extinguished. All 36 employees were exposed to a potential fire during their 8-hour shifts. (Tr. 96-98). Considering these facts and the statutory elements, a penalty of $1,800.00 is assessed.

Repeat Items 8a and 8b—§§ 1910.178(l)(1)(i) and § 1910.178(l)(6)   Items 8a through 10 concern operation of Dierzen's powered industrial trucks. Dierzen did not train employees on the proper operation of powered industrial trucks, in violation of § 1910.178(l)(1)(i) (item 8a). Because of the importance of forklift training to safety in a facility, § 1910.178(l)(6) (item 8b) also requires Dierzen to certify the training, which it did not do. Both the operator and other employees can be exposed to potential broken bones or other injuries or death when untrained operators can strike employees with the forklift or cause material to fall on the operator or others. Ten employees operate forklifts 8 hours a day. Thirty-six other employees worked around the untrained operators. Dierzen's Renee Goff developed a program to conduct training, but the program was never implemented (Tr. 98-101, 122). Considering these facts and the statutory elements, a penalty of $1,800.00 is assessed for the two grouped violations.

Repeat Items 9 and 10—§§ 1910.178(n)(4) and 1910.178(q)(1)   OSHA observed the forklift which did not slow down at a blind intersection and did not sound its horn in violation of § 1910.178(n)(4) (item 9). Significant ambient noise was generated around the elevators and the shears, aggravating the hazard of forklifts speeding through blind intersections without sounding a horn. The potential is for a forklift-to-forklift or a forklift-to-pedestrian collision and resulting broken bones or other serious injury or death. Employees were exposed intermittently during their 8-hour work shifts. Considering these facts and the statutory elements, a penalty of $2,400.00 is assessed for item 9.

Dierzen did not inspect the forklifts in violation of § 1910.178(q)(1) (item 10). On paper Dierzen began an inspection program where unsafe equipment was to be tagged out. However, at the time OSHA arrived, Dierzen had not begun to implement the program (Tr. 101-105, 123, 128). Considering these facts and the statutory elements discussed above, a penalty of $1,800.00 is assessed for item 10.

Repeat Items 11 and 12—§§ 1910.1200(e)(1) and § 1910.1200(h)(1)   Dierzen violated two hazard communication standards when it failed to create or implement an adequate hazard communication program in violation of § 1910.1200(e)(1) (item 11). It did not train employees on how to lessen the impact of the hazardous chemicals on their bodies in violation of § 1910.1200(h)(1) (item 12). Dierzen apparently began to compile some material safety data sheets (MSDSs), but it did secure all of them and did not train employees on how to find or use them. Employees are less likely to understand and protect themselves from the hazards associated with exposure to such substances as paint and solvent fumes, welding fumes, silica, and cylinder gas without access to information and training. Four employees were particularly affected by the failure to train because they worked directly in areas where they were exposed to the hazardous substances (Tr. 106-112, 124-125). Considering these facts, the statutory elements, and the existence of some overlap with other violations, a penalty of $1,500.00 each is assessed for items 11 and 12.

CONCLUSION

It is inescapable that Dierzen considered OSHA and requirements of the OSH Act to be a mere bother and that delays in compliance would work to its benefit. The Act established monetary penalties to counter such attitudes and to encourage employers to be proactive in addressing the safety and health hazards in their facilities.

Findings of Fact and Conclusions of Law

The foregoing decision constitutes the findings of fact and conclusions of law in accordance with Rule 52(a), Fed. R. Civ.P.

Order

Based on the foregoing decision, it is ORDERED:

A total penalty of $133,100.00 is assessed for Docket Nos. 07-0675 and 07-0676. (Note: This case was heard in 2009. OSHA's penalty schedule has changed).

2 OSHA Enforcement

CHAPTER OBJECTIVES

1. Acquire an understanding of the OSHA monetary penalties.
2. Acquire an understanding of the OSHA criminal penalties.
3. Acquire an understanding of the components of an OSHA compliance inspection.
4. Acquire an understanding of how to prepare for an OSHA compliance inspection.

INTRODUCTION

Under the Federal Occupational Safety and Health (OSH) Act, Congress provided civil and criminal penalties for employers who failed to comply with the promulgated standards and regulations. Over the years, the monetary penalties have been modified and the criminal sanctions have been seldom utilized. Many employers who initially addressed the requirements of the OSH Act and the Occupational Safety and Health Administration's (OSHA) standards to avoid the OSHA penalties have found that compliance makes "good business sense." These employers have moved to a higher level in the safety and loss prevention hierarchy, where simply avoiding penalties by OSHA for noncompliance is no longer the objective, but a "given." These employers have moved to a level where safeguarding their human assets pays dividends, not only in personnel but also in terms of dollars saved as well.

Employers who have not heeded the warning of Congress and OSHA have found that failure to comply with the OSHA standards and create a safe and healthful workplace for their employees can be extremely costly. The OSH Act provided OSHA the power to issue monetary penalties, often reaching several million dollars, and, in egregious cases, the ability to pursue criminal sanctions.

MONETARY FINES AND PENALTIES

The OSH Act provides a wide range of penalties, from a simple notice with no fine to criminal prosecution. The Omnibus Budget Reconciliation Act of 1990 multiplied maximum penalties sevenfold. Violations are categorized and penalties may be assessed. In November 2015, Congress enacted legislation requiring federal agencies to adjust their civil penalties to account for inflation. The Department of Labor is adjusting penalties for its agencies, including the Occupational Safety and Health Administration (OSHA). OSHA's maximum penalties, which were last adjusted in

1990, increased by 78%. OSHA will continue to adjust its penalties for inflation each year based on the Consumer Price Index. The new penalties took effect after August 1, 2016, and there were adjustments for subsequent years.

Each alleged violation is categorized, and the appropriate fine is issued by the OSHA area director. It should be noted that each citation is separate and may carry with it a monetary fine. The gravity of the violation is the primary factor in determining penalties. In assessing the gravity of a violation, the compliance officer or area director must consider the severity of the injury or illness that could result and the probability that an injury or illness could occur as a result of the violation. Specific penalty assessment tables assist the area director or compliance officer in determining the appropriate fine for the violation.

After selecting the appropriate penalty table, the area director or compliance officer determines the degree of probability that the injury or illness will occur by considering several factors:

- The number of employees exposed
- The frequency and duration of the exposure
- The proximity of employees to the point of danger
- Factors, such as the speed of the operation, that require work under stress
- Other factors that might significantly affect the degree of probability of an accident

OSHA has defined a serious violation as "an infraction in which there is a substantial probability that death or serious harm could result … unless the employer did not or could not with the exercise of reasonable diligence, know of the presence of the violation." Section 17(b) of the OSH Act requires that a penalty of up to $12,934,000 be assessed for every serious violation cited by the compliance officer. In assembly-line enterprises and manufacturing facilities with duplicate operations, if one process is cited as possessing a serious violation, it is possible that each of the duplicate processes or machines may be cited for the same violation. Thus, if a serious violation is found in one machine, and there are many other identical machines in the enterprise, a very large monetary fine for a single, serious violation is possible.

Currently, the greatest monetary liabilities are for repeat violations, willful violations, and failure to abate cited violations. A repeat violation is a second citation for a violation that was cited previously by a compliance officer. OSHA maintains records of all violations and must check for repeat violations after each inspection. A willful violation is the employer's purposeful or negligent failure to correct a known deficiency. This type of violation, in addition to carrying a large monetary fine, exposes the employer to a charge of an "egregious" violation, and the potential for criminal sanctions under the OSH Act or state criminal statutes if an employee is injured or killed as a direct result of the willful violation. Failure to abate a cited violation has the greatest cumulative monetary liability of all. OSHA may assess a penalty of up to $1,000 per day, per violation, for each day in which a cited violation is not brought into compliance.

In assessing monetary penalties, the area or regional director must consider the good faith of the employer, the gravity of the violation, the employer's past history of compliance, and the size of the employer. OSHA has adopted and is utilizing the egregious case policy. Under the egregious violation policy, when violations are determined to be conspicuous, penalties are cited for each violation, rather than combining the violations into a single, smaller penalty.

Type of Violation	Old Maximum Penalty	New Maximum Penalty
Other-than-Serious	$7,000 per violation	$12,934 per violation – Maximum
Failure to Abate	$7,000 per day	$12,934 per day – Maximum
Willful Violation	$70,000 per violation	$129,336 per violation – Maximum
Repeat Violation	$7,000 per violation	$129,336 per violation – Maximum
Serious Violation	$7,000 per violation	$12,934 per violation – Maximum
Posting Violation	$7,000 per violation	$12,934 per violation – Maximum
(Adjusted civil monetary penalties as of January 2, 2018)		

Safety and loss prevention professionals should be aware that OSHA established the "egregious violation" policy in the mid-1980s. The egregious violations policy instructs OSHA inspectors to cite employers for multiple violations of the same standard where the employer has demonstrated one or more of the following characteristics: (1) persistently high rates of illness/injury or fatalities; (2) extensive history of prior violations; (3) intentional disregard of health and safety responsibilities; or (4) bad faith (a plain indifference to standards or requirements).

Additionally, safety and loss prevention professionals should be aware of the OSHA Severe Violators Enforcement Program. Initiated in 2010, the SVEP program is intended to focus on employers that endanger workers by committing willful, repeat, or failure-to-abate violations in one or more of the following circumstances: a fatality or catastrophe; industry operations or processes that expose workers to severe occupational hazards; exposure to hazards related to the potential releases of highly hazardous chemicals; and all instance-by-instance enforcement actions. In general, the SVEP program subjects employers to mandatory follow-up inspections; increased company/corporate awareness of OSHA enforcement; and, where appropriate, corporate-wide agreements, enhanced settlement provisions, and federal court enforcement under Section 11(b) of the OSH Act. SVEP usually required a fatality, non-fatal catastrophic conditions, or egregious conditions.

INSPECTIONS, CITATIONS, AND PROPOSED PENALTIES

In addition to the potential civil or monetary penalties that could be assessed, OSHA regulations may be used as evidence in negligence, product liability, workers' compensation, and other actions involving employee safety and health issues. OSHA standards and regulations are the baseline requirements for safety and health that must be met, not only to achieve compliance with the OSHA regulations but also to safeguard an organization against other potential civil actions.

CRIMINAL LIABILITY

The OSH Act provides criminal penalties in four circumstances. In the first situation, anyone inside or outside of the Department of Labor or OSHA who gives advance notice of an inspection, without authority from the secretary, may be fined up to $1,000, imprisoned for up to six months, or both. Second, any employer or person who intentionally falsifies statements or OSHA records that must be prepared, maintained, or submitted under the OSH Act, may, if found guilty, be fined up to $10,000, imprisoned for up to six months, or both. Third, any person responsible for a violation of an OSHA standard, rule, order, or regulation that causes the death of an employee may, upon conviction, be fined up to $10,000, imprisoned for up to six months, or both. If convicted for a second violation, punishment may be a fine of up to $20,000, imprisonment for up to one year, or both. Finally, if an individual is convicted of forcibly resisting or assaulting a compliance officer or other Department of Labor personnel, a fine of $5,000, three years in prison, or both can be imposed. Any person convicted of killing a compliance officer or other OSHA or Department of Labor personnel, acting in his or her official capacity, may be sentenced to prison for any term of years or life.

OSHA does not have the authority to impose criminal penalties directly; instead, it refers cases for possible criminal prosecution to the U.S. Department of Justice. Criminal penalties must be based on a violation of a specific OSHA standard; they may not be based on a violation of the general duty clause. Criminal prosecutions are conducted like any other criminal trial, with the same rules of evidence, burden of proof, and rights of the accused. A corporation may be found criminally liable for the acts of its agents or employees. The statute of limitations for possible criminal violations of the OSH Act, as for other federal noncapital crimes, is five years.

Under federal criminal law, criminal charges may range from murder to manslaughter to conspiracy. Several charges may be brought against an employer for various separate violations under one federal indictment.

After a criminal conviction for a federal felony, the sentence to be imposed is defined under the Federal Sentencing Guidelines. Under these guidelines, judges have very little leeway in determining the sentence. Each felony has an offense level. A sentencing table is used to factor in the criminal history, and a fine table provides a minimum and maximum fine range. Deduction from or addition to the sentencing structure based on numerical values is permitted depending on the situation. Departures from the range provided in the guidelines are rare. Of particular interest to safety and health professionals facing this type of situation is the fact that the Federal Sentencing Guidelines provide very little flexibility through which the court can consider factors involved in a work-related injury or fatality. In essence, murder is murder, regardless of whether it happened on the street or in the workplace.

With the Sentencing Reform Act of 1984, which standardized penalties and sentences for federal offenses, the criminal penalty for willful violations of the OSH Act causing loss of human life was amended to be punishable by fines up to $250,000 for individuals and $500,000 for organizations. However, where a willful violation of an OSHA standard causes the death of an employee, Section 17 of the OSH Act is utilized. A violation of Section 17 is not a felony but identified as a "Class B" misdemeanor. And although Section 17(e) carries with it the potential of a prison term, prison sentences rarely occur for individual managers and criminal monetary fines are often utilized.

However, safety and loss prevention professionals should be aware that criminal convictions with prison terms do occur. For a conviction under Section 17(e) of the OSH Act to be obtained, the prosecutor must be able to prove *beyond a reasonable doubt* that the individual or company:

1. knew there was an OSHA regulatory standard and the standard was violated. (Note: A violation of OSHA's "General Duty Clause" may not serve as the basis for an OSH Act criminal charge);
2. the known violation of a specific OSHA standard was committed by the employer;
3. the violation of the standard was the direct cause of an employee's death; and
4. the violation was committed willfully by the employer.

From the U.S. Department of Justice's website located at www.justice.gov, below are some of the willful OSHA violations that caused an employee fatality and employers (and employees) can face criminal sanctions:

1. **Falsifying OSHA documents**
2. **Advance notice of an OSHA inspection**
3. **Perjury during OSHA proceedings**
4. **State criminal laws** The OSH Act does not preempt prosecution under state criminal laws, such as manslaughter or negligent homicide for work-related deaths.
5. **Environmental statutes** Safety and loss prevention professionals should be aware that in catastrophic situations, environmental laws carrying broader criminal provisions can be used in addition to the OSH Act.

Safety and loss prevention professionals should also be aware that although environmental statutes have been used as a vehicle for OSHA criminal enforcement for years, a recent Department of Justice initiative can result in the pursuit of criminal charges under environmental statutes where workers' health and safety is being threatened. In 2015, the U.S. Department of Justice and the U.S. Department of Labor entered into a Memorandum of Understanding regarding the use of environmental standards for prosecution of workplace safety and health violations. The use of environmental standards for workplace safety violations appears to be expanding and safety and loss prevention professionals should be aware of the expanded potential criminal penalties under the environmental laws which far exceed punishment under the OSH Act.

RIGHTS AND RESPONSIBILITIES UNDER THE OSH ACT

The OSHA Inspection

OSHA performs all enforcement functions under the OSH Act. Under section 8(a) of the OSH Act, OSHA compliance officers have the right to enter any workplace of a covered employer without delay, inspect and investigate a workplace during regular hours and at other reasonable times, and obtain an inspection warrant if access to a facility or operation is denied. Upon arrival at an inspection site (any company facility), the compliance officer must present his or her credentials to the owner or designated representative of the employer before starting the inspection. An employer representative and an employee and/or union representative may accompany the compliance officer on the inspection. Compliance officers can question the employer and employees and inspect required records, such as the OSHA Form 300, which records injuries and illnesses. Most compliance officers cannot issue on-the-spot citations; they have only the authority to document potential hazards and report or confer with the OSHA area director before issuing a citation.

A compliance officer or other employee of OSHA may not provide advance notice of the inspection under penalty of law. The OSHA area director is, however, permitted to provide notice under the following circumstances:

1. In cases of apparent imminent danger, to enable the employer to correct the danger as quickly as possible
2. When the inspection can most effectively be conducted after regular business hours or where special preparations are necessary
3. To ensure the presence of employee and employer representatives or appropriate personnel needed to aid in inspections
4. When the area director determines that advance notice would enhance the probability of an effective and thorough inspection

Compliance officers can also take environmental samples and take or obtain photographs related to the inspection. Additionally, compliance officers can use other "reasonable investigative techniques," including personal sampling equipment, dosimeters, air sampling badges, and other equipment. Compliance officers must, however, take reasonable precautions when using photographic or sampling equipment to avoid creating hazardous conditions (i.e., a spark-producing camera flash in a flammable area) or disclosing a trade secret.

An OSHA inspection has four basic components:

1. The opening conference
2. The walk-through inspection
3. The closing conference
4. The issuing of citations, if necessary

In the opening conference, the compliance officer may explain the purpose and type of inspection to be conducted, request records to be evaluated, question the employer, ask for appropriate representatives to accompany him or her during the walk-through inspection, and ask additional questions or request more information. The compliance officer may, but is not required to, provide the employer with copies of the applicable laws and regulations governing procedures and health and safety standards. The opening conference is usually brief and informal; its primary purpose is to establish the scope and purpose of the walk-through inspection.

After the opening conference and review of appropriate records, the compliance officer, usually accompanied by a representative of the employer and a representative of the employees, conducts a physical inspection of the facility or worksite. The general purpose of this walk-through inspection is to determine whether the facility or worksite complies with OSHA standards. The compliance officer must identify potential safety and health hazards in the workplace, if any, and document them to support issuance of citations.

The compliance officer uses various forms to document potential safety and health hazards observed during the inspection. The most commonly used form is the OSHA-1 Inspection Report, wherein the compliance officer records information gathered during the opening conference and walk-through inspection, including the following data:

- the establishment's name
- the inspection number
- the type of legal entity ✓
- the type of business or plant
- any additional citations
- the names and addresses of all organized employee groups
- the name of the authorized representative of employees
- the name of the employee representative contacted
- the names of other persons contacted
- coverage information (state of incorporation, type of goods or services in interstate commerce, etc.)
- the date and time of entry
- the date and time that the walk-through inspection began
- the date and time that the closing conference began
- the date and time of exit
- the recommendation for a follow-up inspection, if needed
- the compliance officer's signature and date
- the names of other compliance officers
- an evaluation of safety and health programs (checklist)
- the closing conference checklist
- any additional comments

Two additional forms are usually attached to the OSHA Inspection Report. The OSHA-1A form, known as the narrative, is used to record information gathered during the walk-through inspection: names and addresses of employees, management officials, and employee representatives accompanying the compliance officer on the inspection and other information. A separate worksheet, known as OSHA-1B, is used by the compliance officer to document each condition that he or she believes could be an OSHA violation. One OSHA-1B worksheet is completed for each potential violation noted by the compliance officer.

When the walk-through inspection is complete, the compliance officer conducts an informal meeting with the employer or the employer's representative to "informally advise (the employer) of any apparent safety or health violations disclosed by the inspection." The compliance officer informs the employer of the potential hazards observed and indicates the applicable section of the allegedly violated standards, advises that citations may be issued, and informs the employer or representative of the appeal process and the employer's rights under the act. Additionally, the compliance officer advises the employer that the OSH Act prohibits discrimination against employees or others for exercising their rights.

In an unusual situation, the compliance officer may issue one or more citations on the spot. When this occurs, the compliance officer informs the employer of the abatement period in addition to the other information provided at the closing conference. In most circumstances, the compliance officer will leave the workplace and file a report with the proper area director (through the secretary of labor) to decide whether a citation should be issued, compute any penalties to be assessed, and set the abatement date for each alleged violation. The area director, under authority from the secretary, must issue the citation with "reasonable promptness." Citations must be issued in writing and must describe with precision the alleged violation, including the relevant standards and regulation. The citation must be issued or vacated within a six-month statute of limitations. OSHA must serve notice of any citation and proposed penalty to an agent or officer of the employer by certified mail, if there has not been personal service.

After the citation and notice of proposed penalty is issued, but before any notice of contest by the employer is filed, the employer may request an informal conference with the OSHA area director. The general purpose of this conference is to clarify the basis for the citation, modify abatement dates or proposed penalties, seek withdrawal of a cited item, or otherwise attempt to settle the case. This conference, as its name implies, is an informal meeting between the employer and OSHA. Employee representatives must have an opportunity to participate if they so request. Safety and health professionals should note that the request for an informal conference does not stay (delay) the 15-working-day period to file a notice of contest to challenge the citation.

Under the OSH Act, an employer, employee, or authorized employee representative (including a labor organization) is given 15 working days, from the date that the citation is issued, to file a notice of contest. If a notice of contest is not filed within 15 working days, the citation and proposed penalty become a final order of the Occupational Safety and Health Review Commission (OSHRC) and are not subject to review by any court or agency. If a timely notice of contest is filed in good faith, the abatement requirement is tolled (temporarily suspended or delayed) and a hearing is scheduled. The employer also has the right to file a petition for modification of the abatement period (PMA) if he or she is unable to comply with the abatement period provided in the citation. If OSHA contests the PMA, a hearing is scheduled to determine whether the abatement requirements should be modified.

When the employer files a notice of contest, the secretary must immediately forward the notice to the OSHRC in order to schedule a hearing before its administrative law judge (ALJ). The secretary of labor is labeled the complainant, and the employer is the respondent. The ALJ may affirm, modify, or vacate the citation, any penalties, or the abatement date. Either party can appeal the ALJ's decision by filing a petition for discretionary review (PDR). Additionally, any member of the OSHRC may direct review any decision by an ALJ, in whole or in part, without a PDR. If a PDR is not filed, and no member of the OSHRC directs a review, the decision of the ALJ becomes final in 30 days. Any party may appeal a final order of the OSHRC by filing a petition for review in the U.S. Court of Appeals for the circuit in which the violation is alleged to have occurred, or in the U.S. Court of Appeals for the District of Columbia Circuit. This petition for review must be filed within 60 days from the date of the OSHRC's final order.

OSHA Inspection Checklist

The following recommended checklist can be used by the safety and loss prevention professional or any employer representative to prepare for an OSHA inspection:

1. Assemble a team from the management group and identify specific responsibilities, in writing, for each team member. The team members should be given appropriate training and education. This team should include, but not be limited to, the following individuals.
   a. an OSHA inspection team coordinator
   b. a document control individual
   c. individuals to accompany the OSHA inspector
   d. a media coordinator
   e. an accident investigation team leader (where applicable)
   f. a notification person
   g. a legal advisor (where applicable)
   h. a law enforcement coordinator (where applicable)
   i. a photographer
   j. an industrial hygienist
2. Define and develop a company policy and procedures to provide guidance to the OSHA inspection team.
3. Prepare an OSHA inspection kit, including all equipment necessary to properly document all phases of the inspection. The kit should include equipment such as a digital camera (or camera with extra film and batteries), a tape recorder (with extra batteries), a video camera, pads of paper, pens, and other appropriate testing and sampling equipment (i.e., a noise-level meter, an air-sampling kit, and so on).
4. Prepare basic forms to be used by the inspection team members during and after the inspection.
5. When notified that an OSHA inspector has arrived, assemble the team members with the inspection kit.
6. Identify the inspector. Check his or her credentials and determine the reason for the inspection and the type of inspection to be conducted.

7. Confirm the reason for the inspection with the inspector (targeted, routine inspection, accident, or in response to a complaint).
   a. Ask the following questions for a random or target inspection.
      i. Did the inspector check the OSHA 300 Form?
      ii. Was a warrant required?
   b. Ask the following questions for an employee complaint inspection.
      i. Does the inspector have a copy of the complaint? If so, obtain a copy.
      ii. Do allegations in the complaint describe an OSHA violation?
      iii. Was a warrant required?
      iv. Was the inspection protested in writing?
   c. Ask the following questions for an accident investigation inspection.
      i. How was OSHA notified of the accident?
      ii. Was a warrant required?
      iii. Was the inspection limited to the accident location?
   d. Ask the following questions if a warrant is presented.
      i. Were the terms of the warrant reviewed by local counsel?
      ii. Did the inspector follow the terms of the warrant?
      iii. Was a copy of the warrant acquired?
      iv. Was the inspection protested in writing?
8. Record information about the opening conference.
   a. Who was present?
   b. What was said?
   c. Was the conference taped or otherwise documented?
9. Record information about the records involved.
   a. What records were requested by the inspector?
   b. Did the document control coordinator number the photocopies of the documents provided to the inspector?
   c. Did the document control coordinator maintain a list of all photocopies provided to the inspector?
10. Record information about the facility inspection.
    a. What areas of the facility were inspected?
    b. What equipment was inspected?
    c. Which employees were interviewed?
    d. Who was the employee or union representative present during the inspection?
    e. Were all the remarks made by the inspector documented?
    f. Did the inspector take photographs?
    g. Did a team member take similar photographs?

There is no replacement for a well-managed safety and loss prevention program. Employers realize that they cannot get by on a shoestring safety program because every aspect of the safety program is important, including preparing for an OSHA inspection. The preceding checklist was provided as an example of what can be done prior to an OSHA inspection. Please remember, OSHA inspectors are human too; if you treat them with respect and courtesy, they will generally be fair and even helpful. This is not to say that you will not be cited for violations, but you may avoid an inspector who is overzealous.

TYPES OF VIOLATIONS

The OSHA monetary penalty structure is classified according to the type and gravity of the particular violation. Violations of OSHA standards or the general duty clause are categorized as de minimis, other (nonserious), serious, repeat, and willful. Monetary penalties assessed by the secretary vary according to the degree of the violation. Penalties range from no monetary penalty to 10 times the imposed penalty for repeat or willful violations. Additionally, the secretary may refer willful violations to the U.S. Department of Justice for imposition of criminal sanctions.

DE MINIMIS VIOLATIONS

When a violation of an OSHA standard does not immediately or directly relate to safety or health, OSHA either does not issue a citation or issues a de minimis citation. Section 9 of the OSH Act provides that "[the] Secretary may prescribe procedures for the issuance of a notice in lieu of a citation with respect to de minimis violations which have no direct or immediate relationship to safety or health."

A de minimis notice does not constitute a citation and no fine is imposed. Additionally, there usually is no abatement period; therefore, there can be no violation for failure to abate. The *OSHA Compliance Field Operations Manual* (*OSHA Manual*) provides two examples of when de minimis notices are generally appropriate:

1. "In situations involving standards containing physical specificity wherein a slight deviation would not have an immediate or direct relationship to safety or health."
2. "Where the height of letters on an exit sign is not in strict conformity with the size requirements of the standard."

OSHA has found de minimis violations in cases where employees, as well as the safety records, are persuasive in exemplifying that no injuries or lost time have been incurred. Additionally, in order for OSHA to conserve valuable resources to produce a greater impact on safety and health in the workplace, it is highly likely that the secretary will encourage the use of the de minimis notice in marginal cases as well as other situations where the possibility of injury is remote, and potential injuries would be minor.

OTHER OR NONSERIOUS VIOLATIONS

Other or nonserious violations are issued if a violation could lead to an accident or occupational illness, but the probability that it would cause death or serious physical harm is minimal. Such a violation, however, does possess a direct relationship to the safety and health of workers. Potential penalties for this type of violation range from no fine up to $7,000 per violation.

In distinguishing between a serious and a nonserious violation, the OSHRC has stated that "a non-serious violation is one in which there is a direct and immediate relationship between the violative condition and occupational safety and health but no such relationship that a resultant injury or illness is death or serious physical harm."

The *OSHA Manual* provides guidance and examples for issuing nonserious violations. It states that:

> An example of non-serious violation is the lack of guardrail at a height from which a fall would more probably result in only a mild sprain or cut or abrasion, i.e., something less than serious harm.

A citation for serious violation may be issued or a group of individual violations (which) taken by themselves would be nonserious, but together would be serious in the sense that in combination they present a substantial probability of injury resulting in death or serious physical harm to employees.

A number of nonserious violations (which) are present in the same piece of equipment which, considered in relation to each other, affect the overall gravity of possible injury resulting from an accident involving the combined violations ... may be grouped in a manner similar to that indicated in the preceding paragraph, although the resulting citation will be for a non-serious violation.

The difference between a serious and a nonserious violation hinges on subjectively determining the probability of injury or illness that might result from the violation. Administrative decisions have usually turned on the particular facts of the situation. The OSHRC has reduced serious citations to nonserious violations when the employer was able to show that the probability of an accident, and the probability of a serious injury or death, was minimal.

SERIOUS VIOLATIONS

Section 17(k) of the OSH Act defines a serious violation as one where:

> There is a substantial probability that death or serious physical harm could result from a condition which exists, or from one or more practices, means, methods, operations or processes which have been adopted or are in use, in such place of employment unless the employer did not, and could not with exercise of reasonable diligence, know of the presence of the violation.

To prove that a violation is within the serious category, OSHA must only show a substantial probability that a foreseeable accident would result in serious physical harm or death. Thus, contrary to common belief, OSHA does not need to show that a violation would create a high probability that an accident would result. Because substantial physical harm is the distinguishing factor between a serious and a nonserious violation, OSHA has defined serious physical harm as "permanent, prolonged, or temporary impairment of the body in which part of the body is made functionally useless or is substantially reduced in efficiency on or off the job." Additionally, an occupational illness is defined as "illness that could shorten life or significantly reduce physical or mental efficiency by inhibiting the normal function of a part of the body."

After determining that a hazardous condition exists and employees are exposed or potentially exposed to the hazard, the *OSHA Manual* instructs compliance officers to use a four-step approach to determine whether the violation is serious:

1. Determine the type of accident or health hazard exposure that the violated standard is designed to prevent in relation to the hazardous condition identified.
2. Determine the type of injury or illness that is reasonably predictable as a result of the type of accident or health hazard exposure identified in step 1.
3. Determine that the type of injury or illness identified in step 2 includes death or a form of serious physical harm.
4. Determine that the employer knew, or with the exercise of reasonable diligence could have known, of the presence of the hazardous condition.

The *OSHA Manual* provides examples of serious injuries, including amputations, fractures, deep cuts involving extensive suturing, disabling burns, and concussions. Examples of serious illnesses include cancer, silicosis, asbestosis, poisoning, and hearing and visual impairment.

Safety and loss prevention professionals should be aware that OSHA is not required to show that the employer actually knew that the cited condition violated safety or health standards. The employer can, however, be charged with constructive knowledge of the OSHA standards. OSHA also does not have to show that the employer could reasonably foresee that an accident would happen, although it does have the burden of proving that the possibility of an accident was not totally unforeseeable. OSHA does need to prove, however, that the employer knew, or should have known, of the hazardous condition and that it knew there was a substantial likelihood that serious harm or death would result from an accident. If the secretary cannot prove that the cited violation meets the criteria for a serious violation, the violation may be cited in one of the lesser categories.

WILLFUL VIOLATIONS

The most severe monetary penalties under the OSHA penalty structure are for willful violations. A willful violation can result in penalties of up to $129,336 per violation. Although the term "willful" is not defined in OSHA regulations, courts generally have defined a willful violation as "an act voluntarily with either an intentional disregard of, or plain indifference to, the Act's requirements." Furthermore, the OSHRC defines a willful violation as "action taken knowledgeably by one subject to the statutory provisions of the OSH Act in disregard of the action's legality. No showing of malicious intent is necessary. A conscious, intentional, deliberate, voluntary decision is properly described as willful."

The major distinctions between civil and criminal willful violations are the due process requirements for a criminal violation and the fact that a violation of the general duty clause cannot be used as the basis for a criminal willful violation. The distinction is usually based on the circumstances and the fact that a criminal willful violation results from a willful violation that caused an employee's death.

According to the *OSHA Manual*, the compliance officer "can assume that an employer has knowledge of any OSHA violation condition of which its supervisor has knowledge; he can also presume that, if the compliance officer was able to discover a violative condition, the employer could have discovered the same condition through the exercise of reasonable diligence."

Courts and the OSHRC have agreed on three basic elements of proof that OSHA must show for a willful violation. OSHA must show that the employer (1) knew or should have known that a violation existed, (2) voluntarily chose not to comply with the OSH Act to remove the violative condition, and (3) made the choice not to comply with intentional disregard of the OSH Act's requirements or plain indifference to them, properly characterized as reckless.

Although these elements of proof appear fairly straightforward and clear, several unresolved issues continue to be litigated, such as the supervisor's role in identifying and correcting the hazardous condition, what the employer actually knew regarding the hazardous condition, and the good faith of the employer.

Regarding the role of a first-line supervisor or other member of the management team, an employer may not be responsible for its supervisor's actions if they are contrary to consistently and adequately enforced work regulations or rules. Conversely, many courts have upheld willful violations on the basis of the supervisor's knowledge of the hazardous condition and his or her subsequent inaction. Additionally, hazards within the plain view of the supervisor have been found to be within the "knew or should have known" category and are potentially willful violations.

Inaction can constitute a willful violation as well as an overt disregard for OSHA standards. In *Georgia Electric Co. v. Marshall*, the Fifth Circuit held that "it is precisely because the Company made no effort whatsoever to make anyone with supervisory authority at the job site aware of the OSHA regulations that the Company can be said to have acted with plain indifference and thereby acted willfully." Additionally, in *Donovan v. Williams Enterprises*, the court upheld a willful violation in finding that "employee safety was never discussed with the company president or any of its supervisory personnel until OSHA inspection of the project began."

Imputed knowledge to the employer has been the basis for willful violation findings by courts and the OSHRC. In *Bergin Corp.*, the OSHRC found a willful violation because of the employer's poor judgment in hiring a supervisor. The employer hired a person experienced in excavation work and instructed him to provide safety instruction to his employees. When a trench caved in on an employee who was being trained, the OSHRC found that the employer's reliance on the supervisor and his instruction to provide safety training were not adequate to remove the willful violation. Courts and the OSHRC have affirmed findings of willful violations in many other circumstances, ranging from deliberate disregard of known safety requirements to fall protection equipment not being provided. Other examples of willful violations include cases where safety equipment was ordered, but employees were permitted to continue work until the equipment arrived; where inexperienced and untrained employees were permitted to perform a hazardous job; and where an employer failed to correct a situation that had been previously cited as a violation.

REPEAT AND FAILURE TO ABATE VIOLATIONS

Repeat and failure to abate violations are often quite similar and can be confusing to safety and health professionals. When a violation of a previously cited standard is found on re-inspection by OSHA, but the violation does not involve the same machinery, equipment, process, or location, this would constitute a repeat violation. If a violation of a previously cited standard is found on re-inspection by OSHA, but evidence indicates that the violation continued uncorrected since the original inspection, this would constitute a failure to abate violation.

The most costly civil penalty under the OSH Act is assessed for repeat violations. The OSH Act authorizes a penalty of up to $12,934 per day per violation, but permits a maximum penalty of 10 times the maximum authorized for the first instance of the violation. Repeat violations can also be grouped within the willful category (i.e., a willful repeat violation) to acquire maximum civil penalties.

In certain cases where an employer has more than one fixed establishment and citations have been issued, the *OSHA Manual* states:

> The purpose for considering whether a violation is repeated, citations issued to employers having fixed establishments (e.g., factories, terminals, stores), will be limited to the cited establishment... . For employers engaged in businesses having no fixed establishments, repeated violations will be alleged based upon prior violations occurring anywhere within the same Area Office Jurisdiction.

When a previous citation has been contested, but a final OSHRC order has not yet been received, a second violation is usually cited as a repeat violation. The *OSHA Manual* instructs the compliance officer to notify the assistant regional director and to indicate on the citation that the violation is contested. If the first citation never becomes a final OSHRC order (i.e., the citation is vacated or otherwise dismissed), the second citation for the repeat violation will be removed automatically.

As noted previously, a failure to abate violation occurs when, on re-inspection, the compliance officer finds that the employer has failed to take necessary corrective action and thus the violation has continued uncorrected. The penalty for a failure to abate violation can be up to $7,000 per day to a maximum of $70,000. Safety and loss prevention professionals should also be aware that citations for repeat violations, failure to abate violations, or willful repeat violations can be issued as violations of the general duty clause. The *OSHA Manual* instructs compliance officers that citations under the general duty clause are restricted to serious violations or to willful or repeat violations that are of a serious nature.

FAILURE TO POST VIOLATION NOTICES

The failure to post violation notices carries a penalty of up to $7,000 for each violation. A failure to post violation occurs when an employer fails to post notices required by the OSHA standards, including the OSHA poster, a copy of the year-end summary of the OSHA 300 form, a copy of OSHA citations when received, and copies of other pleadings and notices. OSHA has recently initiated a program whereby the compliance officer will provide a copy of the required poster to the employer, and if the employer immediately posts this required notice, no citation will be issued.

CRIMINAL PENALTIES

In addition to civil penalties, the OSH Act provides criminal penalties of up to $10,000 and/or imprisonment for up to six months. A repeated willful violation causing an employee's death can double the criminal sanction to a maximum of $20,000, one year of imprisonment, or both. Given the increased use of criminal sanctions by OSHA, safety and loss prevention professionals should advise their employers of the possibility that these sanctions will be used when the safety and health of employees is disregarded or placed on the back burner.

OSHA, as an agency, does not prosecute employers for criminal violations of the OSH Act. In fact, the secretary of labor does not even have authority under the OSH Act to impose criminal sanctions. Instead, the secretary of labor refers all cases meeting criminal sanction criteria to the U.S. Department of Justice for prosecution. The Justice Department prosecutes an OSH Act case as it would any other criminal action. The same rules of evidence and rights of the accused apply, but the rules require a substantially different burden of proof than an OSHRC hearing. The statute of limitations for the Justice Department to file criminal charges against an employer for OSHA violations is the same as any other noncapital federal crime—five years.

Criminal sanctions cannot be brought by OSHA for violations of the general duty clause. For OSHA to refer a case to the Justice Department for possible criminal prosecution, the employer must be alleged to have willfully violated a specific OSHA standard and the willful act must fall within the criminal categorization as well as be defined as serious in nature.

OSHA can refer violations to the Justice Department for criminal prosecution under the following circumstances:

1. Someone inside or outside the Department of Labor or OSHA gives advance notice of an inspection without authority from the secretary. A fine of up to $1,000, imprisonment for up to six months, or both may be imposed on conviction.
2. An employer or other person intentionally falsifies statements or OSHA records that must be prepared, maintained, or submitted under the OSH Act. A fine of up to $10,000, imprisonment for up to six months, or both may be imposed on conviction.
3. Any person willfully violates an OSHA standard, rule, order, or regulation and the violation causes the death of an employee. A fine of up to $10,000, imprisonment for up to six months, or both may be imposed on conviction. If convicted for a second violation, a fine of up to $20,000, imprisonment for up to one year, or both may be imposed.
4. An individual forcibly resists or assaults a compliance officer or other Department of Labor personnel. On conviction, a penalty of $5,000 and/or three years in prison can be imposed.

5. Any person kills a compliance officer or other OSHA or Department of Labor employee acting in his or her official capacity. A prison sentence for any term of years, or life, may be imposed on conviction.

Criminal liability for a willful OSHA violation can attach to an individual or a corporation. In addition, corporations may be held criminally liable for the actions of their agents or officials. Safety and loss prevention professionals, and other corporate officials, may also be subject to criminal liability under a theory of aiding and abetting the criminal violation in their official capacity with the corporation.

Safety and loss prevention professionals should also be aware that an employer could face two prosecutions for the same OSHA violation without the protection of double jeopardy. The OSHRC can bring an action for a civil willful violation using the monetary penalty structure described previously; the case may then be referred to the Justice Department for criminal prosecution of the same violation.

Prosecution of willful criminal violations by the Justice Department has been rare in comparison to the number of inspections performed and violations cited by OSHA on a yearly basis. However, the use of criminal sanctions has increased substantially in the last few years. With workplace accidents and deaths generating adverse publicity and Congress emphasizing reform, a decrease in criminal prosecutions is unlikely.

The law regarding criminal prosecution of willful OSH Act violations is still emerging. Although few cases have actually gone to trial, in most situations the mere threat of criminal prosecution has encouraged employers to settle cases with the assurance that criminal prosecution would be dismissed. Many state plan states are using criminal sanctions permitted under their state OSH regulations more frequently. State prosecutors have also allowed the use of state criminal codes for workplace deaths.

Safety and loss prevention professionals should exercise caution when faced with an on-the-job fatality. The potential for criminal sanctions and criminal prosecution is substantial if a willful violation of a specific OSHA standard is directly involved in the death. The OSHA investigation may be conducted from a criminal perspective in order to gather and secure the appropriate evidence to later pursue criminal sanctions. A prudent safety and loss prevention professional facing a workplace fatality investigation should address the OSHA investigation with legal counsel present and reserve all rights guaranteed under the U.S. Constitution. Obviously, under no circumstances should a health and safety professional condone or attempt to conceal facts or evidence in an attempt to cover up violations.

SELECTED CASE STUDY

OSHA Enforcement
From OSHRC website (OSHRC.gov) and modified for the purposes of this text.
United States Court of Appeals, Sixth Circuit.

Elaine L. CHAO, Secretary of Labor, Petitioner, v.
OCCUPATIONAL SAFETY AND HEALTH REVIEW
COMMISSION, Manganas Painting Co., Inc., Respondents

No. 07-3810.Argued: July 22, 2008.
Decided and Filed: Aug. 29, 2008.

Background
Secretary of Labor petitioned for review of final order of the Occupational Safety
and Health Review Commission, 2007 WL 2285345, which reversed in part and
affirmed in part a decision of Administrative Law Judge, who had affirmed majority
of citations issued to painting contractor for unguarded scaffolding violations.

Holdings
The Court of Appeals, Circuit Judge, held that:

1. statute providing for penalties for failure to correct "a violation for which a
   citation has been issued" was ambiguous as to whether citations issued at
   different worksite covered same condition;
2. deference was owed to statutory construction of Secretary of Labor, not that
   of Occupational Safety and Health Review Commission;
3. Skidmore deference was owed; and
4. Secretary of Labor was not barred from issuing citations for violation of
   the same unguarded scaffolding regulation if new violations occurred at a
   different work site.

Reversed and remanded.
Before: Circuit Judges; District Judge.
The Honorable, United States District Judge for the Southern District of Ohio, sitting
by designation.

Opinion

Circuit Judge.
The Secretary of Labor petitions this court for review of a final order of the
Occupational Safety and Health Review Commission. The Commission affirmed in
part and reversed in part a decision by an Administrative Law Judge, who affirmed
the majority of citations issued to respondent Manganas Painting Co., Inc. by the
Occupational Safety and Health Administration following a 1994 inspection of a
worksite on the southbound structure of the Jeremiah Morrow Bridge. Although the
Commission's order adjudicated numerous citations issued to Manganas Painting,
the Secretary's petition appeals only three citations for unguarded scaffolds that
were vacated by the Commission.

In a 2–1 decision, the Commission held that these citations were barred by § 10(b)
of the Occupational Safety and Health Act of 1970 ("the Act"), 29 U.S.C. § 659(b),
because a 1993 citation for the same unguarded scaffold condition, arising out of an

inspection of the northbound structure of the Morrow Bridge, was pending before the Commission at the time these 1994 citations were issued for the southbound bridge. Commissioner Rogers dissented on the basis that because the citations issued in 1993 and 1994 arose at separate worksites and at different times, § 10(b) did not bar the 1994 unguarded scaffold citations. We agree with the rationale advocated by the dissent and therefore grant the petition for review, reverse the Commission, and remand for further proceedings regarding the merits of the citations at issue.

I.

Manganas Painting began work removing lead-based paint on the Jeremiah Morrow Bridge in Lebanon, Ohio, in 1993, after it entered into a contract with the Ohio *522 Department of Transportation. The Morrow Bridge consists of two parallel bridges: one structure running northbound; the other, southbound. In April 1993, OSHA performed an inspection of the project while Manganas Painting was working on the northbound bridge. Following the inspection, OSHA issued several citations to Manganas Painting, including, inter alia, a citation alleging that Manganas Painting had failed to install guardrails on platforms that were located more than 10 feet above the ground level, in violation of 29 C.F.R. § 1926.451(a)(4) (repealed). Manganas Painting timely appealed the citation, and it was ultimately affirmed by the Commission in 2000. *Sec'y of Labor v. Manganas Painting Co.*, 19 O.S.H. Cas. (BNA) 1102 (2000), aff'd by *Manganas Painting Co. v. Sec'y of Labor*, 273 F.3d 1131 (D.C.Cir.2001).

In December 1994, while Manganas Painting was working on the southbound bridge, OSHA performed another inspection. At the conclusion of this inspection, OSHA issued several new citations, including, inter alia, three alleged instances of unguarded scaffolds, in violation of 29 C.F.R. § 1926.451(a)(4). These citations alleged the following violations:

> Item 13a. Located under and along the east side of the south bound bridge deck, approximate panel point between U38-L38, an employee was observed working from a pic scaffold spray painting a column and the upper cord or steel area without standard guardrails or equivalent, exposing the employee to perimeter exterior falls in excess of 100′ and interior falls of approximately 30′.
> Item 13b. Employees were exposed to a fall in excess of 140′ while using the scaffold pic adjacent to the ladder suspended over the side of the bridge outside the containment area south of pier 4 in that there were no guard rails on the pic.
> Item 13c. Located under and along the east side of the south-bound bridge deck approximate panel point U34, employees were working from a pic scaffold without standard guardrails or equivalent exposing employees to perimeter exterior falls in excess of 100′ and interior falls in excess of 30′.

Manganas Painting timely appealed, resulting in a decision by an administrative law judge vacating the citations on the basis that these violations were duplicative of other citations issued during the 1994 inspection of the southbound bridge.

On review, the Commission affirmed the ALJ's decision, but on different grounds. The Commission majority held that section 10(b) of the Act barred the Secretary from citing Manganas Painting for failing to guard pic scaffolds at the bridge worksite because a 1993 citation for the same condition relating to the northbound bridge was pending before the Commission at the time these alleged violations were cited in December 1994. The Commission reasoned:

> As a result of the April 1993 inspection of the bridge worksite, OSHA cited Manganas for a violation of 523 § 1926.451(a)(4), the same scaffolding standard cited here. The 1993 citation was based on Manganas' failure "to install guardrails on a painter's pick." *Manganas Painting Co.*, 19 BNA OSIIC at 1103, 2000 CCH OSHD at p. 48,767. It is undisputed that at the time OSHA initiated the 1994 inspection and issued the resulting citations, the 1993 citation had been timely contested by Manganas and a hearing in the matter had yet to commence. In fact, the judge who presided over the 1993 matter did not issue his decision until after a decision was issued in the current cases, and his decision did not become a final order of the Commission until 2000.

While the alleged scaffolding violations cited in 1993 and 1994 were observed at what we find to be essentially two different worksites, the citations "covered the same condition" in that each item was based on Manganas' failure to guard the same type of pic scaffold used throughout the bridge worksite during both painting seasons.

The Secretary timely filed a petition for review with this court, limited to the Commission's decision regarding these citations. Neither Manganas Painting nor the Commission has filed a responsive brief in opposition.

II.

In reviewing an agency's interpretation of a statute that it is charged with administering, we apply the familiar two-step process announced by the Supreme Court in *Chevron U.S.A., Inc. v. NRDC*, 467 U.S. 837, 104 S.Ct. 2778, 81 L.Ed.2d 694 (1984). "The initial question under step one of the Chevron framework is 'whether Congress has directly spoken to the precise question at issue' by employing precise, unambiguous statutory language." *Alliance for Community Media v. F.C.C.*, 529 F.3d 763, 776-77 (6th Cir.2008) (citing *Chevron*, 467 U.S. at 842, 104 S.Ct. 2778). If the text of the statute is unambiguous and, therefore, the "intent of Congress is clear, that is the end of the matter; for the court, as well as the agency, must give effect to the unambiguously expressed intent of Congress." *Jewish Hosp., Inc. v. Sec'y of Health & Human Servs.*, 19 F.3d 270, 273 (6th Cir.1994) (citing *Chevron*, 467 U.S. at 842-43, 104 S.Ct. 2778). If, however, "we determine that Congress has not directly addressed the precise question at issue, that is, that the statute is silent or ambiguous on the specific issue, we must determine 'whether the agency's answer is based on a permissible construction of the statute.'" *Battle Creek Health Sys. v. Leavitt*, 498 F.3d 401, 408-09 (6th Cir.2007) (quoting *Jewish Hosp., Inc.*, 19 F.3d at 273). Where the agency's interpretation of the statute does not take the form of a regulation issued following notice and comment rulemaking, but rather is offered through an informal medium—such as an opinion letter, policy statement, or agency manual—Chevron-style deference is not warranted, and we apply the less deferential Skidmore review to the agency's interpretation. Id. at 409. See also infra at III.C.

[1] The threshold issue, then, is whether § 10(b) is ambiguous. Section 10(b) of the Act provides in pertinent part:

> If the Secretary has reason to believe that an employer has failed to correct a violation for which a citation has been issued within the period permitted for its correction (which period shall not begin to run until the entry of a final order by the Commission in the case of *524 any review proceedings under this section initiated by the employer in good faith and not solely for delay or avoidance of penalties), the Secretary shall notify the employer by certified mail of such failure and of the penalty proposed to be assessed under section 666 of this title by reason of such failure, and that the employer has fifteen working days within which to notify the Secretary that he wishes to contest the Secretary's notification or the proposed assessment of penalty.
> 29 U.S.C. § 659(b)

We conclude that the statute is ambiguous regarding the meaning of "a violation for which a citation has been issued." § 659(b). We have explained previously that "[l]anguage is ambiguous when 'to give th[e] phrase meaning requires a specific factual scenario that can give rise to two or more different meanings of the phrase.'" *Alliance*, 529 F.3d at 777 (quoting *Beck v. City of Cleveland*, 390 F.3d 912, 920 (6th Cir.2004)). Here, there are at least two possible constructions of the pertinent phrase. It could refer, as the Secretary insists, to each instance in which an OSHA regulation is breached. The statute's use of the term "violation" could also be read plausibly, however, to apply only to the regulation that was allegedly transgressed, rather than to each individual act. For example, under the former construction of the statute, three separate, individual violations of 29 C.F.R. § 1926.451(a)(4) occurring at different locations on different days would result in three "violations for which a citation has been issued." Under the latter reading, however, one "violation for which a citation has been issued" would result, as only a single regulation would have been alleged to have been breached.

Because § 10(b) permits at least two possible interpretations, we conclude that the statute is ambiguous regarding the meaning of a "violation for which a citation has been issued." Accordingly, under our two-step analysis required by *Chevron*, we turn to the agency's interpretation of the statute for guidance.

III.

A.

Finding "that personal injuries and illnesses arising out of work situations impose a substantial burden upon ... interstate commerce in terms of lost production, wage loss, medical expenses, and disability compensation payments," Congress enacted the Occupational Safety and Health Act in order to "provide for the general welfare [by] assur[ing] so far as possible every working man and woman in the Nation safe and healthful working conditions and to preserve our human resources," 29 U.S.C. § 651(a)-(b). To carry out the objectives of the Act, Congress allotted responsibility for executing the Act to two administrative actors: the Secretary and the Commission, a three-member board appointed by the President with the advice and consent of the Senate, assigned to "carry out adjudicatory functions" under the Act. 29 U.S.C. §§ 651(b)(3). See *Martin v. Occupational Safety and Health Review Comm'n*, 499 U.S. 144, 147-48, 111 S.Ct. 1171, 113 L.Ed.2d 117 (1991) (describing administrative framework); *Chao v. Russell P. Le Frois Builder, Inc.*, 291 F.3d 219, 221-22 (2d Cir.2002) (same).

The Secretary is authorized "to set mandatory occupational safety and health standards applicable to businesses affecting interstate commerce," § 651(b)(3), and to enforce these standards by "inspect[ing] and investigat[ing] during regular working hours and at other reasonable times ... *525 any such place of employment and ... question[ing] privately any such employer, owner, operator, agent or employee." 29 U.S.C. § 657(a)(2). If, during such an inspection, the Secretary (or his or her representative) believes that an employer has violated an OSHA regulation, the Secretary is empowered to issue a citation to the employer, describing "with particularity the nature of the violation, including a reference to the provision of the [Act] ... alleged to have been violated." Id. at § 658(a). The employer then has fifteen days, running from the date of receipt of the citation, to appeal the citation. Id. at § 659(a).

If the employer files a timely notice of its intent to challenge the citation, "the Commission shall afford an opportunity for a hearing" and "shall thereafter issue an order, based on findings of fact, affirming, modifying, or vacating the Secretary's citation or proposed penalty ...," id. at § 659(c). The initial report by the ALJ becomes the final order of the Agency unless any member of the Commission grants discretionary review of the ALJ's decision. Id. at § 661(j). Either party—the Secretary or the employer charged with citations—may appeal the order of the Commission to "any United States court of appeals for the circuit in which the violation is alleged to have occurred" or "in the Court of Appeals for the District of Columbia Circuit," id. at § 660(a).

B.

[2] [3] It is well-settled that appellate courts owe deference to an agency's interpretation of its authorizing statute when the statute is ambiguous. *Ramirez-Canales v. Mukasey*, 517 F.3d 904, 907 (6th Cir.2008) (citing *Chevron*, 467 U.S. at 837, 104 S.Ct. 2778). OSHA's dual-actor framework raises the question of which OSHA entity—the Secretary or the Commission—is entitled to deference, and, similarly, how much deference that entity is entitled. In *Martin*, the Supreme Court considered whether a reviewing court should defer to the Secretary or the Commission when the two actors provided "reasonable but conflicting interpretations" of an ambiguous regulation under the Act. *Martin*, 499 U.S. at 149, 111 S.Ct. 1171. Reasoning that because the Act assigns the Secretary the duty to promulgate and enforce OSHA regulations, while the Commission is delegated "nonpolicy-making adjudicatory powers," the Court concluded

that the Secretary was in the "better position than is the Commission to reconstruct the purpose" of OSHA regulations. *Id.* at 152-54, 111 S.Ct. 1171. Ultimately, the Court held that where both the Commission and the Secretary offer competing reasonable interpretations of an OSHA regulation, "the reviewing court may not prefer the reasonable interpretations of the Commission to the reasonable interpretations of the Secretary... ." *Id.* at 158, 111 S.Ct. 1171.

[4] Left undecided by *Martin*, however, is to whom does a reviewing court defer when the Secretary and Commission offer conflicting interpretations of a provision of the Act. The Second Circuit Court of Appeals' opinion in *Russell P. Le Frois Builder* ("*Russell*") is instructive on this question. In *Russell*, the Second Circuit faced competing interpretations of the Act concerning the Commission's ability to exercise jurisdiction over an employer's untimely challenge to OSHA citations. *Russell*, 291 F.3d at 221. In deciding which agency actor's interpretation warranted deference, the Second Circuit observed the Supreme Court's holding in *United States v. Mead Corporation*, 533 U.S. 218, 226-27, 121 S.Ct. 2164, 150 L.Ed.2d 292 (2001), that "administrative implementation of a particular statutory provision qualifies for *526 Chevron deference when it appears that Congress delegated authority to the agency generally to make rules carrying the force of law, and that the agency interpretation claiming deference was promulgated in the exercise of that authority." *Russell*, 291 F.3d at 226. Reasoning that, as explained by the Court in *Martin*, Congress delegated rule-making authority under the Act to the Secretary, rather than the Commission, the Russell court held that "the Secretary not the Commission[ ] has authority to interpret the statute ... and we should therefore defer to the views of the Secretary rather than the Commission." *Russell*, 291 F.3d at 226-27.

[5] We agree with the Second Circuit that the Secretary, rather than the Commission, warrants deference in her interpretation of the Act. In *Martin*, the Court pointed out that Congress did not invest the Commission with the power to make law or policy, 499 U.S. at 154, 111 S.Ct. 1171, and repeatedly emphasized the Commission's role as limited to serving as a "neutral arbiter." Id. at 152, 111 S.Ct. 1171 (quoting *Cuyahoga Valley Ry. Co. v. United Transp. Union*, 474 U.S. 3, 7, 106 S.Ct. 286, 88 L.Ed.2d 2 (1985)). In contrast, the Court observed that the Secretary is empowered with the ability to promulgate OSHA rules and standards. *Martin*, 499 U.S. at 152, 111 S.Ct. 1171. Moreover, as the Martin Court noted, because she is empowered to write and enforce OSHA standards, the Secretary "comes into contact with a much greater number of regulatory problems than does the Commission" and, as a result, "the Secretary is more likely to develop the expertise relevant to assessing the effect of a particular regulatory interpretation." Id. at 152-53, 111 S.Ct. 1171. Thus, applying the Court's opinion in *Mead* and *Martin*, we choose to follow *Russell* and defer to the Secretary, rather than the Commission, in her interpretation of the Act.

C.

[6] Next, we must decide the degree of deference owed the Secretary's interpretation of § 10(b)—specifically, whether we should give full Chevron deference or rather the more limited Skidmore deference. In contrast to the deferential review required by *Chevron*, under *Skidmore*, we give an agency interpretation deference that "depend[s] upon the thoroughness evident in its consideration, the validity of its reasoning, its consistency with earlier and later pronouncements, and all those factors which give it power to persuade, if lacking power to control." 323 U.S. at 140, 65 S.Ct. 161. Because the Secretary's interpretation of § 10(b) is not the product of notice-and-comment rulemaking, we conclude*527 that the less-deferential Skidmore level of review is warranted.

[7] [8] In *Christensen v. Harris County*, 529 U.S. 576, 587, 120 S.Ct. 1655, 146 L.Ed.2d 621 (2000), the Supreme Court explained that, although *Chevron* requires a reviewing court to give effect to an agency's reasonable interpretation (offered in the form of an agency regulation) of an ambiguous statute, such deference is not required where the interpretation is offered via an informal medium—such as an opinion letter, agency manual, policy statement, or enforcement guideline—that lacks the force of law. Here, the Secretary has offered her interpretation of § 10(b) only in her litigation position; she has not pointed to any regulation or any other format that carries the force of law which reflects her interpretation of § 10(b). Accordingly, under *Christensen*, we hold that the Secretary's interpretation of § 10(b) is entitled to deference only to the extent that it has the power to persuade.

IV.

Although the Secretary is entitled to only Skidmore deference regarding her position, we conclude that her interpretation of § 10(b) is a compelling, reasonable construction of the statute. As the administrative actor charged with enforcing the Act, the Secretary is "'in the best position' to develop 'historical familiarity and policy-making expertise'" in applying § 10(b). *Russell*, 291 F.3d at 228 (quoting *Martin*, 499 U.S. at 153, 111 S.Ct. 1171). On this issue, the Secretary has repeatedly advanced the interpretation that § 10(b) does not prohibit her from alleging multiple violations of the same regulation where the violations occurred at different places or different times. See, e.g., *Sec'y of Labor v. MJP Constr. Co., Inc.*, 19 O.S.H. Cas. (BNA) 1638 (2001) (upholding multiple violations under same regulation where citation items alleged violations occurring on different dates); *Sec'y of Labor v. J.A. Jones Constr. Co.*, 15 O.S.H. Cas. (BNA) 2201 (1993) (holding that the Secretary may appropriately cite separate violations for each individual instance of improper fall protection where each alleged instance of violation involves either a different floor or a different location on each floor). See also *Good Samaritan Hosp. v. Shalala*, 508 U.S. 402, 417, 113 S.Ct. 2151, 124 L.Ed.2d 368 (1993) ("The consistency of an agency's position is a factor in assessing the weight that position is due").

Most importantly, the Secretary's position is consistent with the text of the Act. The text of § 10(b) speaks in terms of a singular, discrete violation—providing that "[i]f the Secretary has reason to believe that an employer has failed to correct a violation for which a citation has been issued … ." § 659(b). The citations at issue in this case were likewise written in singular terms, referring to individual violations of 29 C.F.R. § 1926.451(a)(4) by specific Manganas Painting employees. Had Congress intended for the expansive interpretation of "a violation" that the Commission applied, it is likely that it would have referred to a "type of violation for which a citation has been issued" or a "practice for which a citation has been issued." See *Alden Leeds, Inc. v. Occupational Safety and Health Review Comm'n*, 298 F.3d 256, 262-63 (3d Cir.2002) (distinguishing citations alleging discrete violations and citations alleging violative practices, and pointing*528 out that Secretary could have cited employer for categorical practices, rather than for discrete, individual violations). Instead, the statute refers only to "a violation," and we "must presume that a legislature says in a statute what it means and means in a statute what it says there." *Conn. Nat'l Bank v. Germain*, 503 U.S. 249, 253-54, 112 S.Ct. 1146, 117 L.Ed.2d 391 (1992). Thus, we hold that the Secretary's interpretation of § 10(b) is consistent with the text of the statute.

> [9] Finally, we emphasize the limits of our opinion. The Commission found, and substantial evidence supports the factual finding, that the 1993 and 1994 citations occurred at two separate worksites, one year apart from each other. In this regard, we review for clear error the factual findings made by the Commission. *Fields Excavating, Inc. v. Sec'y of Labor*, 383 F.3d 419, 420 (6th Cir.2004); *CMC Elec., Inc. v. OSHA*, 221 F.3d 861, 865 (6th Cir.2000). We conclude the Commission's finding of two different worksites is supported by substantial evidence. See *Nat'l Eng'g & Contracting Co. v. Herman*, 181 F.3d 715, 721 (6th Cir.1999) (explaining that "[s]ubstantial evidence is such relevant evidence as a reasonable mind might accept as adequate to support a conclusion.") [internal quotations omitted].
> [10] Thus, we are not called upon to decide, and express no opinion on the question, whether § 10(b) bars successive penalties for the same violation at the same worksite. Rather, we decide only that where an employer challenges a citation issued by the Secretary, the Secretary is not barred by § 10(b) from issuing a second citation for a violation of the same regulation that occurs at a different worksite at a subsequent time.

V.

For these reasons, we grant the petition for review, reverse the decision of the Occupational Safety and Health Review Commission with respect to citation items 13a–13c, and remand for further proceedings consistent with this opinion.

3 OSHA Requirements

CHAPTER OBJECTIVES

1. Acquire an understanding of the OSHA Recordkeeping standard.
2. Acquire an understanding of a variance and its use.
3. Acquire the reporting requirements for fatalities and multiple injury situations.
4. Acquire an understanding of the amputation reporting requirements.

INTRODUCTION

The Occupational Safety and Health Administration (OSHA) has promulgated numerous standards addressing a wide variety of specific hazards in the workplace. Additionally, OSHA has established specific procedures that require compliance, such as notification of OSHA in fatality or multiple injury situations, which are especially important to safety and loss prevention professionals.

ACCIDENT INVESTIGATION AND OSHA NOTIFICATION

Accident investigation procedures are a fundamental element of any safety and loss prevention program. They identify deficiencies in equipment, the environment, machine processes, and human acts in order to prevent the recurrence of accidents and injuries. Virtually all safety and loss prevention programs have some form of an accident investigation procedure in order to collect vital information and meet minimum regulatory requirements. Many companies have very sophisticated procedures to identify trends and areas of potential exposures, to spot deficiencies in equipment or processes, and to evaluate the effectiveness of safety and loss prevention efforts.

Completion of an accident investigation following a work-related injury or illness is required under the OSH Act to meet the recordkeeping requirements. Many states' workers' compensation laws also require investigation of accidents that result in compensable injuries and illnesses. Major components of most accident investigation programs include: providing first aid and transport of the victim, securing the accident scene, analyzing the facts to determine root causes, interviewing witnesses and/or the victim, documenting and analyzing the evidence, and initiating corrective action. Most accident investigation programs are designed to find facts rather than determine fault.

SEVERE INJURY REPORTS

OSHA requires employers to report all severe work-related injuries, defined as an amputation, in-patient hospitalization, or loss of an eye. The requirement began on January 1, 2015. Information required includes a description of the incident and the name and address of the establishment where it happened. Injuries are coded using the Occupational Injury and Illness Classification System.

Three areas within the accident investigation process pose the greatest danger of liability for safety and loss prevention professionals:

1. Recordkeeping
2. Required notification in fatality and multiple injury situations
3. Fatality or serious injury investigations

In each area, safety and loss prevention professionals should exercise extreme caution while ensuring compliance with regulations and preserving their rights in case of future prosecution or litigation.

RECORDKEEPING

Under OSHA recordkeeping requirements, covered companies must maintain specific records that identify injuries and illnesses that meet the recordable standard. From 1970 to 1980, the recordkeeping requirements were not overly burdensome to most employers, and were of little concern to safety and loss prevention professionals. In 1980, OSHA promulgated a requirement that employers must retain all monitoring, exposure, and medical records relating to more than 40,000 allegedly toxic or hazardous substances for a period of 30 years or longer. Employees or their representatives were also given the right to access these records, and the right to obtain copies of them. These requirements, which augment the previous OSHA posting, logging, and other recordkeeping requirements, have created new areas of potential liability for safety and loss prevention professionals.

In 2002, OSHA made major changes to the old 1904 recordkeeping rule, which included new forms, new recording criteria, a new accounting system, new reporting requirements, and new privacy requirements, among the numerous other changes to the rule.

The first area of potential liability is employee access to personal medical records and OSHA-required records. Under the Federal Occupational Safety and Health Act of 1970 (OSH Act), employees and their representatives have the right to view and copy specific records. Safety and loss prevention professionals who refuse access to these records expose themselves and their companies to potential litigation.

The second area of potential liability, and possibly the most costly in terms of OSHA's monetary penalties, is the failure to record information regarding work-related injuries and illnesses accurately and appropriately on the OSHA 300 log or other OSHA-required forms. One major area of danger is discrepancies that can arise when determining whether the injury or illness "fits" within the recordable, non-recordable, or first aid category. Recordable occupational injuries or illnesses are defined as:

Any occupational injuries or illnesses which result in

1. fatalities, regardless of time between the injury and death, or the length of the illness; or
2. lost workday cases, other than fatalities, that result in lost workdays; or
3. nonfatal cases without lost workdays which result in transfer to another job or termination of employment, or require medical treatment (other than first aid) or involve loss of consciousness or restriction of work or motion. This category also includes any diagnosed occupational illnesses which are reported to the employer but are not classified as fatalities or lost workday cases.

Injuries or illnesses that fall within a gray area, where interpretation by the safety and loss prevention professional is required, present the greatest potential danger. If, for example, the safety and loss prevention professional interprets a gray-area injury as non-recordable and this interpretation is repeated throughout the year by medical personnel, and, if during an OSHA inspection the injury is viewed as a recordable injury, then each gray-area interpretation on the OSHA 300 log could be viewed as an OSHA violation carrying a multiplying monetary penalty. Gray-area interpretations are often required when evaluating early cumulative trauma disorders, minor cuts and lacerations, and sprain or strain type injuries. In 1992, recordkeeping was the fourth highest category of citation by OSHA, and the vast majority of these citations involved violations of the OSHA 300 log. The fines levied for these violations ranged from no penalty to a proposed penalty of $3.3 million. In addition to potential monetary penalties, safety and loss prevention professionals should be aware that falsifying or failing to keep records or reports required by the OSH Act and standards is prohibited. Under the regulations and section 17 of the Act, "Whoever knowingly makes any false statement, representation, or certification in any application, record, report, plan or other document filed or required to be maintained pursuant to this Act shall, upon conviction, be punished by a fine of not more than $10,000.00 or by imprisonment for not more than 6 months or both."

FAILURE TO REPORT AN INJURY OR FATALITY

Another area of potential liability is failure to report a fatality or injury situation involving five or more employees within the eight-hour time limitation. The OSHA standard requires that "within 8 hours after the occurrence of an employment accident which is fatal to one or more employees or which results in hospitalization of 4 or more employees, the employer of any so injured or killed shall report the accident orally or in writing to the nearest office of the Area Director." The report must include the circumstances of the accident, the extent of the injuries, and the number of fatalities. Although most cases involving violation of this requirement resulted in minor citations and minimal monetary penalties by OSHA, this is still a high-risk area of potential liability because state prosecutors can use the state criminal codes to prosecute in fatality situations.

Potential liability following a fatality or multiple serious injury situation has changed dramatically since 1989 because the OSHA penalties that can be assessed have increased sevenfold. In addition, OSHA has increased its use of criminal sanctions provided under the Act, and a fatality may be investigated by a state or local prosecutor's office as a homicide.

Under some states' criminal codes, monetary penalties and criminal sanctions have been a liability risk for safety and prevention professionals since 1970; however, the use of state criminal codes to prosecute company officials for murder and manslaughter is a new phenomenon. Not only can an OSHA inspector investigate an accident, but police and local prosecutors can do so as well.

For example, in 1985, the Los Angeles County District Attorney's office created a special section, the first of its kind in the United States, specifically to prosecute corporate managers "whose disregard of safety standards [cause] on-the-job deaths and injuries." This office has created a roll-out program where an attorney and an investigator are on call around the clock; they respond to the scene of a workplace accident and conduct an investigation. In the past seven years, this program has prosecuted more than 50 criminal cases against corporate managers and it now serves as the model for a possible federal program.

INVESTIGATING A FATALITY OR MULTIPLE INJURY SITUATION

Safety and loss prevention professionals should know what to expect in a fatality or multiple injury situation and be prepared to exercise their constitutional and other rights to protect themselves and their companies. As noted previously, the character of accident investigations has changed dramatically, and safety and loss prevention professionals should adapt and be prepared for all contingencies.

The expertise and experience of OSHA investigators or investigation teams, the local prosecutors, or any other individuals investigating an accident should not be taken for granted. In most cases the individuals investigating an accident are highly educated, have specialized training, and are well-schooled in every party's rights at the scene. In fatality or multiple injury situations, safety and loss prevention professionals may want legal counsel present during the investigation and they certainly should adhere to legal counsel's advice. If statements are provided to an OSHA inspector or prosecutor, they should be cleared with legal counsel or corporate officials first. In such a situation, safety and loss prevention professionals have the right to say nothing, and everything they say can and will likely be used against them in a court of law. In most circumstances, safety and health professionals should not volunteer opinions, conjecture, or theories, but should simply "stick to the facts." If the company terminates the employment of the loss prevention and safety professional after the accident, but before litigation, the safety professional can still be made a party to the action or criminal prosecution. In many of these cases, the safety professional is left to fend for him- or herself and cover the cost of a legal defense.

Preparations should also be made for addressing the media, the other site employees, and possibly family members of those injured or killed who may arrive at the scene. The accident scene should be secured, and the investigation should be conducted as soon as feasible. All information should be well-documented so that it may be provided to legal counsel for protection under the work product rule.

The accident scene is not the place for a confrontation or disagreement with OSHA inspectors or other individuals investigating an accident. The safety and loss prevention professional should gather the same information and documentation as the OSHA inspector or other investigator. All photographs taken by the OSHA inspector should be immediately duplicated by taking identical photographs from the same position and angle. If a video recording is made, a correlating video recording should be made. This procedure should be used for all evidence gathering, including any type of sampling.

If an OSHA inspector wishes to interview employees, the safety and loss prevention professional or some other representative of the employer may be present during those interviews. However, in some cases, the OSHA inspector will deny the company representative's request to be present at the interview. If the interviews would interrupt the investigation or operation, the names, addresses, and telephone numbers of employees may be provided to the OSHA inspector for later contact away from the worksite.

Although a safety and loss prevention professional's life's work is preventing injuries and fatalities in the workplace, a prudent professional may want to develop a plan of action in the unlikely event that a fatality or multiple injury situation should occur. Planning for such a situation might include preparing documentation kits (i.e., camera, sampling equipment, video camera, and so on), preparing individuals to accompany the investigative agencies, establishing a media area, and choosing individuals to notify family members of affected employees. When a serious accident happens, those involved may not be fully capable of rational thought, and it is often left to the safety and loss prevention professional to steer the ship through the rough waters.

Safety and loss prevention professionals should be prepared for allegations of liability (either civil or criminal) after a fatality or multiple injury situation. Preventing the accident in the first place is the best method of eliminating potential liability, but after an accident has occurred, damage control is essential. Safety and loss prevention professionals should know their rights and responsibilities and should not be afraid to exercise them for fear of offending someone. After the accident has happened, liability has already attached itself; it is now a matter of determining who is responsible and how much it is going to cost.

VARIANCES

An alternative that is often overlooked by safety and loss prevention professionals for achieving and maintaining compliance with OSHA standards and directives is the use of permanent or temporary variances. Using variances is not a new method of achieving compliance in unusual circumstances. In fact, variance actions were originally provided under the OSH Act in 1970.

Variance actions have been rarely used by safety and loss prevention professionals due to the fact that they were widely perceived as complicated and costly. Safety and loss prevention professionals should take a second look at this option, however, because new and complex standards are being promulgated by OSHA.

Under section 6(b)(6) of the OSH Act, an employer may apply to the secretary of labor for permission to use means other than those prescribed in the OSHA standard to protect its employees. This application, known as a variance request, asks the secretary of labor to permit the employer to use other methods to safeguard employees in particular circumstances or situations that are not permitted under OSHA standards.

There are two basic types of variances: permanent and temporary. Permanent and temporary variances apply only to particular worksites or pieces of equipment; they are not blanket exemptions. Individual variances must be obtained for each worksite or piece of equipment and are nontransferable. Employers in state plan states may apply for a variance with the individual state plan program. Where the workplace involves several states or a single (non- state plan) state, the employer must apply for OSHA for variance protection. The OSHA regulations provide that actions taken on variances and interim orders will be "deemed prospectively an authoritative interpretation of the employer or employers' compliance obligation" regarding the state plan standard as long as the state plan standard is the same as the federal standard.

Two other types of variances also are permitted under the OSH Act, although they are seldom used: experimental variances and national defense variances. Under section 6(b)(6)(C), experimental variances are available when the employer is participating in an experiment approved by OSHA or the National Institute for Occupational Safety and Health (NIOSH) "designed to demonstrate or validate new and improved techniques to safeguard the health or safety of workers." The OSH Act also authorizes "variances, tolerances, and exemptions" where necessary to "avoid serious impairment of the national defense." The procedures for applying for these variances are the same as the procedure for permanent variances. Employers have the option of applying for a variance with either the state plan program or OSHA, but not both. Under OSHA regulations, the election of either the state or federal remedy is binding, and after a variance application is filed with either the state or OSHA, application to the other is barred. To assist employers, OSHA has established the Office of Variance Determination, for managing and processing variance applications.

TEMPORARY VARIANCES

A temporary variance can be applied for when an employer cannot comply with a specific OSHA standard because technical personnel or materials are lacking, equipment needed to achieve compliance is not available, or necessary modifications or alterations cannot be made to achieve compliance within the specified time period required under a new OSHA standard. Applying for a temporary variance is an option that safety and health professionals should consider when major modifications of equipment or a facility are necessary, or in other circumstances when additional time is needed in order to achieve compliance. The completion of a modification variance can protect the employer from the potential of OSHA citations for noncompliance.

Under section 6(b)(6)(A) of the OSH Act, OSHA can grant temporary variances when the employer cannot comply with a standard because specific personnel are unavailable or equipment or facilities' modifications cannot be completed within the specified time frame. As part of the variance application, the employer must state that all available steps are being taken to protect employees and that it has developed and implemented an effective program for reaching compliance with the OSHA standard as quickly as possible. A temporary variance may remain in effect for a maximum of one year, with the possibility of two renewals of no more than 180 days per renewal.

When preparing an application for a temporary variance, safety and health professionals should provide the following information:

- the employer's name and address
- the address of the worksite involved
- the specific OSHA standard or section thereof for which the variance is sought
- a detailed statement, supported by representations from qualified persons, explaining why compliance with the standard is not attainable
- the steps taken or to be taken to protect employees against the hazard covered by the OSHA standard
- a statement identifying when the applicant expects to be able to come into compliance with dates, and the steps taken or to be taken
- a statement showing why the applicant cannot come into compliance by the OSHA standard's effective date, explaining that all steps are being taken to safeguard employees from the hazards covered by the OSHA standard, and demonstrating that the applicant has an effective program for coming into compliance
- any request for a hearing on the temporary variance
- a statement that affected employees have been notified of the variance request and its contents
- a description of how affected employees have been notified
- for state plan states, information concerning the state OSHA standard, any variance applications filed with the state plan program, and identification of any state OSHA citation involving the comparable state OSHA standard

The process for obtaining a temporary variance is the same as that for a permanent variance, which is described in the following section. Safety and health professionals may want to further protect the worksite by applying for an interim order to protect the workplace from possible inspections and citations while the temporary variance application is being reviewed. Temporary variances may also be used while solutions to difficult problems are being evaluated or while the employer prepares an application for a permanent variance.

Permanent Variances

A permanent variance can be obtained by an employer through showing, by a preponderance of evidence, that the safety procedures, practices, or equipment modifications in question, although not in compliance with the specific OSHA standard, are as safe and healthful as the practices required under the standard. Safety and loss prevention professionals should consider obtaining a permanent variance when an OSHA standard seems impracticable as applied to the particular workplace or when they believe that the alternative safeguards achieve equivalent or superior protection for employees. In some circumstances, applying for a variance may be the only alternative, given the facts of the situation. For example, in *General Electric Co. v. Secretary of Labor*, the court ruled that an employer cannot raise the defense to an OSHA citation that the practices or procedures used by the employer are as safe as or safer than those required under the OSHA standard unless it has first filed a variance application or can show that a variance application would be inappropriate. Additionally, an employer should not wait to be cited for alleged violations before applying for a variance because OSHA regulations state that OSHA may decline to entertain a variance application until the citation has been resolved.

On the application form for a permanent variance, the following items should be included:

- the employer's name and address
- the address of the worksite involved
- a description of the conditions, practices, means, methods, operations, or processes used or proposed to be used in lieu of the specified OSHA standard
- a statement explaining how the alternative measures will provide employees with a work environment that is as safe and healthful as that provided by the OSHA standard
- certification that the employees have been informed of the application
- any request for a hearing
- a description of how employees have been informed of both the application and their right to petition for a hearing
- in state plan states, information concerning the state OSHA standard, any variance applications filed with the state plan program, and identification of any state OSHA citation involving the comparable state OSHA standard

Applicants are entitled to a hearing on the application, and employees and labor organizations, if any, are also entitled to a hearing. Variance application hearings are formal in nature and are normally documented by a court reporter or other methods. Under the OSHA regulations, the hearing examiner can make a finding and render a decision regarding the application. The decision is final unless an appeal is taken to the assistant secretary of OSHA. In applications for variances, an informal process is used by the parties. Any disputes regarding the application are usually negotiated in one or more informal conferences held between the employer and the Office of Variance Determination. In most circumstances, a representative of the Office of Variance Determination will inspect the area or equipment involved before any determination is made. When the informal conferences and site inspection are completed, if it is believed that the variance application should be granted, a notice of the application will be published in the *Federal Register*. If there are no strong objections or responses to the application, a final order is published in the *Federal Register* a few months later.

In the informal conferences or at a formal hearing, specific terms and conditions may be attached to the variance application to ensure that the alternative system incorporates adequate protective measures. The OSH Act requires that a variance "prescribe the conditions the employer must maintain, and the practices, means, methods, operations, and processes that he must adopt and utilize to the extent they differ from the OSHA standard in question." Additionally, in lieu of granting the requested variance, OSHA may decide to clarify the particular standard in a manner that grants substantial relief to the employer. This is normally done on an informal basis by letter or directive to the regional OSHA offices, with a copy sent to the employer.

INTERIM ORDERS

OSHA regulations provide interim orders that grant temporary relief from inspection or citation pending the outcome of a formal hearing regarding a variance application. Employers can include an application for an interim order with an application for a temporary or permanent variance. The assistant secretary of OSHA reviews all applications for interim orders and can issue interim orders on an ex parte basis. Normally, when an interim order is granted by the assistant secretary, a notice of the application for the interim order, with the application for a permanent or temporary variance and a request for public comment, is published in the *Federal Register*.

APPENDIX 3: A SAMPLE VARIANCE FORM

Petition for Permanent Variance
    Lockout/Tagout during the Operation

    1. Petitioner:
       XYZ Manufacturers' Address
       Counsel for Petitioners:
       Thomas D. Schneid
       Anywhere Street
       Richmond, Kentucky 40475
       (888) XXX-XXXX
    2. Place of Employment Involved:
       XYZ Manufacturers
       Address
       Anywhere City, USA

3. A Description of the Conditions, Practices, Means, Methods, Operations, or
   Processes Used or Proposed to be Used by the Petitioner:
   XYZ Manufacturers, Inc. makes widgets at their facility in Anywhere,
   USA. The primary function of XYZ Manufacturers is to mold the widget
   materials into a finished widget.

   XYZ Manufacturers are required to comply not only with the Kentucky
   Labor Cabinet, Division of Occupational Safety and Health standards,
   but also with the Federal Meat Inspection Act (21 U.S.C. § 1621 et seq.).
   The United States Department of Agriculture has promulgated specific
   cleaning and sanitation regulations governing meat processing facilities
   (21 U.S.C. § 608 and 9 C.F.R. § 308.1). On-site inspectors for the United
   States Department of Agriculture possess the ability to test and approve all
   chemicals utilized in the cleaning process, the water utilized, conduct bac-
   teriological testing, and perform other testing to insure the cleanliness of
   the machinery and operations within the facility (21 U.S.C. § 695). Under a
   Memorandum of Agreement between the Federal Occupational Safety and
   Health Administration and the Department of Agriculture in 1993, federal
   meat inspectors are empowered to evaluate and inspect for appropriate conditions not only in the areas specified in the Federal Meat Inspection Act
   but also for safety and health conditions. The United States Department
   of Agriculture (USDA) inspectors may enter and inspect the machinery at
   any time day or night (21 U.S.C. § 606 and 609) and if the machinery or
   operations do not meet the specific sanitation standard, the inspector may
   withdraw the inspection services and thus prohibit the plant from operating
   until appropriate safety and sanitation requirements are achieved (21 U.S.C.
   § 608 and 671; 9 C.F.R. § 305.5).
4. A Statement Showing How the Conditions, Practices, Means, Methods,
   Operations, or Processes Used or Proposed to Be Used Would Provide
   Employment and Places of Employment to Employees Which Are as Safe
   and Healthful as Those Required by the Standard from Which a Variance
   Is Sought:
   In analyzing the "widget making" phase of the operations without the
   proposed procedures addressed in this Variance Petition, the authorized
   employees would be required to lockout and tagout the specified machinery
   utilizing the prescribed procedures, walk to the hose connection area, con-
   nect the sanitizer hose, spray the machinery, turn off the water hose, turn
   off the sanitizer, walk back to the machinery (varied distance), remove the
   lockout and tagout utilizing the prescribed procedure, energize the machinery, turn the belt or screw 25 degrees to 33 degrees, perform the lockout
   and tagout procedure, walk to the hose, and repeat the above sequence.
   This procedure would be required to be repeated 2–5 times in order that the
   high-pressure water and sanitizing agent to all angles can be applied on the
   screw or belt in order to comply with the sanitation requirements specified
   by the United States Department of Agriculture regulations. In this procedure, the authorized employee may de-energize and re-energize the specific
   equipment 2–5 times while standing in the residual water from the water

spray and walk continuously over slippery floors. Once the "sanitizing" phase is completed, the authorized employee would perform the lockout and tagout procedure prior to initiating the other steps of the clean-up operation. XYZ Manufacturers contends that this repeated interaction with the specific machinery and the lockout process creates a greater risk of potential harm to the employee than the process requested in this variance petition.

Under the current standard, the employee is required to continuously repeat the lockout and tagout process during the "sanitizing" phase, thus creating a greater potential exposure of the employee removing the guards from the specific machine to be able to manipulate the hose nozzle to an angle to spray water under the screw or belt. Additionally, given human nature, there is a greater potential hazard of human error in the fact that the employee may not wish to continuously lockout and tagout the machine and return to the designated spray area thus he/she may circumvent the lockout and tagout procedures in order to avoid the walking and repeated steps.

In analyzing the potential ergonomic hazards, the authorized employee may be required, absent the proposed procedures specified in this Variance Petition, to stand or lay in awkward angles in order to focus the water spray at the underside of the de-energized equipment.

As specified in 29 CFR 1910.147(D)(2)(ii)(A) & (B) (as codified in 803 KAR 2.300-2.320), normal production operations are not covered under the Control of Hazardous Energy standard; however, service and/or maintenance would be covered if: (A) "An employee is required to remove or bypass a guard or other safety device." In the proposed process, the "sanitizing" procedure should not be construed as either service or maintenance of the equipment. The "widget making" process consists of simply spraying a water and sanitizer mix on the specified equipment through the use of a high-pressure stream of water connected to a sanitizing agent hose to provide a mixture of sanitizing agent and water. In this widget making process, the employee does not bypass any safeguard (such as the emergency stop systems) or other safety controls.

In order to insure complete clarity, XYZ Manufacturers requires all equipment to be locked and tagged out whenever employees are working on or within the plane of any equipment. If there is any doubt, employees are instructed to lock and tag the equipment. All equipment remains locked and tagged until such time as the process is completed. This Variance Petition is submitted only to address the specific time period of short duration which is known in the industry as the "widget making" period. All other operations, processes, and procedures performed by the XYZ Manufacturers at the designated location will continue to require the equipment to be locked and tagged whenever cleaning, repair, or other activity is to be performed.

5. A Statement That the Petitioner Has Informed His Affected Employees of the Application by Giving a Copy Thereof to Their Authorized Representative, Posting a Statement, Giving a Summary of the Application and Specifying

Where a Copy May Be Examined, at the Place or Places Where Notices to Employees Are Normally Posted, and by Other Appropriate Means:

Mr. Barnie Widget, President of XYZ Manufacturers, hereby certifies that he has informed the employees currently working in the affected facility of this petition for a permanent variance. As set forth in this petition, a notice will be posted on the bulletin board of the lunchroom (canteen) and a copy of the full petition will be available in the office area for review by all employees.

Mr. Barnie Widget _____ Date _____ President

6. A Description of How Employees Have Been Informed of the Petition and of Their Right to Petition the Commissioner of Labor for a Hearing.

A posting informing all XYZ Manufacturers' employees of this petition will be posted on the bulletin board in the canteen area of the facility for review by all employees. This is the normal and routine area for all employees to review company policies and other postings. A copy of the notice is attached for your review and evaluation.

A copy of this entire petition, including the supplemental information, will be located in the office area of this facility for review by all employees. This full text will be maintained in the office area because the USDA sanitation requirements mandate a complete washdown of the facility at the end of each working day.

7. Request for a Hearing:

XYZ Manufacturers hereby formally request that the secretary of labor grant a formal hearing on this Petition for Permanent Variance if necessary.

Signed this _____ day of _____, 1996.

Mr. Barnie Widget _____

4 Employer's Rights

CHAPTER OBJECTIVES

1. Acquire an understanding of OSHA enforcement.
2. Acquire an understanding of the employer's rights under the OSH Act.
3. Acquire an understanding of the *Marshall v. Barlow's* case.
4. Acquire an understanding of the use of OSHA information in civil actions.

INTRODUCTION

All enforcement functions under the OSH Act rest with OSHA, which is under the direction of the Department of Labor. All OSHA compliance officers can, under Section 8 of the Act, inspect any public- or private-sector workplace covered by the Act. The compliance officer is required to present his or her credentials to the owner, operator, or agent in charge before proceeding with the inspection tour. The employer and a union or employee representative have the right to accompany the compliance officer during the inspection. Upon completion of the inspection, a closing conference is usually held in which the compliance officer and the employer discuss safety and health conditions as well as possible violations. Most compliance officers cannot issue on-the-spot citations; they must first confer with the regional or area director.

When a compliance officer observes a violation in an employer's workplace and notes this observation on his or her report, the area director, after the completion of the on-site inspection, usually decides whether or not to issue a citation. The area director normally computes any penalties and sets abatement dates for each violation. The citation is mailed by means of the U.S. Postal Service (usually certified mail) to the employer as soon as possible after the inspection, but in no event can it be sent more than six months after the alleged violation occurred. All citations must be in writing and must describe with particularity the violation alleged, including the relevant standard and regulation.

The OSH Act enforcement scheme includes both civil and criminal penalties for violations. Violators of specific standards or of the general duty clause may face civil penalties. Penalties may be assessed only within the range set forth under the Act.

The Act currently allows imprisonment of up to six months for willful violations that cause the death of an employee. OSHA normally reserves the use of criminal sanctions for the most serious and egregious circumstances. OSHA usually relies on the monetary fines to rectify workplace violations. The good faith of the employer, the gravity of the violation, the employer's past history of compliance, and the size of the employer are usually considered in assessing the penalty. The area director has the authority to compromise, reduce, or remove a violation.

After a citation is issued, the employer, any employee, or any authorized union representative has 15 working days to file a notice of contest. If the employer does not contest the violation, abatement date, or proposed penalty, the citation becomes final and is not subject to review by any court or agency. If a timely notice of contest is filed in good faith, the abatement requirement is tolled, and a hearing is scheduled. An employer may also file a petition of modification of the abatement period (PMA) if it cannot comply with any abatement that has become a final order. If the secretary of labor or an employer contests the PMA, a hearing is held to determine whether any abatement requirement, even if part of an uncontested citation, should be modified.

The secretary of labor must immediately forward any notice of contest to the Occupational Safety and Health Review Commission (OSHRC). In cases before the OSHRC the secretary of labor is usually referred to as the complainant and has the burden of proving the violation. Conversely, the employer is called the respondent. The hearing is presided over by an ALJ, who renders a decision affirming, modifying, or vacating the citation, penalty, or abatement date. The ALJ's decision automatically goes before the OSHRC for review. The aggrieved party may file a petition for discretionary review of the ALJ's decision, but even without this discretionary review, any OSHRC member may call for review of any part or all of the ALJ's decision. If, however, no member of the OSHRC calls for a review within 30 days, the ALJ's decision is final. Through either review method the OSHRC may reconsider the evidence and issue a new decision. After this review, any party adversely affected by the OSHRC's final order may file, within 60 days of the decision, a petition for review in the U.S. Court of Appeals for the circuit in which the alleged violation had occurred or in the U.S. Court of Appeals for the District of Columbia Circuit.

The inspection, violation, and appeal procedures in virtually all state programs are identical to those of OSHA; however, the names of the reviewing commission may be different (e.g., Kentucky Labor Cabinet, Occupational Safety and Health Review Commission). After exhausting the state's administrative route, an adversely affected employer may usually file, within 60 days of the decision, a petition for review in the state supreme court or the state court of appeals for the circuit in which the employer is located.

When addressing the enforcement mechanism of either OSHA or a state plan program, the employer must be prepared in advance and be completely aware of its rights and duties under the OSH Act or correlating state laws. Preplanning and preparation of an efficient and effective safety and loss prevention program is vital to ensuring compliance with OSHA standards. In addition to physical preparation to ensure compliance, corporations should preplan their strategy for dealing with OSHA and state plan enforcement agencies.

EMPLOYER'S RIGHTS DURING AN OSHA INSPECTION

When a compliance officer or other Department of Labor representative enters a facility to perform an inspection, the employer has proscribed rights. First, the employer is entitled to know the purpose of the inspection, e.g., whether it is based on an employee complaint or it is a routine inspection. Second, the employer also has the right to accompany the compliance officer during the inspection. This can be helpful or harmful—helpful in the sense that the employer can avoid certain areas, but harmful if a major violation is found and the employer, in trying to explain, talks itself into more trouble and ends up with a higher fine and a more serious violation.

Corporate officials should know their rights under the OSH Act and U.S. Constitution. Because of the decision in *Marshall v. Barlow's, Inc.*, specific avenues for addressing OSHA enforcement efforts have been developed and can be efficiently used, depending on the circumstances.

In *Barlow's*, the Supreme Court held that section 8(a) of the OSH Act, which empowered OSHA compliance officers to search the work areas of any employment facility within the OSH Act's jurisdiction without a search warrant or other process, was unconstitutional. The Court concluded that "the concerns expressed by the Secretary (of Labor) do not suffice to justify warrantless inspections under OSHA or vitiate the general constitutional requirement that for a search to be reasonable a warrant must be obtained." This decision opened the door to one avenue of approach—namely, requiring OSHA and state enforcement officers to acquire a warrant before entering a facility to conduct an inspection. This approach should be carefully evaluated with the help of legal counsel, given the potential pitfalls and the possibility of sanctions against the employer for bad faith.

Another successful approach is limiting the scope and inspection techniques used by the OSHA or state inspection officer. This is normally an informal process whereby the safety and loss prevention professional can contact the regional or area director before a voluntary compliance inspection to reach an agreement regarding the specific area to be inspected, or to place limitations on photographing or videotaping in order to protect trade secrets. If an agreement cannot be reached before the inspection, a court order may be acquired to protect the confidentiality of a trade secret. An additional approach, which is most often utilized, is to permit the inspection of the facility without a warrant.

Safety and loss prevention professionals should analyze their situation and facility and develop a policy and plan of action that advises their management team regarding the specific approach to be utilized when addressing OSHA and state compliance officers. This plan should include specific individuals who will represent the employer and detailed steps to be followed when OSHA or state compliance officers attempt entry, during the inspection, at the closing conference, and after the inspection as well as forms to assist the management team in this procedure. In addition, contingencies, such as when a member of a team is on vacation or unavailable, should also be addressed. In essence, this plan should prepare the management team for every contingency that could happen during a compliance inspection.

OSHA and most state plan programs offer free consultation services. While OSHA or a state plan is assisting the employer in this voluntary compliance effort, the agency will only cite violations under very limited and life-threatening circumstances.

EMPLOYER RESPONSIBILITIES

Under the OSH Act, employers have a responsibility to provide a safe workplace. This is a short summary of key employer responsibilities:

- Provide a workplace free from serious recognized hazards and comply with standards, rules, and regulations issued under the OSH Act.
- Examine workplace conditions to make sure they conform to applicable OSHA standards.
- Make sure employees have and use safe tools and equipment and properly maintain this equipment.
- Use color codes, posters, labels, or signs to warn employees of potential hazards.
- Establish or update operating procedures and communicate them so that employees follow safety and health requirements.
- Employers must provide safety training in a language and vocabulary workers can understand.
- Employers with hazardous chemicals in the workplace must develop and implement a written hazard communication program and train employees on the hazards they are exposed to and proper precautions (and a copy of safety data sheets must be readily available). See the OSHA page on Hazard Communication.
- Provide medical examinations and training when required by OSHA standards.
- Post, at a prominent location within the workplace, the OSHA poster (or the state-plan equivalent) informing employees of their rights and responsibilities.
- Report to the nearest OSHA office all work-related fatalities within 8 hours, and all work-related inpatient hospitalizations, all amputations, and all losses of an eye within 24 hours. Call our toll-free number: 1-800-321-OSHA (6742); TTY 1-877-889-5627. [Employers under federal OSHA's jurisdiction were required to begin reporting by Jan. 1, 2015. Establishments in a state with a state-run OSHA program should contact their state plan for the implementation date].
- Keep records of work-related injuries and illnesses. (Note: Employers with 10 or fewer employees and employers in certain low-hazard industries are exempt from this requirement.)
- Provide employees, former employees, and their representatives access to the Log of Work-Related Injuries and Illnesses (OSHA Form 300). On February 1, and for three months, covered employers must post the summary of the OSHA log of injuries and illnesses (OSHA Form 300A).
- Provide access to employee medical records and exposure records to employees or their authorized representatives.
- Provide to the OSHA compliance officer the names of authorized employee representatives who may be asked to accompany the compliance officer during an inspection.

- Not discriminate against employees who exercise their rights under the Act. See our "Whistleblower Protection" webpage.
- Post OSHA citations at or near the work area involved. Each citation must remain posted until the violation has been corrected, or for three working days, whichever is longer. Post abatement verification documents or tags.
- Correct cited violations by the deadline set in the OSHA citation and submit required abatement verification documentation.
- OSHA encourages all employers to adopt an Injury and Illness Prevention Program. Injury and Illness Prevention Programs, known by a variety of names, are universal interventions that can substantially reduce the number and severity of workplace injuries and alleviate the associated financial burdens on U.S. workplaces. Many states have requirements or voluntary guidelines for workplace Injury and Illness Prevention Programs. Also, numerous employers in the United States already manage safety using Injury and Illness Prevention Programs, and we believe that all employers can and should do the same. Most successful Injury and Illness Prevention Programs are based on a common set of key elements. These include: management leadership, worker participation, hazard identification, hazard prevention and control, education and training, and program evaluation and improvement. OSHA's Injury and Illness Prevention Programs topics page contains more information including examples of programs and systems that have reduced workplace injuries and illnesses.

APPEAL RIGHTS AND PROCEDURES

Under section 9(a) of the OSH Act, if the secretary of labor believes that an employer "has violated a requirement of Section 5 of this Act, of any standard, rule or order promulgated pursuant to Section 6 of this Act, or of any regulations prescribed pursuant to this Act, he shall within reasonable promptness issue a citation to the employer." Reasonable promptness has been defined as within six months of the occurrence of a violation.

Section 9(a) also requires that citations be in writing and "describe with particularity the nature of the violation, including a reference to the provision of the Act, standard, rule, regulation, or order alleged to have been violated." The OSHRC has adopted a fair notice test that is satisfied if the employer is notified of the nature of the violation, the standard allegedly violated, and the location of the alleged violation.

The OSH Act does not specifically provide a method of service for citations. Section 10(a) authorizes service of notice of proposed penalties by certified mail, and in most instances, the written citations are attached to the penalty notice. Regarding the proper party to be served, the OSHRC has held that service is proper if it "is reasonably calculated to provide an employer with knowledge of the citation and notification of proposed penalty and an opportunity to determine whether to contest or abate."

Under Section 10 of the Act, after a citation is issued, the employer, any employee, or any authorized union representative has 15 working days to file a notice of contest.

If the employer does not contest the violation, abatement date, or proposed penalty, the citation becomes final and, therefore, not subject to review by any court or agency. If a timely notice of contest is filed in good faith, the abatement requirement is met and a hearing is scheduled. An employer may contest any part or all of the citation, proposed penalty, or abatement date. Employee contests are limited to the reasonableness of the proposed abatement date. Employees also have the right to elect party status after an employer has filed a notice of contest.

An employer may also file a PMA if it cannot comply with any abatement that has become a final order. If the secretary of labor or an employer contests the PMA, a hearing is held to determine whether any abatement requirement, even if part of an uncontested citation, should be modified.

The notice of contest does not have to be in any particular form and it is sent to the area director who issued the citation. (Note: Several state plan programs offer fill-in-the-blank forms to assist the employers in filing a notice of contest.) The area director is required to forward the notice to the OSHRC, which is required to docket the case for hearing.

After pleading, discovery, and other preliminary matters, a hearing is scheduled before an ALJ. Witnesses testify and are cross-examined under oath, and a verbatim transcript is usually made. The Federal Rules of Evidence apply to this hearing.

After closure of the hearing, parties may submit briefs to the ALJ. The ALJ's decision contains findings of fact and conclusions of law and affirms, vacates, or modifies the citation, proposed penalty, and abatement requirements. The ALJ's decision is filed with the OSHRC and may be directed for review by any OSHRC member *sua sponte* or in response to a party's petition for discretionary review. Failure to file a petition for discretionary review precludes subsequent judicial review.

The secretary of labor has the burden of proving each cited violation. Through either review route, the OSHRC may reconsider the evidence and issue a new decision.

The factual determinations of the ALJ, especially regarding credibility findings, are often afforded great weight by the OSHRC. Briefs may be submitted to the OSHRC, but oral argument can be permitted by the OSHRC, although extremely rare.

In this administrative phase of the OSH Act's citation adjudication process, the employer's good faith, the gravity of the violation, the employer's past history of compliance, and the employer's size are all considered in the penalty assessment. The area director can compromise, reduce, or remove a violation. Many citations can be compromised or reduced at this stage.

Although the OSHRC's rules mandate the filing of a complaint by the secretary, and an answer by the employer, pleadings are liberally construed and easily amended. Approximately 90% of the cases filed are resolved without a hearing, either through settlement, withdrawal of the citation by the secretary, or withdrawal of the notice of contest by the employer.

As permitted under section 11(a) of the Act, any person adversely affected by the OSHRC's final order may file, within 60 days of the decision, a petition for review in the U.S. Court of Appeals for the circuit in which the alleged violation occurred or in the U.S. Court of Appeals for the District of Columbia Circuit. Under section 11(b), the secretary may seek review only in the circuit in which the alleged violation occurred or where the employer has its principal office.

The courts apply the substantial evidence rule to factual determinations made by the OSHRC and its ALJs, but courts vary on the degree of deference afforded the OSHRC's interpretations of the statutes and standards. The burden of proof is placed on the secretary of labor at this hearing. The rules of civil procedure, rules of evidence, and all other legal requirements apply, as with any trial before the federal court.

Safety and loss prevention professionals should be aware of their rights and responsibilities under the law. Although they should not fear inspections by OSHA, they should prepare for them to ensure that the legal rights of the employer are protected. In most circumstances, simple communication with the OSHA inspector or area director can correct most difficulties during an inspection. After a citation is issued, safety and health professionals should at least consider contesting the citation at the area director level in order to discuss reduction of monetary penalties. If an amicable solution cannot be reached at the regional level, safety and health professionals should be certain that the time limitations are met in order to preserve the right to appeal the decision. Meeting the specific time limitations set forth in the Act is of utmost importance. If the time limitation is permitted to lapse, the opportunity to appeal is lost.

SEARCH WARRANTS IN OSHA INSPECTIONS

OSHA's authority to conduct inspections and investigations is derived from Section 8(a) of the OSH Act, which states:

> In order to carry out the purpose of this Act, the Secretary, upon presenting appropriate credentials to the owner, operator, or agent in charge is authorized:
>
> 1. to enter without delay and at reasonable times any factory, plant, establishment, construction site, or other area, workplace, or environment, where work is performed by an employee of an employer; and
> 2. to inspect and investigate during regular working hours and at other reasonable times, and within reasonable limits and in a reasonable manner, any such place of employment and all pertinent conditions, structures, machines, apparatus, devices, equipment, and materials therein, and to question privately any such employer, owner, operator, agent, or employee.

As noted previously, OSHA may not provide advance notice of its intent to inspect a particular worksite under penalty of law, except when the secretary approves such notice. By regulation, the secretary can provide advance notice:

- in cases of apparent imminent danger, to enable the employer to abate the danger as quickly as possible
- when the inspection can most efficiently be conducted after regular business hours and special preparations are necessary for the inspection

- to ensure the presence of employer and employee representatives or the appropriate personnel needed to aid in the inspection
- when the area director determines that advance notice would enhance the probability of an effective and thorough investigation

In virtually all OSHA inspections, no advance notice to the employer is provided. The decision in *Marshall v. Barlow's, Inc.* settled the issue regarding OSHA's ability to conduct warrantless inspections, but opened the door to many related issues. *Barlow's* was not, however, the first case to address the requirement of administrative search warrants. In *Camara v. Municipal Court* and *See v. City of Seattle* (companion cases), the Supreme Court first required a search warrant for nonconsensual administrative inspections. These cases laid the foundation for the Barlow's decision and also specified and defined four exceptions to the warrant requirement in administrative inspections. The first three exceptions, namely the consent, plain view, and emergency inspection exceptions, were drawn directly from the law of search and seizure. The fourth, known as the Colonnade-Biswell or licensing exception, is the most controversial.

THE CONSENT EXCEPTION

The usual exception to the search warrant requirement for OSHA that is used by the vast majority of employers is consent to the safety and health inspection. Valid consent by the employer waives the employer's Fourth Amendment rights and protection. In the administrative setting of an OSHA inspection, consent can be provided by the employer by simply failing to object to the inspection. This form of consent differs greatly from the criminal investigation requirement that the consent be knowing and voluntary. Additionally, OSHA compliance officers are not required to inform employers of their right to demand a warrant or even to ask for the employer's consent.

Employers must affirmatively exercise their right to require a search warrant before an inspection. OSHA has instructed compliance officers to answer employers' questions regarding search warrants in a straightforward manner, but they are not required to volunteer information and are not allowed to mislead, coerce, or threaten an employer.

A major issue regarding the consent exception occurs whether the individual providing the consent has the authority to do so on behalf of the employer. Courts and the OSHRC have provided a broad interpretation to this question, permitting plant managers, foremen, and even senior employees to provide consent. OSHA has also found that general contractors may provide consent to inspect a common worksite where other subcontractors are working.

THE PLAIN VIEW EXCEPTION

A search warrant is not required for equipment, apparatus, or worksites that are in the plain view of the compliance officer or open to public view. For a compliance officer to issue a citation for a workplace hazard that is within the plain view exception, he or she must meet two criteria:

1. Be in a place or location where he or she possesses a right to be.
2. Observe (or smell, hear, or acquire through other senses) what is visible or held out to public view.

In the past, simply being the target of the OSHA inspection was usually sufficient for being able to complain of a search or seizure violation by OSHA. The U.S. Supreme Court has set up a significant hurdle for challenging the plain view exception by limiting the right to challenge search and seizure claims to those individuals, companies, or other entities who have an actual and legitimate expectation of intrusion by a government action. This interpretation, as applied to the plain view exception, would permit a compliance officer to observe and issue a citation for workplace violations from a distance away from company property (such as from an adjacent hill or another building), and if the employer challenged the citation on grounds of search or seizure, the court would not permit the claim because of lack of standing. The plain view exception is often used when a compliance officer observes a violation in a public area, such as a trenching site located on a public street. Safety and health professionals should be aware of the areas surrounding the facility and operation that are open to public view and should keep the plain view exception in mind.

THE EMERGENCY EXCEPTION

The third exception to the search warrant requirement is an emergency situation. When an urgent threat to human life exists and a delay in acquiring a search warrant might increase the hazard, or consent cannot readily be obtained from the employer because of the emergency situation and the emergency need outweighs the individual's right to privacy, the compliance officer may enter the facility without acquiring a search warrant.

In these rare circumstances, the compliance officer's duty to safeguard the employees who are in imminent danger far outweighs the employer's right to privacy. There are no specific OSHA regulations defining what would constitute an emergency situation; however, it is highly likely that a court would find that the emergency situation would require an extreme life-threatening situation likely to cause death or severe injury if the compliance officer did not intervene immediately.

THE COLONNADE-BISWELL EXCEPTION

The fourth and most controversial exception to the search warrant requirement is the Colonnade-Biswell or licensure exception. Before the *Barlow's* decision in 1978, the Supreme Court held in *Colonnade Catering Corp. v. United States* that warrantless nonconsensual searches of licensed liquor stores were permitted. Additionally, in *United States v. Biswell*, the Court permitted nonconsensual warrantless searches of pawn shops under the Gun Control Act of 1968. The Court found that a business in a regulated industry, such as a liquor store or pawn shop, provided an implied waiver of its Fourth Amendment rights by engaging in these industries.

From 1970 through 1978, several courts determined that the Colonnade-Biswell exception applied to the OSH Act and OSHA compliance inspections; conversely, others found that the *Camara* and *See* decisions required OSHA to acquire a warrant. The Barlow's decision settled the issue regarding OSHA's requirement to obtain a search warrant for routine compliance inspections, but it also opened the door to peripheral issues involving search warrant requirements.

Marshall v. Barlow's, Inc. and Probable Cause

In basic terms, the Barlow's decision utilized the Camara and See standards requiring that the probable-cause standard be applied to OSHA and other administrative searches. Issues involving whether the Barlow's decision applies to nonroutine inspections and whether or not OSHA may acquire an ex parte warrant are still unresolved. The Court in *Barlow's* did conclude, however, that the OSH Act authorized or intended to authorize issuance of search warrants and suggested that OSHA could amend its regulations to authorize the issuance of ex parte search warrants.

Safety and loss prevention professionals should be aware that the probable-cause requirement for a search warrant is significantly different from the criminal probable-cause standard. In *Barlow's*, the Court defined the administrative probable-cause standard as a warrant showing that a specific business has been chosen for an OSHA search on the basis of a general administrative plan for the enforcement of the OSH Act derived from neutral sources. For example, dispersion of employees in various types of industries across a given area and the desired frequency of searches in any of the lesser divisions of an area would protect an employer's Fourth Amendment rights.

Probable cause for an OSHA or other administrative search can be developed from any of three basic categories:

1. General information about the employer's industry
2. General information about the individual employer
3. Specific information about the employer's workplace

The general information regarding the employer's specific industry can be acquired by OSHA from many sources, including industry-by-industry data regarding workplace injuries and illnesses, days lost from work, and other data. General and specific information regarding the employer and workplace is usually acquired through employee complaints. Other sources of information include data acquired during past fatality or accident investigations, plain view observations by the compliance officer, and the employer's past history of citations.

When probable cause is shown and an administrative search warrant is issued, the permitted scope of the inspection is usually broad enough to encompass the entire operation. This is often referred to as a wall-to-wall inspection. Courts have generally permitted wall-to-wall inspections so that compliance officers, who normally are unfamiliar with the operation or facility under inspection, are provided great latitude in meeting the intent and purposes of the OSH Act in locating and identifying workplace hazards.

Although employers now have a constitutional right to require a search warrant before an OSHA inspection, few have exercised this right. OSHA inspections conducted under administrative search warrants continue to be rare; "approximately 97 to 99 percent of all employers voluntarily consent to inspection when the compliance officer knocks at the door." Most employers have found that it is fairly easy for OSHA to acquire an administrative search warrant, and creating an adversarial relationship with the compliance officer might do greater harm than good. Although most employers prefer to maintain good working relationships with compliance officers, some employers, due to individual circumstances, such as a belief that OSHA is harassing them or where impostors have posed as OSHA inspectors soliciting bribes, have exercised their constitutional right to require a search warrant before entry.

CHALLENGING A SEARCH WARRANT

In most circumstances, employers should carefully consider their option to require an OSHA compliance officer to obtain a search warrant. Many courts have sanctioned employers seeking a search warrant for contempt of court and other Rule 11 sanctions for frivolous actions. Additionally, the likelihood of successfully challenging an OSHA inspection warrant unless unusual circumstances are present is minimal at best. The decision to require an administrative search warrant by OSHA should be carefully and extensively evaluated. Given the potential risks involved in requiring a search warrant, this decision should involve the board of directors, officers, and legal counsel of the organization.

If an employer chooses to require a search warrant, careful planning and preparation should take place before the actual inspection:

- A policy statement or other directive should be distributed to other management team members informing them of the decision.
- On-site personnel who will be in charge when the compliance officer arrives at the scene should be trained in all aspects of the pre-inspection and inspection processes as well as the potential risks involved.
- All appropriate statements, forms, and other documents should be (a) provided to the compliance officer, (b) used for training the team members responsible for documenting the inspection, and (c) used during the actual inspection by team members.
- On-site team members should be provided with necessary equipment to document all aspects of the inspection (i.e., cameras and noise dosimeters) and must be properly trained to use, maintain, and calibrate the equipment and to document test results.

In short, the management team responsible for performance of the company directive should be well prepared to address any and all issues or circumstances that might arise while the compliance officer is on site.

Four routes for challenging an administrative search warrant are normally available to the employer, depending on the court and the circumstances:

1. Seek to enjoin issuance of the administrative warrant in federal district court.
2. Refuse to permit the inspection after a warrant is issued and then move to quash the warrant or civil contempt proceedings brought by OSHA.
3. Seek to enjoin enforcement of citations in federal district court after the inspection has taken place under protest.
4. Contest the validity of the warrant after the inspection has taken place before the OSHRC.

Attempting to have the court enjoin OSHA from acquiring an ex parte warrant is normally difficult, given the fact that the employer seldom knows that the compliance officer is attempting to acquire an ex parte warrant. In 1980, OSHA reintroduced its regulation authorizing ex parte warrant applications. Under this regulation, an employer's demand for a search warrant for a previous inspection is one of the factors that OSHA will consider in finding that an ex parte warrant is "desirable and necessary" for subsequent inspections, even before seeking the employer's consent. Refusal to admit a compliance inspector without a search warrant is only one of the situations where an ex parte warrant can be sought. Conversely, OSHA has not always elected to pursue ex parte warrants in certain circumstances, such as the denial of entry, where grounds existed for pursuing such a warrant.

Defying the search warrant after the compliance officer has obtained it from the magistrate or court is potentially the most dangerous route for challenging a search warrant. As in *Barlow's*, the employer may move the court to quash the search warrant, or wait and defend a refusal to comply with the search warrant in civil contempt proceedings initiated by OSHA. Although the Third Circuit in *Babcock & Wilcox Co. v. Marshall* found that the federal court had authority to test the validity of a search warrant before an OSHA inspection, other courts have found that a motion to quash is not the proper method to challenge a warrant. Selection of this route carries many other potential dangers. In addition to assessing other penalties, courts have found that some employers lacked good faith in defying the OSHA warrant and placed those employers in civil contempt of court.

The route for judicially challenging OSHA's enforcement authority is usually taken after the compliance officer completes the inspection under a search warrant. This option has not been successful. Most courts have found that, even if the inspection is completed under a search warrant and under protest, jurisdiction lies with the OSHRC and not with the courts. The leading case in this area is *In re Quality Products, Inc.* The court found that the magistrate who issued the warrant had no authority to "stay and recall" the administrative search warrant. In addition, the court had no jurisdiction to consider the warrant's validity in a separate action while OSHA enforcement proceedings were pending (i.e., the employer would have to exhaust all administrative remedies with the OSHRC first). In the few decisions that have permitted a motion to quash after an inspection, the motion to quash was treated as a motion to suppress the evidence obtained during the warrant required and acquired during the protested inspection.

The most frequently used route for challenging a search warrant is to challenge the validity of the warrant with the OSHRC after an inspection. The OSHRC does not have the authority to question the validity of an administrative search warrant before an inspection; however, after the inspection takes place, the OSHRC does have the authority and jurisdiction to rule on the constitutionality of an administrative search warrant obtained for the purpose of conducting an OSHA inspection. The employer must affirmatively present any challenges to the administrative search warrant to the OSHRC for review. The OSHRC's authority is consistent with the holding of several courts; the employer must exhaust all administrative remedies before seeking judicial relief.

Safety and loss prevention professionals may want to consider alternatives to requiring an administrative search warrant of OSHA. In many circumstances, understanding the scope of the inspection to be performed, or a simple telephone call to the regional director can solve most inspection concerns or conflicts. Another alternative is the use of protective orders. These can modify the terms of the inspection, the time of the inspection, and other conditions to make the inspection more reasonable to the employer.

In short, requiring an administrative search warrant should be the last alternative considered when addressing an OSHA inspection situation. In most circumstances, the chances of prevailing or preventing the compliance officer from conducting an inspection of the worksite are minimal. In addition, the cost in management time, expected effort, legal fees, potential loss of goodwill in the community, or the creation of ill will with the local OSHA office generally outweighs the potential benefits of preventing an inevitable OSHA inspection.

DISCRIMINATION PROTECTION UNDER THE OSH ACT

WHO IS PROTECTED?

The OSH Act provides employees working for covered employers basic rights to file complaints with OSHA, to be protected against discrimination for reporting violations of the OSH Act, and even the right to refuse unsafe work without retaliation. In addition to the OSH Act, employees have additional protection under other federal laws, such as the National Labor Relations Act and state and local laws.

What Type of Employee Is Afforded Protection under the OSH Act?

By regulation, the secretary of labor has interpreted the term person as defined by section 3(4) of the OSH Act, and section 11(c) "no person shall discharge or in any manner discriminate against any employee ..." and the discrimination prohibitions "are not limited to actions taken by employers against their own employees." Thus, the term employee must be literally construed and extended to applicants for employment as well as traditional employees. Additionally, the extent of the business relationship "is to be based upon economic realities rather than upon common law doctrines and concepts" due to the "broad remedial nature" of the Act. However, given the statutory definition and interpretations of employer and employee under the Act, public sector employees of states or other political subdivisions are excluded from the discriminatory protection provided under the Act.

Protection under the OSH Act has been extended to labor organization representatives working within the confines of the employer's premises. In *Marshall v. Kennedy Tubular Products*, the court also extended protection against discrimination to a union business agent who was banned from the employer's facility after reporting safety violations to OSHA. The court ordered the employer to allow the business agent to return to the facility and participate in safety meetings. The court determined that the employer's discrimination against the employee's union representative was, in effect, discrimination against the represented employees.

WHAT ACTIVITIES ARE PROTECTED?

The OSH Act prohibits discharging or otherwise discriminating against an employee who has filed a complaint, instituted or testified in any proceeding, or otherwise exercised any right afforded by the Act. The Act also specifically gives employees the right to contact OSHA and request an inspection without retaliation from the employer if the employee believes a violation of a health or safety standard threatens physical harm or creates an imminent danger. Employees exercising the right to contact OSHA with a complaint can also remain anonymous to the employer and to the public under the Act.

Employees are also protected against discrimination under the Act when testifying in proceedings under or related to the Act, including inspections, employee-contested abatement dates, employee-initiated proceedings for promulgating new standards, employee applications for modifying or revoking variances, employee-based judicial challenges to OSHA standards, or employee appeals from decisions by the OSHRC. An employee "need not himself directly institute the proceedings" but may merely set "into motion activities of others which result in proceedings under or related to the Act."

When testifying in any proceeding related to the Act, employees are protected against discrimination by employers. This protection is extended to proceedings instituted or caused to be instituted by the employee, as well as "any statement given in the course of judicial, quasi-judicial, and administrative proceedings, including inspections, investigations, and administrative rule making or adjudicative functions."

The OSH Act also provides protection against discrimination for employees who petition for hearings on variance requests, request inspections, challenge abatement dates, accompany the OSHA inspector during the inspection, participate in and challenge OSHRC decisions and citation contests, and bring actions for injunctive relief against the secretary of labor for imminent danger situations.

There are few reported cases of discrimination and, although employee rights appear to be straightforward under the Act, determining when the protection is relevant to the employee and the situation remains an unresolved area.

In *Dunlop v. Hanover Shoe Farms*, the employer argued that the employee was terminated for just cause before a complaint was filed with OSHA. The court, in rejecting the employer's argument, found that the employee's complaint of unsafe and unhealthful working conditions was filed five days before his termination. Therefore, the employer discriminated against the employee when it discharged him for exercising his rights to notify OSHA.

WAIVER OF RIGHTS

In *Marshall v. N.L. Industries,* the court addressed the issue of an employee waiving discriminatory rights provided under the OSH Act. In this case, the Seventh Circuit Court of Appeals held that an employee's acceptance of an arbitration award did not preclude the secretary of labor from bringing an action against the employer based upon the same facts. Specifically, the employee refused to load metal scraps into a melting kettle because the pay-loader did not have a windshield or enclosed cab to protect him from the molten metal. The employer discharged the employee, and the employee filed a complaint with OSHA and a grievance with his union. An arbitrator awarded the employee reinstatement without back pay, and the employee accepted the award. The lower court found that acceptance of the award constituted a voluntary waiver of the right to statutory relief under the Act. The Seventh Circuit reversed the decision, finding that "the OSHA legislation was intended to create a separate and general right of broad social importance existing beyond the parameters of an individual labor agreement and susceptible of full vindication only in a judicial forum."

FILING A COMPLAINT AGAINST AN EMPLOYER

Specific administrative rules govern the nondiscrimination provisions of the OSH Act. An employee who believes he or she has been discriminated against may file a complaint with the secretary within 30 days of the alleged violation. The purpose of the 30-day limitation is "to allow the Secretary to decline to entertain complaints that have become stale." This relatively short period can be tolled under special circumstances and has no effect on other causes of action. When an employee has filed a complaint, the secretary must notify him or her as to whether or not an action will be filed on his or her behalf in federal court. At least one court has ruled that OSHA may bring discrimination action against corporate officers as individuals as well as against the corporation itself and the officers in their official capacities.

Regarding an employee's right to refuse unsafe or unhealthy work, the Supreme Court, in *Whirlpool Corp. v. Marshall,* stated:

> [C]ircumstances may exist in which the employee justifiably believes that the express statutory arrangement does not sufficiently protect him from death or serious injury. Such circumstances will probably not often occur, but such a circumstance may arise when (1) the employee is ordered by the employer to work under conditions that the employee reasonably believes pose an imminent risk of death or serious bodily injury, and (2) the employee has reason to believe that there is not sufficient time or opportunity either to seek effective redress from the employer or to apprise OSHA of the danger.

In this case, two employees refused to perform routine maintenance tasks that required them to stand on a wire mesh guard approximately 20 feet above the work surface. The mesh screen was designed to catch appliance components that might fall from an overhead conveyor. While performing this activity in the past, several employees had punctured the screen, and one employee died after falling through the mesh guard. The employees refused to perform the task and the employer reprimanded them. The district court denied relief, but the Sixth Circuit reversed the decision. The Supreme Court, in affirming the Sixth Circuit, found the Act's provisions were "designed to give employees full protection in most situations from the risk of injury or death resulting from an imminently dangerous condition at the worksite."

PRIVATE LITIGATION UNDER THE OSH ACT

Although there is no common law basis for actions under the OSH Act, OSHA regulations are used in many tort actions, such as negligence and product liability suits, as evidence of the standard of care and conduct to which the party must comply. Additionally, documents generated in the course of business that are required under the OSH Act are usually discoverable under the Freedom of Information Act (FOIA) and can be used as evidence of a deviation from the required standard of care. According to section 653(b)(4) of the OSH Act:

> Nothing in this Act shall be construed to supersede or in any manner affect any workmen's compensation law or to enlarge or diminish or affect in any other manner the common law or statutory rights, duties, or liabilities of employers and employees under any law with respect to injuries, diseases, or death of employees arising out of, or in the course of, employment.

This language prevents injured employees or families of employees killed in work-related accidents from directly using the OSH Act or OSHA standards as an independent basis for a cause of action (i.e., wrongful death actions). However, many federal and state courts have found that section 653(b)(4) does not bar application of the OSH Act or OSHA standards in workers' compensation litigation or application of the doctrine of negligence or negligence per se to an OSHA violation. These decisions distinguish between use of an OSHA standard as the basis for a standard of care in a state or federal common law action and the use of OSH Act or OSHA standards to create a separate and independent cause of action. *tort – act or admission*

NEGLIGENCE ACTIONS

OSHA standards are most widely used in negligence actions. The plaintiff in a negligence action must prove the four elements: duty, breach of duty, causation, and damages. Black's Law Dictionary defines negligence per se as conduct, whether of action or omission, that may without any argument or proof as to the particular surrounding circumstances, either because it is in violation of a statute or valid municipal ordinance, or because it is so palpably opposed to the dictates of common prudence that it can be said without hesitation or doubt that no careful person would have been guilty of it.

In simpler terms, if a plaintiff can show that an OSHA standard applied to the circumstances and the employer violated the OSHA standard, the court can eliminate the plaintiff's burden of proving the negligence elements of duty and breach through a finding of negligence per se.

The majority of courts have found that relevant OSHA standards and regulations are admissible as evidence of the standard of care, and thus violation of OSHA standards can be used as evidence of an employer's negligence or negligence per se. It should be noted, however, that some courts have prohibited use of OSHA standards and regulations, and evidence of their violation, if the proposed purpose of the OSHA standards' use conflicts with the purposes of the OSH Act, unfairly prejudices a party, or is meant to enlarge a civil cause of action. The Fifth Circuit, reflecting the general application, approved the admissibility of OSHA standards as evidence of negligence, but permits the court to accept or reject the evidence as it sees fit.

In using OSHA standards to prove negligence per se, safety and loss prevention professionals should be aware that numerous courts have recognized the OSHA standards as the reasonable standard of conduct in the workplace. With this recognition, a violation by the employer would constitute negligence per se to the employee. A few other courts have held, however, that violations of OSHA standards can never constitute negligence per se because of section 653(b)(4) of the Act.

In *Walton v. Potlatch Corp.*, the court set forth four criteria to determine whether OSHA standards and regulations could be used to establish negligence per se:

1. The statute or regulation must clearly define the required standard of conduct.
2. The standard or regulation must have been intended to prevent the type of harm that the defendant's act or omission caused.
3. The plaintiff must be a member of the class of persons that the statute or regulation was designed to protect.
4. The violation must have been the proximate cause of the injury.

If the court provides an instruction on negligence per se rather than an instruction on simple negligence, the effect is that the jury cannot consider the reasonableness of the employer's conduct. In essence, the court has already established a violation that constituted unreasonable conduct on the part of the employer and has also established that the conduct was prohibited or required under a specific OSHA standard. Thus, as a matter of law, the jury will not be permitted to address the reasonableness of the employer's actions.

OSHA Standards as a Defense

Under appropriate circumstances, an employer may be able to use OSHA standards and regulations as a defense. Simple compliance with required OSHA standards is not in and of itself a defense, and the use of OSHA standards as a defense has received mixed treatment by the courts. However, at least one court has held that a violation of a state OSHA plan by an employee could be considered in determining the employee's comparative negligence in a liability case. Use of OSHA standards and regulations to demonstrate an appropriate standard of care in third-party product liability actions, workers' compensation litigation, and other actions may be permitted and should be explored by safety and loss prevention professionals in appropriate circumstances.

The use of OSHA citations and penalties in tort actions has also received mixed treatment by the courts. In *Industrial Tile v. Stewart*, the Alabama Supreme Court stated:

> We hold that it was not error to admit the regulation if the regulations are admissible as going to show a standard of care, then it seems only reasonable that the evidence of violation of the standards would also be admissible as evidence that the defendant failed to meet the standards that it should have followed. Clearly, the fact that Industrial Tile had been cited by OSHA for violating the standards, and the fact that Industrial Tile paid the fine, are relevant to the conduct of whether it violated the standards of care applicable to its conduct. It was evidenced from a number of witnesses that the crane violated the 10-foot standards. It seems to us that evidence Industrial Tile paid the fine without objection was properly admitted into evidence as a declaration against interest.

Other courts have found that OSHA citations and fines are inadmissible under the hearsay rule of the Federal Rules of Evidence. However, this usually can be easily overcome by offering a certified copy of the citations and penalties to the court, under the investigatory report exception to the Federal Rules of Evidence.

Investigation records and other documents gathered in the course of an OSHA inspection are normally available under the FOIA. As noted previously, if particular citations are deemed inadmissible, a certified copy of the citations and penalties is normally considered admissible under section 803(8)(c) of the Federal Rules of Evidence and 28 U.S.C. Section 1733 governing admissibility of certified copies of government records. Although the issue of whether OSHA citations and penalties are admissible is determined by the court under the rules of evidence, safety and loss prevention professionals should be prepared for all of the documents collected or produced during an OSHA inspection or investigation to be presented to the court. Given the nature of these government documents and the methods of presenting OSHA documents under the Federal Rules of Evidence, it is highly likely in any type of related litigation that the opposing party will obtain the documents from OSHA and that they will be submitted for use at trial. Other information and documents, such as photographs, recordings, and samples, may also be admissible under the same theory. Thus, safety and loss prevention professionals should maintain as much control as possible over information gathered during an investigation or inspection (i.e., trade secrets and speculation by management team members) and be prepared for the information to become public through the FOIA and used by opponents in litigation or elsewhere.

In addition to direct litigation with OSHA and in tort actions, OSHA standards used as evidence of the standard of care and citations used to show a breach of the duty of care have also been used in product liability.

Within the fifteen (15) working day period to file a notice to contest, it is imperative that safety and loss prevention professionals analyze each and every word of the citation and the alleged violations, as well as the proposed monetary penalties and categorization of the proposed penalties. The safety and loss prevention professional, along with the management team, legal counsel, and other applicable parties, should determine the potential defenses that may be available, the cost effectiveness of appeal, the timeliness of the abatement, and related issues in order to develop a defense strategy for each and every alleged violation. Again, as previously noted, time is of the essence in preparing this strategy and identifying applicable defenses.

One avenue to address the alleged violations is the use of an informal conference, which is very cost effective as well as applicable if the alleged violation has been corrected, additional time is needed for abatement, or related issues are applicable. Within the fifteen (15) working days time period, the safety and loss prevention professional may request an informal conference with the assistant regional director or state plan director to discuss, on an informal basis, any issues regarding the "inspection, citation, notice of proposed penalty, or notice of intention to contest." The informal conference is usually held at the local OSHA office in a conference room or office area. Attending for OSHA is usually the assistant regional director or his/her designee, the compliance officer(s) who conducted the inspection, and other representatives from OSHA. Although the employer may be represented by legal counsel, safety and loss prevention professionals as well as other members of the management team often represent their employer at the informal conference.

The informal conference forum is usually an open discussion with an agenda that is directed by the safety and loss prevention professional. Safety and loss prevention professionals should come to the informal conference prepared with all arguments, ideas, and supporting documentation with the objective of settling the citation(s). Alleged violations cited from the inspection that have been repaired or otherwise corrected should be documented with photographs, video recordings, repair orders, or other documentation. The good faith shown by the employer through the safety and loss prevention professional is often taken into account when requesting reductions in violation categorization and monetary penalties. Safety and loss prevention professionals should be aware that most informal conferences are not adversarial proceedings, but more of a discussion and exchange of ideas focused on improving safety and health in the workplace. This being said, there is no substitute for proper preparation for an informal conference.

In the event that a settlement cannot be reached at the informal conference, safety and loss prevention professionals must protect their employer's right to appeal within the fifteen (15) work day time period. Although legal counsel often represents the employer at the more formal stages of appeal before the Occupational Safety and Health Review Commission, it is vital that the safety and loss prevention professional file the notice of contest in a timely manner or ensure that legal counsel is aware of the time limitations. The notice of contest can take a variety of written forms and, in several state plan states, a notice of contest page is provided in the handout of rights that often accompanies the citation. The safety and loss prevention professional must ensure that the notice of contest document is in writing and filed in a timely manner with the OSHA or state plan office. Safety and loss prevention professionals must be aware that failure to file the notice of contest within the fifteen (15) working days will result in the employer losing all rights to contest the citation(s) and any/all aspect or issues involving the citation(s).

DEFENSES TO AN OSHA CITATION

Safety and loss prevention professionals should start with the fact that OSHA possesses the burden of proof, or in essence, OSHA must prove each and every alleged violation. The preponderance of the evidence standard is utilized in Occupational Safety and Health Review Commission hearings and rules of evidence are utilized throughout. At this level of appeal, most employers are represented by legal counsel. However, the safety and loss prevention professional is essential in assisting legal counsel in preparing the defenses to the alleged violations.

In general, there are several layers of potential defenses, ranging from the broad constitutional challenges to procedural and jurisdictional defenses to the more specific factual defenses. Most safety and loss prevention professionals will not be involved in the constitutional and jurisdictional defenses, but often play a vital role in developing and designing the procedural and factual defenses. Procedural and factual defenses are usually based upon the actions or inactions of OSHA or the state plan program during and after the compliance inspection. Working with legal counsel, safety and loss prevention professionals often examine and analyze each of the possible defenses to ascertain whether or not the defense is applicable to the facts of the given situation. In essence, because OSHA possesses the burden of proving each and every alleged violation, the procedural and factual defenses are attempting to identify issues that may create difficulties in proving the alleged violation or vacate the alleged violation because of other facts or situations.

Within the area of procedural defenses, these technical defenses were often raised early in the history of the OSH Act. However, they are infrequently successful today because the case law regarding these defenses has been established. Procedural defenses are usually very technical in nature, such as a defect in the inspection procedure, and are usually not successful unless the safety and loss prevention professional possesses facts that can show actual prejudice in the ability to defend against the citation or alleged violations. Although more jurisdictional than procedural, two general defenses in this area have been utilized, including the statute of limitation defense and the lack of reasonable promptness defense. The statute of limitation defense is often utilized when the citation is issued beyond six (6) months from the time of the alleged violation. In general, if the citation is not issued within six (6) months from the date of the alleged violation, this is grounds to dismiss the citation. The lack of reasonable promptness defense, although not an absolute defense, generally involves a similar delay in issuing the citation that prejudices the employer in preparing the defense. Safety and loss prevention professionals should be aware that at least one court has found that the six (6) month statute of limitation is not applicable when an alleged criminally willful violation has been issued by OSHA. Other potential procedural defenses can include the failure to timely forward notice of contested defenses (OSHA possesses a seven (7) day requirement to forward the notice of contest to the OSHRC) and improper service defense.

Of particular importance within the category of procedural defenses for safety and loss prevention professionals working on government or military worksites is the preemption defense. In essence, this defense involves whether OSHA possesses the jurisdiction to inspect the particular workplace and issue a citation regarding the alleged violations found during this inspection. In general, the OSHRC has interpreted the preemption clause to determine OSHA's jurisdiction very narrowly by utilizing the test of (1) whether a regulation has actually been promulgated by an agency other than OSHA, and (2) whether that regulation covers the specific working conditions at issue. In general, OSHA does not possess concurrent jurisdiction with other governmental agencies and usually will expand its jurisdiction unless there is another governmental agency with a specific regulation addressing safety and health. This jurisdiction defense can also be applicable within state plan states.

One procedural defense used by safety and loss prevention professionals working within the construction industry and other industries with contractors on site is the lack of an employment relationship defense. This defense may be applicable for general contractors where the alleged hazard was created by and controlled by a subcontractor and only the subcontractor's employees are exposed to the alleged hazard.

Within the factual defense area, safety and loss prevention professionals can be invaluable to legal counsel in preparing the strategy and defenses. The factual defenses are often based on the specific actions, inactions, and circumstances that took place during the actual compliance inspection. The experience and expertise of the safety and loss prevention professional, as well as their documentation during the compliance inspection, is often used to assist legal counsel in preparing these factual defenses.

One of the most utilized factually based defenses, which is used where the safety and loss prevention professional has a well-established safety and health program, is the defense of unpreventable employee misconduct. To prove this defense, safety and loss prevention professionals must gather documentation to assist legal counsel to prove the following circumstances:

1. There is an established work rule or policy to prevent the violation from occurring.
2. The work rule or policy was adequately communicated to all employees, including their supervisors or team leaders.
3. The safety and loss prevention professional or other management team member took reasonable steps to discover the alleged violation.
4. The safety and loss prevention professional or other management team member, as the agent of the employer, effectively enforced the work rule or policy when employees violated the work rule or policy or when the safety and loss prevention professional or other management team member discovered the violation.

Safety and loss prevention professionals can assist legal counsel in developing this defense by providing written compliance programs, written safety policies, documentation of fair and consistent disciplinary action for violations and training and education documentation, as well as safety inspections, safety audits, and related documentation. Safety and loss prevention professionals should be aware that some of the circuit courts require the defense of unpreventable Employee Misconduct to be plead as an affirmative defense. Thus, it is vital that the safety and loss prevention professional communicate this possible defense to legal counsel prior to any filing with the OSHRC.

Another commonly utilized factual defense by safety and loss prevention professionals with established safety and health programs is the Isolated Incident defense. This defense is exactly what the title implies—namely, the employer possesses an adequate safety and health program and the alleged violation observed by the compliance officer was an isolated incident, which is out of the norm within the safety and health program and efforts of the employer. Safety and loss prevention professionals should be aware that in order to establish this defense, documentation and other supporting evidence must support the demonstration of the following circumstances:

1. The alleged violation resulted exclusively from the employee's conduct or misconduct.
2. The alleged violation was not "participated in, observed by or performed with the knowledge and/or consent of any supervisory personnel."
3. The employee's conduct or misconduct contravened a well-established company policy or work rule that was in effect at the time, well-published to the employees, and actively enforced through disciplinary action or other appropriate procedures.

Safety and loss prevention professionals should be aware that the "employer must have a specific program for instructing employees in safe work practices" in place at the time of the alleged violation. Thus, written programs, training documents, training schedules, and other documentation become vital to this defense. As with the unpreventable employee misconduct defense, the isolated incident defense is an affirmative defense. Thus, it is important for safety and loss prevention professionals to communicate the use of this possible defense to legal counsel as soon as possible.

Depending on the circumstances of the alleged violation, safety and loss prevention professionals should be aware that the greater hazard in compliance defense may be applicable. This defense, in essence, challenges the OSHA standard for the particular situation, arguing that compliance with the standard would create a greater hazard for employees than an alternative that is not in compliance with the promulgated standard. Safety and loss prevention professionals should be aware that this defense carries a substantial evidentiary burden to provide the following evidence:

1. The hazards of compliance with the standard are greater than the hazards of noncompliance.
2. Alternative means of protection were used or were unavailable.
3. A variance was unavailable or application for a variance would be inappropriate.

To bolster this defense, safety and loss prevention professionals must have explored and documented all possible alternative methods to protect employees beyond the methods identified in the standard. Safety and loss prevention professionals should be aware that if compliance simply created another or additional hazard, but the other or additional hazard could be controlled, the use of this defense is in jeopardy. Prudent safety and loss prevention professionals encountering this situation may consider pursuing a variance if applicable prior to any compliance inspection.

The correlating defense of impossibility or infeasibility of compliance defense may also be available to safety and loss prevention professionals, depending on the situation. In essence, the basis of this defense is that achieving compliance with the standard is impossible because of the nature of the specific job or work. This defense is very narrow and safety and loss prevention professionals must be aware that the burden shifts to the employer to prove this defense. The safety and loss prevention professional should be aware that there are two levels of proof, with the first being

1. "it would have been technologically or economically infeasible to implement the standard's requirements under the circumstances, or
2. compliance with the standard would have precluded performance of necessary work operations."

Additionally, the safety and loss prevention professional must then prove that

1. "the company used an alternative method of employee protection; or
2. no feasible alternative means of protection was available."

Safety and loss prevention professionals should be aware that this defense carries a heavy burden of proof. Difficulty achieving compliance or the fact that compliance may be inconvenient or expensive is usually not sufficient to sustain this burden. Safety and loss prevention professionals should be prepared to show that all possible alternative forms of protection had been explored and tested, where feasible. Additionally, safety and loss prevention professionals should assemble documentation to prove that required work could not be properly performed if the standard is utilized to achieve compliance. If the economically infeasible argument is to be made, safety and loss prevention professionals should be prepared to demonstrate that the cost is unreasonable in light of the protection afforded and to show any adverse effects the cost would have on the business as a whole.

A relatively easy affirmative defense that safety and loss prevention professionals should bring to the attention of legal counsel, if applicable, is that of the machine was not in use. If the facts of the situation support that the machine cited by the compliance officer during the inspection was not in use and, thus, could not create the alleged violation identified in the citation, the defense that the machine or equipment was not in use may be a viable defense.

Lastly, although technically not a factual defense, safety and loss prevention professionals, working in conjunction with legal counsel, have successfully argued the lack of the employer's knowledge in specific situations. More specifically, within the definition of a serious violation in the OSH Act, is the statement that "unless the employer did not, and could not with the exercise of reasonable diligence, know of the presence of the violation." Thus, with violations categorized as serious, safety and loss prevention professionals may challenge OSHA's proof of the employer's knowledge of the alleged serious violation.

The previously listed defenses are several of the commonly utilized defenses to alleged violations that have been utilized by safety and loss prevention professionals and legal counsel. These defenses are not all-inclusive and creative safety and loss prevention professionals, working with legal counsel, can shape and create new defenses depending on the facts and circumstances. Safety and loss prevention professionals should consider taking proactive advantage of the established variance process prior to an inspection, if feasible. After a citation is issued, most safety and loss prevention professionals, as agents for their company, must work diligently with legal counsel to defend their company utilizing the defenses available.

SELECTED CASE STUDY

Note: Case may be modified for the purpose of this text.
Supreme Court of the United States

Ray MARSHALL, Secretary of Labor, et al., Appellants, v. BARLOW'S, INC.

No. 76-1143.
Argued Jan. 9, 1978.
Decided May 23, 1978.

Action was brought by an employer to enjoin enforcement of inspection provisions of the Occupational Safety and Health Act of 1970. The Three-Judge District Court, 424 F.Supp. 437, held that the Fourth Amendment required a warrant for type of OSHA search involved and that statutory authorization for warrantless inspections was unconstitutional, and the Secretary of Labor appealed. The Supreme Court, Mr. Justice White, held that the Occupational Safety and Health Act, which empowers Secretary's agents to search work area of any employment facility within OSHA's jurisdiction for safety hazards and violations, is unconstitutional insofar as it purports to authorize inspections without warrant or its equivalent. Affirmed.

On the morning of September 11, 1975, an OSHA inspector entered the customer service area of Barlow's, Inc., an electrical and plumbing installation business located in Pocatello, Idaho. The president and general manager, Ferrol G. "Bill" Barlow, was on hand; and the OSHA inspector, after showing his credentials, informed Mr. Barlow that he wished to conduct a search of the working areas of the business. Mr. Barlow inquired whether any complaint had been received about his company. The inspector answered no, but that Barlow's, Inc., had simply turned up in the agency's selection process. The inspector again asked to enter the nonpublic area of the business; Mr. Barlow's response was to inquire whether the inspector had a search warrant. The inspector had none. Thereupon, Mr. Barlow refused the inspector admission to the employee area of his business. He said he was relying on his rights as guaranteed by the Fourth Amendment of the United States Constitution.

Three months later, the Secretary petitioned the United States District Court for the District of Idaho to issue an order compelling Mr. Barlow to admit the inspector. The requested order was issued on December 30, 1975, and was presented to Mr. Barlow on January 5, 1976. Mr. Barlow again refused admission, and he sought his own injunctive relief against the warrantless searches assertedly permitted by OSHA. A three-judge court was convened. On December 30, 1976, it ruled in Mr. Barlow's favor. 424 F. Supp. 437. Concluding that *Camara v. Municipal Court*, 387 U.S. 523, 528-529, 87 S. Ct. 1727, 1730, 1731, 18 L. Ed. 2d 930 (1967), and *See v. City of Seattle*, 387 U.S. 541, 543, 87 S.Ct. 1737, 1739, 18 L.Ed.2d 943 (1967), controlled this case, the court held that the Fourth Amendment required a warrant for the type of search involved here and that the statutory authorization for warrantless inspections was unconstitutional. An injunction against searches or inspections pursuant to § 8(a) was entered. The Secretary appealed, challenging the judgment, and we noted probable jurisdiction. 430 U.S. 964, 98 S.Ct. 474, 54 L.Ed.2d 309.

The Secretary urges that warrantless inspections to enforce OSHA are reasonable within the meaning of the Fourth Amendment. Among other things, he relies on § 8(a) of the Act, 29 U.S.C. § 657(a), which authorizes inspection of business premises without a warrant and which the Secretary urges represents a congressional construction of the Fourth Amendment that the courts should not reject. Regrettably, we are unable to agree.

The Warrant Clause of the Fourth Amendment protects commercial buildings as well as private homes. To hold otherwise would belie the origin of that Amendment,**1820 and the American colonial experience. An important forerunner of the first 10 Amendments to the United States Constitution, the Virginia Bill of Rights, specifically opposed "general warrants, whereby an officer or messenger may be commanded to search suspected places without evidence of a fact committed." The general warrant was a recurring point of contention in the Colonies immediately preceding the Revolution. The particular offensiveness it engendered was acutely felt by the merchants and businessmen whose premises and products were inspected for compliance with the several parliamentary revenue measures that most irritated the colonists. "[T]he Fourth Amendment's commands grew in large measure out of the colonists' experience with the writs of assistance ... [that] granted sweeping power to customs officials and other agents of the King to search at large for smuggled goods." *United States v. Chadwick*, 433 U.S. 1, 7-8, 97 S.Ct. 2476, 2481, 53 L.Ed.2d 538 (1977). *312 See also *G. M. Leasing Corp. v. United States*, 429 U.S. 338, 355, 97 S.Ct. 619, 630, 50 L.Ed.2d 530 (1977). Against this background, it is untenable that the ban on warrantless searches was not intended to shield places of business as well as of residence.

This Court has already held that warrantless searches are generally unreasonable, and that this rule applies to commercial premises as well as homes. In *Camara v. Municipal Court*, *supra*, 387 U.S., at 528-529, 87 S.Ct., at 1731, we held:

[E]xcept in certain carefully defined classes of cases, a search of private property without proper consent is 'unreasonable' unless it has been authorized by a valid search warrant.

On the same day, we also ruled:

> As we explained in *Camara*, a search of private houses is presumptively unreasonable
> if conducted without a warrant. The businessman, like the occupant of a residence, has
> a constitutional right to go about his business free from unreasonable official entries
> upon his private commercial property. The businessman, too, has that right placed in
> jeopardy if the decision to enter and inspect for violation of regulatory laws can be
> made and enforced by the inspector in the field without official authority evidenced by
> a warrant. *See v. City of Seattle, supra*, 387 U.S., at 543, 87 S.Ct., at 1739.

These same cases also held that the Fourth Amendment prohibition against unreasonable searches protects against warrantless intrusions during civil as well as criminal
investigations. *Ibid.* The reason is found in the "basic purpose of this Amendment …
[which] is to safeguard the privacy and security of individuals against arbitrary inva-
sions by governmental officials." *Camara, supra*, 387 U.S., at 528, 87 S.Ct. at 1730. If
the government intrudes on a person's property, the privacy interest suffers whether
the government's motivation is to investigate violations of criminal laws or breaches
of other statutory or *313 regulatory standards. It therefore appears that unless some
recognized exception to the warrant requirement applies, *See v. City of Seattle* would
require a warrant to conduct the inspection sought in this case.

The Secretary urges that an exception from the search warrant requirement has
been recognized for "pervasively regulated business[es]," *United States v. Biswell*,
406 U.S. 311, 316, 92 S.Ct. 1593, 1596, 32 L.Ed.2d 87 (1972), and for "closely
regulated" industries "long subject to close supervision and inspection." **1821
Colonnade Catering Corp. v. United States, 397 U.S. 72, 74, 77, 90 S.Ct. 774, 777, 25
L.Ed.2d 60 (1970). These cases are indeed exceptions, but they represent responses
to relatively unique circumstances. Certain industries have such a history of government oversight that no reasonable expectation of privacy, see *Katz v. United States*,
389 U.S. 347, 351-352, 88 S.Ct. 507, 511, 19 L.Ed.2d 576 (1967), could exist for a
proprietor over the stock of such an enterprise. Liquor (*Colonnade*) and firearms
(*Biswell*) are industries of this type; when an entrepreneur embarks upon such a business, he has voluntarily chosen to subject himself to a full arsenal of governmental
regulation.

Industries such as these fall within the "certain carefully defined classes of
cases," referenced in *Camara*, 387 U.S., at 528, 87 S.Ct., at 1731. The element that
distinguishes these enterprises from ordinary businesses is a long tradition of close
government supervision, of which any person who chooses to enter such a business
must already be aware. "A central difference between those cases [*Colonnade* and
Biswell] and this one is that businessmen engaged in such federally licensed and regulated enterprises accept the burdens as well as the benefits of their trade, whereas
the petitioner here was not engaged in any regulated or licensed business. The businessman in a regulated industry in effect consents to the restrictions placed upon
him." *Almeida-Sanchez v. United States*, 413 U.S. 266, 271, 93 S.Ct. 2535, 2538,
37 L.Ed.2d 596 (1973).

The clear import of our cases is that the closely regulated industry of the type
involved in *Colonnade* and *Biswell* is the exception. The Secretary would make it

the rule. Invoking *314 the Walsh-Healey Act of 1936, 41 U.S.C. § 35 et seq., the Secretary attempts to support a conclusion that all businesses involved in interstate commerce have long been subjected to close supervision of employee safety and health conditions. But the degree of federal involvement in employee working circumstances has never been of the order of specificity and pervasiveness that OSHA mandates. It is quite unconvincing to argue that the imposition of minimum wages and maximum hours on employers who contracted with the Government under the Walsh-Healey Act prepared the entirety of American interstate commerce for regulation of working conditions to the minutest detail. Nor can any but the most fictional sense of voluntary consent to later searches be found in the single fact that one conducts a business affecting interstate commerce; under current practice and law, few businesses can be conducted without having some effect on interstate commerce.

The Secretary also attempts to derive support for a Colonnade-Biswell-type exception by drawing analogies from the field of labor law. In *Republic Aviation Corp. v. NLRB*, 324 U.S. 793, 65 S.Ct. 982, 89 L.Ed. 1372 (1945), this Court upheld the rights of employees to solicit for a union during nonworking time where efficiency was not compromised. By opening up his property to employees, the employer had yielded so much of his private property rights as to allow those employees to exercise § 7 rights under the National Labor Relations Act. But this Court also held that the private property rights of an owner prevailed over the intrusion of nonemployee organizers, even in nonworking areas of the plant and during nonworking hours. *NLRB v. Babcock & Wilcox Co.*, 351 U.S. 105, 76 S.Ct. 679, 100 L.Ed. 975 (1956).

The critical fact in this case is that entry over Mr. Barlow's objection is being sought by a Government agent. Employees are not being prohibited from reporting OSHA violations. What they observe in their daily functions is undoubtedly beyond the employer's reasonable expectation of privacy. The Government inspector, however, is not an employee. Without a warrant he stands in no better position than a member of the public. What is observable by the public is observable, without a warrant, by the Government inspector as well. The owner of a business has not, by the necessary utilization of employees in his operation, thrown open the areas where employees alone are permitted to the warrantless scrutiny of Government agents. That an employee is free to report, and the Government is free to use, any evidence of noncompliance with OSHA that the employee observes furnishes no justification for federal agents to enter a place of business from which the public is restricted and to conduct their own warrantless search.

The Government has asked that Mr. Barlow be ordered to show cause why he should not be held in contempt for refusing to honor the inspection order, and its position is that the OSHA inspector is now entitled to enter at once, over Mr. Barlow's objection.

I.

The Secretary nevertheless stoutly argues that the enforcement scheme of the Act requires warrantless searches, and that the restrictions on search discretion contained in the Act and its regulations already protect as much privacy as a warrant would. The Secretary thereby asserts the actual reasonableness of OSHA searches, whatever the general rule against warrantless searches might be. Because "reasonableness is still the ultimate standard," *Camara v. Municipal Court*, 387 U.S., at 539, 87 S.Ct., at 1736, the Secretary suggests that the Court decide whether a warrant is needed by arriving at a sensible balance between the administrative necessities of OSHA inspections and the incremental protection of privacy of business owners a warrant would afford. He suggests that only a decision exempting OSHA inspections from the Warrant Clause would give "full recognition to the competing public and private interests here at stake."

The Secretary submits that warrantless inspections are essential to the proper enforcement of OSHA because they afford the opportunity to inspect without prior notice and hence to preserve the advantages of surprise. While the dangerous conditions outlawed by the Act include structural defects that cannot be quickly hidden or remedied, the Act also regulates a myriad of safety details that may be amenable to speedy alteration or disguise. The risk is that during the interval between an inspector's initial request to search a plant and his procuring a warrant following the owner's refusal of permission, violations of this latter type could be corrected and thus escape the inspector's notice. To the suggestion that warrants may be issued ex parte and executed without delay and without prior notice, thereby preserving the element of surprise, the Secretary expresses concern for the administrative strain that would be experienced by the inspection system, and by the courts, should ex parte warrants issued in advance become standard practice.

We are unconvinced, however, that requiring warrants to inspect will impose serious burdens on the inspection system or the courts, will prevent inspections necessary to enforce the statute, or will make them less effective. In the first place, the great majority of businessmen can be expected in normal course to consent to inspection without warrant; the Secretary has not brought to this Court's attention any widespread pattern of refusal. In those cases where an owner does insist on a warrant, the Secretary argues that inspection efficiency will be impeded by the advance notice and delay. The Act's penalty provisions for giving advance notice of a search, 29 U.S.C. § 666(f), and the Secretary's own regulations, 29 CFR § 1903.6 (1977), indicate that surprise searches are indeed contemplated. However, the Secretary has also promulgated a regulation providing that upon refusal to permit an inspector to enter the property or to complete his inspection, the inspector shall attempt to ascertain the reasons for the refusal and report to his superior, who shall "promptly take appropriate action, including compulsory process, if necessary." 29 CFR § 1903.4 (1977). The regulation represents a choice to proceed by process where entry is refused; and on the basis of evidence available from present practice, the Act's effectiveness has not been crippled by providing those owners who wish to refuse an initial requested entry with a time lapse while the inspector obtains the necessary process. Indeed, the kind of process sought in this case and apparently anticipated by the regulation provides notice to the business operator. If this safeguard endangers the efficient administration of OSHA, the Secretary should never have adopted it, particularly when the Act does not require it. Nor is it immediately apparent why the advantages of surprise would be lost if, after being refused entry, procedures were available for the Secretary to seek an ex parte warrant and to reappear at the premises without further notice to the establishment being inspected.

Whether the Secretary proceeds to secure a warrant or other process, with or without prior notice, his entitlement to inspect will not depend on his demonstrating probable cause to believe that conditions in violation of OSHA exist on the premises. Probable cause in the criminal law sense is not required. For purposes of an administrative search such as this, probable cause justifying the issuance of a warrant may be based not only on specific evidence of an existing violation but also on a showing that "reasonable legislative or administrative standards for conducting an ... inspection are satisfied with respect to a particular [establishment]." *Camara v. Municipal Court*, 387 U.S., at 538, 87 S.Ct., at 1736. A warrant showing that a specific business has been chosen for an OSHA search on the basis of a general administrative plan for the enforcement of the Act derived from neutral sources such as, for example, dispersion of employees in various types of industries across a given area, and the desired frequency of searches in any of the lesser divisions of the area, would protect an employer's Fourth Amendment rights. We doubt that the consumption of enforcement energies in the obtaining of such warrants will exceed manageable proportions.

Finally, the Secretary urges that requiring a warrant for OSHA inspectors will mean that, as a practical matter, warrantless-search provisions in other regulatory statutes are also constitutionally infirm. The reasonableness of a warrantless search, however, will depend upon the specific enforcement needs and privacy guarantees of each statute. Some of the statutes cited apply only to a single industry, where regulations might already be so pervasive that a Colonnade-Biswell exception to the warrant requirement could apply. Some statutes already envision resort to federal-court enforcement when entry is refused, employing specific language in some cases and general language in others. In short, we base today's opinion on the facts and law concerned with OSHA and do not retreat from a holding appropriate to that statute because of its real or imagined effect on other, different administrative schemes.

Nor do we agree that the incremental protections afforded the employer's privacy by a warrant are so marginal that they fail to justify the administrative burdens that may be entailed. The authority to make warrantless searches devolves almost unbridled discretion upon executive and administrative officers, particularly those in the field, as to when to search and whom to search. A warrant, by contrast, would provide assurances from a neutral officer that the inspection is reasonable under the Constitution, is authorized by statute, and is pursuant to an administrative plan containing specific neutral criteria. Also, a warrant would then and there advise the owner of the scope and objects of the search, beyond which limits the inspector is not expected to proceed. These are important functions for a warrant to perform, functions which underlie the Court's prior decisions that the Warrant Clause applies to inspections for compliance with regulatory statutes. *Camara v. Municipal Court*, 387 U.S. 523, 87 S.Ct. 1727, 18 L.Ed.2d 930 (1967); *See v. City of Seattle*, 387 U.S. 541, 87 S.Ct. 1737, 18 L.Ed.2d 943 (1967). We conclude that the concerns expressed by the Secretary do not suffice to justify warrantless inspections under OSHA or vitiate the general constitutional requirement that for a search to be reasonable a warrant must be obtained.

II.

We hold that Barlow's was entitled to a declaratory judgment that the Act is unconstitutional insofar as it purports to authorize inspections without warrant or its equivalent and to an injunction enjoining the Act's enforcement to that extent.FN23 The judgment of the District Court is therefore affirmed.

So ordered.

Mr. Justice BRENNAN took no part in the consideration or decision of this case.

Mr. Justice STEVENS, with whom Mr. Justice BLACKMUN and Mr. Justice REHNQUIST join, dissenting.

Congress enacted the Occupational Safety and Health Act to safeguard employees against hazards in the work areas of businesses subject to the Act. To ensure compliance, Congress authorized the Secretary of Labor to conduct routine, nonconsensual inspections. Today the Court holds that the Fourth Amendment prohibits such inspections without a warrant. The Court also holds that the constitutionally required warrant may be issued without any showing of probable cause. I disagree with both of these holdings.

The Fourth Amendment contains two separate Clauses, each *326 flatly prohibiting a category of governmental conduct. The first Clause states that the right to be free from unreasonable searches "shall not be violated"; the second unequivocally prohibits the issuance of warrants except "upon probable cause." In this case the ultimate question is whether the category of warrantless searches authorized by the statute is "unreasonable" within the meaning of the first Clause.

In cases involving the investigation of criminal activity, the Court has held that the reasonableness of a search generally depends upon whether it was conducted pursuant to a valid warrant. See, e.g., *Coolidge v. New Hampshire*, 403 U.S. 443, 91 S.Ct. 2022, 29 L.Ed.2d 564. There is, however, also a category of searches which are reasonable within the meaning of the first Clause even though the probable-cause requirement of the Warrant Clause cannot be satisfied. See *United States v. Martinez-Fuerte*, 428 U.S. 543, 96 S.Ct. 3074, 49 L.Ed.2d 1116; *Terry v. Ohio*, 392 U.S. 1, 88 S.Ct. 1868, 20 L.Ed.2d 889; *South Dakota v. Opperman*, 428 U.S. 364, 96 S.Ct. 3092, 49 L.Ed.2d 1000; *United States v. Biswell*, 406 U.S. 311, 92 S.Ct. 1593, 32 L.Ed.2d 87. The regulatory inspection program challenged in this case, in my judgment, falls within this category.

III.

The warrant requirement is linked "textually ... to the probable-cause concept" in the Warrant Clause. *South Dakota v. Opperman, supra*, 428 U.S., at 370 n.5, 96 S.Ct. at 3097. The routine OSHA inspections are, by definition, not based on cause to believe there is a violation on the premises to be inspected. Hence, if the inspections were measured against the requirements of the Warrant Clause, they would be automatically and unequivocally unreasonable.

Because of the acknowledged importance and reasonableness of routine inspections in the enforcement of federal regulatory statutes such as OSHA, the Court recognizes that requiring full compliance with the Warrant Clause would invalidate all such inspection programs. Yet, rather than simply analyzing such programs under the "Reasonableness" Clause of the Fourth Amendment, the Court holds the OSHA program invalid under the Warrant Clause and then avoids a blanket prohibition on all routine, regulatory inspections by relying on the notion that the "probable-cause" requirement in the Warrant Clause may be relaxed whenever the Court believes that the governmental need to conduct a category of "searches" outweighs the intrusion on interests protected by the Fourth Amendment.

The Court's approach disregards the plain language of the Warrant Clause and is unfaithful to the balance struck by the Framers of the Fourth Amendment—"the one procedural safeguard in the Constitution that grew directly out of the events which immediately preceded the revolutionary struggle with England." This preconstitutional history includes the controversy in England over the issuance of general warrants to aid enforcement of the seditious libel laws and the colonial experience with writs of assistance issued to facilitate collection of the various import duties imposed by Parliament. The Framers' familiarity with the abuses attending the issuance of such general warrants provided the principal stimulus for the restraints on arbitrary governmental intrusions embodied in the Fourth Amendment.

IV.

Even if a warrant issued without probable cause were faithful to the Warrant Clause, I could not accept the Court's holding that the Government's inspection program is constitutionally unreasonable because it fails to require such a warrant procedure. In determining whether a warrant is a necessary safeguard in a given class of cases, "the Court has weighed the public interest against the Fourth Amendment interest of the individual ... ." *United States v. Martinez-Fuerte*, 428 U.S., at 555, 96 S.Ct., at 3081. Several considerations persuade me that this balance should be struck in favor of the routine inspections authorized by Congress.

Congress has determined that regulation and supervision of safety in the workplace furthers an important public interest and that the power to conduct warrantless searches is necessary to accomplish the safety goals of the legislation. In assessing the public interest side of the Fourth Amendment balance, however, the Court today substitutes its judgment for that of Congress on the question of what inspection authority is needed to effectuate the purposes of the Act. The Court states that if surprise is truly an important ingredient of an effective, representative inspection program, it can be retained by obtaining ex parte warrants in advance. The Court assures the Secretary that this will not unduly burden enforcement resources because most employers will consent to inspection.

The Court's analysis does not persuade me that Congress' determination that the warrantless-inspection power as a necessary adjunct of the exercise of the regulatory power is unreasonable. It was surely not unreasonable to conclude that the rate at which employers deny entry to inspectors would increase if covered businesses, which may have safety violations on their premises, have a right to deny warrantless entry to a compliance inspector. The Court is correct that this problem could be avoided by requiring inspectors to obtain a warrant prior to every inspection visit. But the adoption of such a practice undercuts the Court's explanation of why a warrant requirement would not create undue enforcement problems. For, even if it were true that many employers would not exercise their right to demand a warrant, it would provide little solace to those charged with administration of OSHA; faced with an increase in the rate of refusals and the added costs generated by futile trips to inspection sites where entry is denied, officials may be compelled to adopt a general practice of obtaining warrants in advance. While the Court's prediction of the effect a warrant requirement would have on the behavior of covered employers may turn out to be accurate, its judgment is essentially empirical. On such an issue, I would defer to Congress' judgment regarding the importance of a warrantless-search power to the OSHA enforcement scheme.

The Court also appears uncomfortable with the notion of second-guessing Congress and the Secretary on the question of how the substantive goals of OSHA can best be achieved. Thus, the Court offers an alternative explanation for its refusal to accept the legislative judgment. We are told that, in any event, the Secretary, who is charged with enforcement of the Act, has indicated that inspections without delay are not essential to the enforcement scheme. The Court bases this conclusion on a regulation prescribing the administrative response when a compliance inspector is denied entry. It provides: "The Area Director shall immediately consult with the Assistant Regional**1830 Director and the Regional Solicitor, who shall promptly take appropriate action, including compulsory process, if necessary." 29 CFR § 1903.4 (1977). The Court views this regulation as an admission by the Secretary that no enforcement problem is generated by permitting employers to deny entry and delaying the inspection until a warrant has been obtained. I disagree. The regulation was promulgated against the background of a statutory right to immediate entry, of which covered employers are presumably aware and which Congress and the Secretary obviously thought would keep denials of entry to a minimum. In these circumstances, it was surely not unreasonable for the Secretary to adopt an orderly procedure for dealing with what he believed would be the occasional denial of entry. The regulation does not imply a judgment by the Secretary that delay caused by numerous denials of entry would be administratively acceptable.

Even if a warrant requirement does not "frustrate" the legislative purpose, the Court has no authority to impose an additional burden on the Secretary unless that burden is required to protect the employer's Fourth Amendment interests. The essential function of the traditional warrant requirement is the interposition of a neutral magistrate between the citizen and the presumably zealous law enforcement officer so that there might be an objective determination of probable cause. But this purpose is not served by the newfangled inspection warrant. As the Court acknowledges, the inspector's "entitlement to inspect will not depend on his demonstrating probable cause to believe that conditions in violation of OSHA exist on the premises ... . For purposes of an administrative search such as this, probable cause justifying the issuance of a warrant may be based ... on a showing that 'reasonable legislative or administrative standards for conducting an ... inspection are satisfied with respect to a particular [establishment].'" *Ante*, at 1824. To obtain a warrant, the inspector need only show that "a specific business has been chosen for an OSHA search on the basis of a general administrative plan for the enforcement of the Act derived *332 from neutral sources ... ." *Ante*, at 1825. Thus, the only question for the magistrate's consideration is whether the contemplated inspection deviates from an inspection schedule drawn up by higher level agency officials.

What purposes, then, are served by the administrative warrant procedure? The inspection warrant purports to serve three functions: to inform the employer that the inspection is authorized by the statute, to advise him of the lawful limits of the inspection, and to assure him that the person demanding entry is an authorized inspector. *Camara v. Municipal Court*, 387 U.S. 523, 532, 87 S.Ct. 1727, 1732, 18 L.Ed.2d 930. An examination of these functions in the OSHA context reveals that the inspection warrant adds little to the protections already afforded by the statute and pertinent regulations, and the slight additional benefit it might provide is insufficient to identify a constitutional violation or to justify overriding Congress' judgment that the power to conduct warrantless inspections is essential.

The inspection warrant is supposed to assure the employer that the inspection is in fact routine, and that the inspector has not improperly departed from the program of representative inspections established by responsible officials. But to the extent that harassment inspections would be reduced by the necessity of obtaining a warrant, the Secretary's present enforcement scheme would have precisely the same effect. The representative inspections are conducted "'in accordance with criteria based upon accident experience and the number of employees exposed in particular industries.'" *Ante*, at 1825 n.17. If, under the present scheme, entry to covered premises is denied, the inspector can gain entry only by informing his administrative superiors of the refusal and seeking a court order requiring the employer to submit to the inspection. The inspector who would like to conduct a nonroutine search is just as likely to be deterred by the prospect of informing his superiors of his intention and of making false representations to the court when he seeks compulsory process as by the prospect of having to make bad-faith representations in an ex parte warrant proceeding.

The other two asserted purposes of the administrative warrant are also adequately achieved under the existing scheme. If the employer has doubts about the official status of the inspector, he is given adequate opportunity to reassure himself in this regard before permitting entry. The OSHA inspector's statutory right to enter the premises is conditioned upon the presentation of appropriate credentials. 29 U.S.C. § 657(a)(1). These credentials state the inspector's name, identify him as an OSHA compliance officer, and contain his photograph and signature. If the employer still has doubts, he may make a toll-free call to verify the inspector's authority. *Usery v. Godfrey Brake & Supply Service, Inc.*, 545 F.2d 52, 54 (CA 8 1976), or simply deny entry and await the presentation of a court order.

The warrant is not needed to inform the employer of the lawful limits of an OSHA inspection. The statute expressly provides that the inspector may enter all areas in a covered business "where work is performed by an employee of an employer," 29 U.S.C. § 657(a)(1), "to inspect and investigate during regular working hours and at other reasonable times, and within reasonable limits and in a reasonable manner ... all pertinent conditions, structures, machines, apparatus,*334 devices, equipment, and materials therein ... ." 29 U.S.C. § 657(a)(2). *See also* 29 CFR § 1903 (1977). While it is true that the inspection power granted by Congress is broad, the warrant procedure required by the Court does not purport to restrict this power but simply to ensure that the employer is apprised of its scope. Since both the statute and the pertinent regulations perform this informational function, a warrant is superfluous.

Requiring the inspection warrant, therefore, adds little in the way of protection to that already provided under the existing enforcement scheme. In these circumstances, the warrant is essentially a formality. In view of the obviously enormous cost of enforcing a health and safety scheme of the dimensions of OSHA, this Court should not, in the guise of construing the Fourth Amendment, require formalities which merely place an additional strain on already overtaxed federal resources.

Congress, like this Court, has an obligation to obey the mandate of the Fourth Amendment. In the past the Court "has been particularly sensitive to the Amendment's broad standard of 'reasonableness' where ... authorizing statutes permitted the challenged searches." *Almeida-Sanchez v. United States*, 413 U.S. 266, 290, 93 S.Ct. 2535, 2548, 37 L.Ed.2d 596 (WHITE, J., dissenting). In *United States v. Martinez-Fuerte*, 428 U.S. 543, 96 S.Ct. 3074, 49 L.Ed.2d 1116, for example, respondents challenged the routine stopping of vehicles to check for aliens at permanent checkpoints located away from the border. **1832 The checkpoints were established pursuant to statutory authority and their location and operation were governed by administrative criteria. The Court rejected respondents' argument that the constitutional reasonableness of the location and operation of the fixed checkpoints should be reviewed in a Camara warrant proceeding. The Court observed that the reassuring purposes of the inspection warrant were adequately served by the visible manifestations of authority exhibited at the fixed checkpoints.

Moreover, although the location and method of operation of the fixed checkpoints were deemed critical to the constitutional reasonableness of the challenged stops, the Court did not require Border Patrol officials to obtain a warrant based on a showing that the checkpoints were located and operated in accordance with administrative standards. Indeed, the Court observed that "[t]he choice of checkpoint locations must be left largely to the discretion of Border Patrol officials, to be exercised in accordance with statutes and regulations that may be applicable ... [and] [m]any incidents of checkpoint operation also must be committed to the discretion of such officials." 428 U.S., at 559-560, n.13, 96 S.Ct., at 3083. The Court had no difficulty assuming that those officials responsible for allocating limited enforcement resources would be "unlikely to locate a checkpoint where it bears arbitrarily or oppressively on motorists as a class." *Id.* at 559, 96 S.Ct. at 3083.

The Court's recognition of Congress' role in balancing the public interest advanced by various regulatory statutes and the private interest in being free from arbitrary governmental intrusion has not been limited to situations in which, for example, Congress is exercising its special power to exclude aliens. Until today, we have not rejected a congressional judgment concerning the reasonableness of a category of regulatory inspections of commercial premises. While businesses are unquestionably entitled to Fourth Amendment protection, we have "recognized that a business by its special nature and voluntary existence, may open itself to intrusions that would not be permissible in a purely private context." *G. M. Leasing Corp. v. United States*, 429 U.S. 338, 353, 97 S.Ct. 619, 629, 50 L.Ed.2d 530. Thus, in *Colonnade Catering Corp. v. United States*, 397 U.S. 72, 90 S.Ct. 774, 25 L.Ed.2d 60, the Court recognized the reasonableness of a statutory authorization to inspect the premises of a caterer dealing in alcoholic beverages, noting that "Congress has broad power to design such powers of inspection under the liquor laws as it deems necessary to meet the evils at hand." *Id.*, at 76, 90 S.Ct. at 777. And in *United States v. Biswell*, 406 U.S. 311, 92 S.Ct. 1593, 32 L.Ed.2d 87, the Court sustained the authority to conduct warrantless searches of firearm dealers under the Gun Control Act of 1968 primarily on the basis of the reasonableness of the congressional evaluation of the interests at stake.

> We have little difficulty in concluding that where, as here, regulatory inspections further urgent federal interest, and the possibilities of abuse and the threat to privacy are not of impressive dimensions, the inspection may proceed without a warrant where specifically authorized by statute. 406 U.S., at 315, 317, 92 S.Ct., at 1596.

The Court, however, concludes that the deference accorded Congress in *Biswell* and *Colonnade* should be limited to situations where the evils addressed by the regulatory statute are peculiar to a specific industry and that industry is one which has long been subject to Government regulation. The Court reasons that only in those situations can it be said that a person who engages in business will be aware of and consent to routine, regulatory inspections. I cannot agree that the respect due the congressional judgment should be so narrowly confined.

In the first place, the longevity of a regulatory program does not, in my judgment, have any bearing on the reasonableness of routine inspections necessary to achieve adequate enforcement of that program. Congress' conception of what constitute urgent federal interests need not remain static. The recent vintage of public and congressional awareness of the dangers posed by health and safety hazards in the workplace is not a basis for according less respect to the considered judgment of Congress. Indeed, in *Biswell*, the Court upheld an inspection program authorized by a regulatory statute enacted in 1968. The Court there noted that "[f]ederal regulation of the interstate traffic in firearms is not as deeply rooted in history as is governmental control of the liquor industry, but close scrutiny of this traffic is undeniably" an urgent federal interest. 406 U.S., at 315, 92 S.Ct. at 1596. Thus, the critical fact is the congressional determination that federal regulation would further significant public interests, not the date that determination was made.

In the second place, I see no basis for the Court's conclusion that a congressional determination that a category of regulatory inspections is reasonable need only be respected when Congress is legislating on an industry-by-industry basis. The pertinent inquiry is not whether the inspection program is authorized by a regulatory statute directed at a single industry, but whether Congress has limited the exercise of the inspection power to those commercial premises where the evils at which the statute is directed are to be found. Thus, in *Biswell*, if Congress had authorized inspections of all commercial premises as a means of restricting the illegal traffic in firearms, the Court would have found the inspection program unreasonable; the power to inspect was upheld because it was tailored to the subject matter of Congress' proper exercise of regulatory power. Similarly, OSHA is directed at health and safety hazards in the workplace, and the inspection power granted the Secretary extends only to those areas where such hazards are likely to be found.

Finally, the Court would distinguish the respect accorded Congress' judgment in *Colonnade* and *Biswell* on the ground that businesses engaged in the liquor and firearms industry "'accept the burdens as well as the benefits of their trade ... .'" *Ante*, at 1821. In the Court's view, such businesses consent to the restrictions placed upon them, while it would be fiction to conclude that a businessman subject to OSHA consented to routine safety inspections. In fact, however, consent is fictional in both contexts. Here, as well as in *Biswell*, businesses are required to be aware of and comply with regulations governing their business activities. In both situations, the validity of the regulations depends not upon the consent of those regulated, but on the existence of a federal statute embodying a congressional determination that the public interest in the health of the Nation's work force or the limitation of illegal firearms traffic outweighs the businessman's interest in preventing a Government inspector from viewing those areas of his premises which relate to the subject matter of the regulation.

The case before us involves an attempt to conduct a warrantless search of the working area of an electrical and plumbing contractor. The statute authorizes such an inspection during reasonable hours. The inspection is limited to those areas over which Congress has exercised its proper legislative authority. The area is also one to which employees have regular access without any suggestion that the work performed or the equipment used has any special claim to confidentiality. Congress has determined that industrial safety is an urgent federal interest requiring regulation and supervision, and further, that warrantless inspections are necessary to accomplish the safety goals of the legislation. While one may question the wisdom of pervasive governmental oversight of industrial life, I decline to question Congress' judgment that the inspection power is a necessary enforcement device in achieving the goals of a valid exercise of regulatory power.

I respectfully dissent.

5 Managing an Effective Safety and Loss Prevention Program

CHAPTER OBJECTIVES

1. Acquire the knowledge to develop and manage the safety and loss prevention function.
2. Acquire the knowledge to manage OSHA compliance programs.
3. Acquire a level of training and education knowledge.
4. Acquire knowledge as to how accidents happen and how to stop accidents.

INTRODUCTION

The ultimate goal for every safety and loss prevention professional is to safeguard employees from harm in the workplace. A secondary goal, although vitally important, is the achievement and maintenance of compliance with the OSHA standards and requirements. To achieve these important goals, a comprehensive management approach and all-inclusive strategy to direct and control the completion of the required tasks must be developed to comply with the OSHA standards and regulations and to safeguard employees.

Incorporate a management philosophy to serve as the foundation of the safety and loss prevention function. It should provide the necessary style through which to manage the safety and loss prevention function. The selection of the management philosophy and style is an individual decision based upon the background and personality of the safety and loss prevention professional, type of industry, employee population, and numerous other factors. The key to achieving and maintaining compliance with the OSHA standards is to manage the safety and loss prevention function in the same manner and style as you would manage production, quality, or other functions. Safety and loss prevention must be managed effectively to be successful.

The principles that virtually all management team members use in daily supervision of production, quality control, or any other operation are the same as principles used to manage safety in the workplace. In production you plan, organize, direct, and control your operation to produce a product, whereas in safety you plan, organize, direct, and control the safety and health of the employees in the workplace.

PLAN OF ACTION

Creating a written plan of action is the initial step when developing the appropriate mechanisms to manage OSHA compliance in the workplace. A written plan of action, not unlike a battle plan in military terms, sets forth the objective of each activity; delineates the activity into smaller, manageable elements; names the responsible parties for each element of the activity; and provides target dates, at which time the responsible party will be held accountable for the achievement of the particular element or activity. In order to manage this planning phase, safety and loss prevention professionals can use a planning document that permits easy evaluation of the progress toward the objective on a daily basis, as well as holds the appropriate party accountable for the achievement of the particular element or activity. This type of planning document can be computerized or simply completed in written form.

MANAGEMENT TEAM MEMBERS

All levels of the management team should be involved in developing priorities and scheduling each component of the plan of action. Team involvement and interaction allows each management team member input into the process and leads to "buy-in" by the team member into the overall safety and loss prevention effort. Additionally, team involvement permits individual input regarding potential obstacles that could be encountered as well as the development of a realistic, targeted schedule, given the OSHA time requirements and worksite pressures. Ranking of the various mandated safety and loss prevention programs, as well as programs not required by OSHA, should be carefully analyzed and prioritized so that the programs are developed to provide maximum protection to employees over and above the OSHA requirements.

Management team members should be advised that each team member will be held accountable for the successful and timely completion of his or her specifically assigned tasks and duties as set forth under the plan of action. With the required commitment to the safety and loss prevention goals and objectives by top management and management team members' participation during the development of the plan of action, each management team member should be well aware of his or her specific duties and responsibilities within the framework of the overall safety and loss prevention effort of the organization (i.e., his or her piece of the overall safety and loss prevention pie). Appropriate positive or negative reinforcement can be utilized to motivate this achievement of specific goals and objectives.

COMPLIANCE PROGRAMS

Safety and loss prevention professionals should be cautious when developing written safety and loss prevention programs to meet compliance requirements. The methods used in the development and documentation of OSHA compliance programs are a direct reflection on the safety and loss prevention professional and the safety and health efforts of the organization. Safety and loss prevention professionals should always develop written safety and health compliance programs in a professional manner. Safety and loss prevention professionals should be aware that when compliance officers are evaluating the written compliance programs, the professional quality of the written program could set the tone for the entire inspection or investigation. In general, safety and loss prevention professionals should consider the following:

- Acquire and read all applicable standards.
- Re-read the standard and identify the key elements.
- Visualize the program in action and identify additional elements needed for effective and efficient operations.
- Draft the written compliance program and acquire input as to changes, modifications, and additions.
- Finalize the written compliance program and acquire all required approvals.

Compliance programs should also be developed in a defensive manner. Every element of the OSHA standard should be addressed in written form and all training and education requirements should be documented. A work-related accident or incident places the written safety and loss prevention programs on trial. The written program will be placed under a microscope for every detail to be scrutinized from every angle. Using proper preparation, evaluation, and scrutiny when developing a written compliance program can help to avoid substantial embarrassment, cost, and possible liabilities in the future.

When developing and writing a compliance program, safety and loss prevention professionals should attempt to visualize the compliance program in action. Each step within the compliance program should be carefully examined and scrutinized, looking specifically for issues and items that may impact the operations of the program but were not specifically addressed within the OSHA standard. For example, a visitor's personal protection equipment (PPE) or prescription glasses should be considered in an eye protection program. A visualization of the program in a step-by-step manner can often identify these issues during the development stage and eliminate many headaches during the implementation stage.

TRAINING AND EDUCATION

Documentation is vital in the area of required education and training elements mandated under a particular OSHA standard. Safety and loss prevention professionals should closely evaluate the compliance program to ensure that all required training and education mandated under the standard is being completed in a timely manner. In addition, the training documentation should confirm, beyond a shadow of a doubt, that a particular employee attended the required training. This documentation should show that an individual employee not only attended but also understood the information provided during the training and education session. A written examination may be helpful to prove comprehension and an adequate level of competency. If a required training and education element is not documented, there is no conclusive proof, outside verbal statements, to substantiate that employees obtained a certain type of training.

In the area of training and education elements involved with a compliance program, safety and loss prevention professionals are reminded of the educational maxim of "tell them, show them, and tell them again." Safety and loss prevention training should be conducted, when feasible, in an atmosphere conducive to learning and at a time when employees are mentally alert. The individuals performing the training should be competent and enthusiastic. Hands-on training has been found to be the best method to provide the greatest understanding and the method that allows employees to retain the information better. Audiovisual aids are an exceptional method of increasing the retention level; however, safety and loss prevention professionals are cautioned not to rely solely on video recordings for the total training experience.

Additionally, although safety and loss prevention is a serious matter, training does not have to be a sober and boring endeavor that employees are required to endure on a periodic basis. Safety and health professionals should strive to make training an experience that employees will remember and enjoy. There is no rule that safety and loss prevention training cannot be mentally stimulating or even fun. Remember, the information that is provided in a safety and loss prevention training session may be the difference between an employee going home or not going home at the end of the day—do everything possible to ensure that the employee accesses, understands, and retains the information from the training session.

PERSONAL PROTECTIVE EQUIPMENT

Purchasing the appropriate personal protective equipment in order to achieve the objectives of the OSHA standard is vitally important. Safety and loss prevention professionals should be actively involved in the selection, purchase, monitoring, inspection, and replacement of personal protective equipment. Although cost is always a factor, the safety parameters, comfort levels, and approval or certifications, as well as other factors, need to be scrutinized to ensure that the personal protective equipment meets or exceeds the requirements that are mandated under the OSHA standard. Safety and loss prevention professionals should also become familiar with the new Personal Protective Equipment Final Rule, addressed in greater detail later in this text, which addresses the payment for personal protective equipment. The personal protective equipment must be of a type and quality that will not cause employees difficulties in everyday use. Many safety and loss prevention professionals provide the initial evaluation and selection of broad types of personal protective equipment and then permit the individuals who will be required to wear the personal protective equipment to make the final selection. This type of employee and management team member involvement in the selection process often leads to greater compliance and participation in the program.

RECEIVING ASSISTANCE

If a safety and loss prevention professional is unsure about the requirements of a particular standard, it is imperative that he or she acquire a definite answer or clarification. OSHA often provides clarification of particular issues or problems without identifying the caller, or OSHA may transfer the call to the state education and training division. In many state plan states and in federal states (usually through a state agency), a separate section of OSHA has been established to assist employers in achieving compliance. Upon request, the education and training section can assist employers with a wide variety of compliance issues, ranging from program development to the acquisition of pertinent information, at no cost. Safety and loss prevention professionals should be aware that this section of OSHA does possess the ability to issue citations, but normally issues citations only in situations involving imminent harm or where the employers have failed to follow prescribed advice.

AUDIT INSTRUMENT

Safety and loss prevention professionals should adopt a strategy to effectively manage a number of compliance programs simultaneously. The use of a safety and health audit can effectively identify deficiencies within a compliance program and permit timely correction of identified deficiencies. Although there are various types of safety and health audit instruments, all audits possess the basic methods for identification of the required elements of a compliance program, i.e., track the current level of performance, identify deficiencies, and identify potential corrective actions. A safety and health audit mechanism can provide numeric, letter, grade, or other methods of scoring, so that the management team can ascertain the current level of performance and identify areas in need of improvement.

To achieve compliance with an OSHA standard, it is imperative to check that every element and aspect of the standard has been evaluated and is in compliance. The following questions form a basic evaluation instrument to assist the safety and loss prevention professional in addressing potential areas that may have been overlooked:

1. Is the employer covered under the OSH Act?
2. Has the facility or worksite been evaluated to ascertain which specific OSHA standards are applicable?
3. Has the management group been educated regarding the requirements of the OSH Act and standards? Has the management acquired the necessary support and funding for the programs?
4. Is there a copy of the OSHA standards (29 C.F.R. 1910 et seq and other standards) at the worksite?
5. Is the *Federal Register* or other appropriate sources for new standards or emergency standards being reviewed to see whether standards are applicable to the worksite?
6. For new programs or new standards promulgated by OSHA, is the OSHA standard applicable to the workplace, situation, or industry?
7. Is there an OSHA guideline for particular situations or hazards in the workplace?
8. If there is no applicable OSHA standard, will the situation qualify under the general duty clause as an unsafe or unhealthful situation or hazard?

9. If there is no applicable OSHA standard and the situation is deemed to qualify under the general duty clause, have the National Advisory Committee on Occupational Safety and Health (NIOSH) publications, the American National Standards Institute (ANSI) standards, or other applicable journals and texts been reviewed to acquire guidance?

10. If there is no applicable OSHA standard or there is an OSHA standard that conflicts with other government agency regulations, there are other options:
    • Contact the regional OSHA office and request consultation assistance.
    • Employ an outside consultant possessing the specific expertise to assist.
    • Pursue a variance action.

11. If there is an applicable OSHA standard, has each and every word of the standard been read so that the requirement of the standard is completely understood?

12. Is there a written program to ensure compliance with all of the requirements of the OSHA standard? Remember, if the program is not in writing, there is no evidence to prove the existence of the program during an OSHA inspection or in litigation. Is the original version of this written program in a secure location?

13. Is the program written in a defensive manner? Can the written program be scrutinized by OSHA or a court of law without the identification of flaws in the program? Is the program written in a neutral and nondiscriminatory language?

14. Is the documentation of every purchase and equipment modification included in the program in order to ensure compliance with the OSHA standard? Is the documentation in the written program or in a secure location?

15. Does the written program possess the purpose of this program? Is there a copy of the applicable OSHA standard for easy reference? Have the responsibilities been delineated, and are they specific for each level of the management team and each position within the levels of management?

16. Have the OSHA standards and other applicable information been closely scrutinized and evaluated? Has each and every requirement in the standard been achieved or exceeded? Have any steps or elements that are required in the OSHA standard been omitted? If the standard is vague, make sure the program is clear, concise, and to the point.

17. Has all training been documented as required under the OSHA standard? (Remember, the OSHA standard is only a minimum requirement. Every program can be better than, but not less than, the OSHA standard requirements.)

18. Is there detailed documentation regarding each and every phase of the training? Is this documentation in the written program? Is documentation provided in the written program to prove the use of audiovisual aids, the instructor's qualifications, and other pertinent information?

19. Is there a schedule of the classroom and hands-on training sessions in the written program?

20. Have all employees who have completed the required training signed a document showing the exact training completed (in detail), the instructor's name, the date of the training, and so on? Have auxiliary aids and other accommodations for individuals with disabilities been provided in the training programs? (For employees who cannot write, a thumbprint can be used.) A video recording of the training is also an acceptable method of documentation. Remember to maintain the individual recordings on file, like other documents, for future use as evidence of this training.

21. Is the training offered in the languages used by the employees? Are the documents and the written program interpreted into the languages read and spoken by the employees and are the interpreted programs provided for use by these employees?

22. Is there a posting requirement under the applicable OSHA standard? Is the necessary poster from OSHA in the facility and has it been posted in an appropriate location by the required date? Have posters in the language of the employees been acquired and are they posted in an appropriate location by the required date?

23. Are there any other requirements, such as labeling of containers and material safety data sheets (MSDS)? Does the OSHA standard require support or information from an outside vendor or agency? Have all requests to outside vendors or agencies requesting the necessary information (e.g., MSDS) been documented and placed in the written program?

24. Is there a disciplinary procedure in the written program informing employees of the potential disciplinary action for failure to follow the written program?

25. Has the written program been reviewed prior to publication? Does it meet or exceed each and every step required under the applicable OSHA standard?

26. Has the written program been reviewed by legal counsel or the upper management group prior to publication? After acquiring necessary approvals of the written program, have copies of the program been made for distribution to strategic locations in the facility? Are there translated copies available for use by the employees? Does the upper management group possess individual copies of the program?

27. Is there a plan to review and critically evaluate the effectiveness of the program at least one time per month for the first six months? When deficiencies are identified, are plans prepared to make the necessary changes or modifications while ensuring compliance with the OSHA standard?

28. Is there a safety and health audit assessment procedure and instrument? Is there a plan for auditing the programs on a periodic basis? Is the audit scheduled yet?

MANAGING AN EFFECTIVE SAFETY AND LOSS PREVENTION PROGRAM

The principles that managers use in their daily supervision of production, quality control, or any operation are the same principles that should be used when managing the safety and health function in the workplace. In production, managers utilize the basic management principles of planning, organizing, directing, and controlling the operation to produce a product. Safety and loss prevention professionals should utilize the same basic management principles to plan, organize, direct, and control the safety and loss prevention function in the workplace.

For many years, safety compliance was a secondary job function, or in many cases, an afterthought. The safety and health function was often managed utilizing a "squeaky wheel" theory. That is, the only time that the management team paid any attention to the safety and loss prevention function was when the wheel squeaked (i.e., after an accident or incident had already occurred). Today, with increasing costs of work-related injuries and illnesses, increasing compliance requirements and liability, increasing insurance costs, and other increasing costs in the area of safety and health, a proactive stance should be taken to ensure that a safe and healthful environment is created and maintained in the workplace.

DIRECT AND INDIRECT COSTS

In order for most management groups to embrace the concept of a proactive safety and loss prevention program, it is imperative that the management group be thoroughly educated regarding the cost effectiveness of safety and loss prevention. Professionals in the field are often able to show the monetary, as well as the humanitarian, benefits of a proactive safety and loss prevention program through the use of a cost-benefit analysis.

One of the basic examples used for direct and indirect costs of an accident is an iceberg with direct costs being above the waterline and indirect costs being below the waterline. This example is often used to exemplify the actual costs of accidents, injuries, and illnesses in the workplace. When most individuals think of accident costs, the first thoughts that cross their minds are the direct costs. Direct costs include the cost of maintaining a medical facility at the worksite, the medical costs, time loss benefits provided under workers' compensation, and the premium costs of insurance. In most organizations, direct cost figures are easily identified for use by the management group. Direct costs, in most organizations, are substantial and normally result in a percentage of the profits being utilized to pay for these costs (e.g., 4% of the fiscal year 2009 profit was paid in workers' compensation benefits). When management team members actually take the time to understand the amount of money (lost profit) that is being spent on the direct costs of accidents, safety and loss prevention professionals usually obtain the immediate and intense attention of the management group.

The safety and loss prevention professional can express to upper management the actual costs of work-related accidents by combining the above-water direct costs with the more elusive indirect costs shown below the water line. When the management group understands that the indirect costs of an accident can sometimes be as much as 50 times the direct costs of an accident, the safety and loss prevention professional is beginning to acquire the management's buy-in necessary for their commitment to the proactive safety and loss prevention effort. As the model shows, indirect costs can arise from equipment damage, replacement costs, quality losses, production losses, and many other areas. This model can be customized for a particular organization by using actual dollars to provide a greater impact to the presentation. This visual demonstration works especially well with the financial number crunchers in the organization.

DOMINO THEORY

When the management groups fully understand the cost factors involved in work-related accidents, safety and loss prevention professionals should also be prepared to show the dividends, in both monetary and humanitarian terms, that can be acquired through a comprehensive and systematic management approach to safety and loss prevention. To ensure that the management group fully understands the concepts involved in a proactive program, the management group should understand how accidents happen and how accidents can be prevented. Utilizing the simple domino theory, safety and loss prevention professionals can easily explain the causal factors leading up to an accident to management as well as the negative impacts and costs following an accident.

Picture seven (7) dominos lined up in a row. Each domino identifies an underlying factors that could lead to an accident. The safety and loss prevention professional should emphasize that these underlying causes for workplace injuries and illnesses can be identified and corrected through the use of a proactive safety and health program. If the underlying factors leading to an accident are not identified and corrected, the dominos begin to fall. After the dominos begin to fall, it is almost impossible to prevent an accident from happening. The key is to ensure that the management group realizes that to prevent an accident, the underlying risk factors must be minimized or eliminated.

Safety and loss prevention professionals often use the following progression model to drive home the point that near-misses and other underlying factors, if not addressed, will ultimately lead to an accident. In this model, for every 300 equipment damage accidents or near-misses an employer may experience, there will be 29 minor injuries. If the deficiencies and underlying risk factors are not identified and corrected, the 300 near-misses will ultimately lead to one major injury or fatality. The key is to ensure the management team's complete understanding that they must take a proactive approach to the safety and loss prevention function rather than simply reacting when an incident or accident happens.

MANAGEMENT THEORIES

There are several management theories and approaches, including but not limited to management control system management, management by objectives, group dynamic and human approach management, and total safety management. These approaches have successfully been utilized in organizations to manage the safety and loss prevention function. The particular management theory selected for use within any given organization must meet the needs and management style of the organization. There is no single right or wrong management theory for any given organization, as long as the management system that is selected provides a consistent and systematic approach that proactively addresses the underlying reasons and risk factors that may ultimately lead to an accident.

MANAGEMENT BY OBJECTIVES

Many organizations have found that the management by objectives (MBO) theory is a simple but effective systematic and practical methodology for the management of their safety and loss prevention function. This style provides a "stair step," long-term approach to achieving the ultimate safety goals or objectives. Using MBO, each element within a safety and health program can be assigned an achievable objective or goal. When all objectives from each element within a specific safety and loss prevention program are achieved, the overall objective of the program will be achieved concurrently. When all of the individual safety program objectives are achieved, the larger, overall objective of safety and loss prevention effort will also be achieved. In simplistic terms, MBO provides a series of "building block" objectives upon which other objectives are based; achievement of the smaller objectives will ultimately lead to the achievement of the larger objectives or goals.

Zero Accident Goal Theory

In developing specific safety and loss prevention goals or objectives for an organization, all levels of the management team and employees should be provided the opportunity to interject their ideas and opinions into the goal development process. There are two basic schools of thought in this area; namely the zero accident goal theory and the progressional accident goal theory.

Under the zero accident goal theory, the ultimate goal is zero accidents. To attain less than this goal is to permit employees to incur injuries and illnesses on the job. Using the ultimate goal of zero, the entire organization team possesses a common goal that is at the pinnacle of the safety and loss prevention. The downside of this theory is the possibility that organization team members may view the zero accident goal as unrealistic and unattainable and thus lose interest and momentum in striving to achieve their safety and health goals.

Under the progressional goal theory, organizations will continuously phase in the reduction of safety and loss prevention goals over a period of time in order to achieve the ultimate goal of zero accidents (e.g., 2009—25% reduction from 2008 accident total; 2010—50% reduction from 2008 accident total, etc., ultimately reaching the zero accident goal over a number of months or years). The downside of this theory is the fact that the organization will be accepting a certain number of accidents, and thus injuries and illnesses, while the organization strives to achieve its ultimate goal.

First-Line Supervisor/Team Leader

Although every organization team member is important in any safety and loss prevention program, the key management position is the first-line supervisor or team leader. This particular management level normally provides the communications link between upper management and the employees. In most organizations, the first-line supervisor or team leader is the individual who will have daily interaction with the employees within his or her department or area, direct the activities of the employees in the department or area, perform disciplinary functions, and perform the training function and other related activities. This management level embodies the commitment of the organization to safety and loss prevention and relays it to the employees. If the first-line supervisor or team leader has been properly educated and adopts the goals and objectives of the safety and loss prevention program, and effectively communicates these goals and objectives to his or her employees, the employees will normally embrace the safety and loss prevention effort or, at the very least, adhere to the safety and loss prevention policies and procedures.

The first-line supervisor or team leader should be educated, trained, and motivated to make safety and loss prevention part of his or her everyday activities. First-line supervisors and team leaders must be provided the tools to effectively manage the safety and loss prevention function. Upper-level management commitment and motivation, in combination with the necessary education and training, provides the equipment to get the job done for supervisors or team leaders to manage effectively; however, upper management must hold the supervisor or team leader accountable for the safety performance or achievement of the goals or objectives in his or her identified areas or job responsibilities.

One of the first questions asked by first-line supervisors and team leaders is, "Where am I going to find the time to manage safety and loss prevention when I don't have enough hours in the day to do my job now?" With a proactive approach to safety and loss prevention, the first-line supervisor or team leader will be provided the skills to effectively manage the safety and loss prevention function within his or her department or area rather than reacting to problems. The management skills taught for the effective management of the safety and loss prevention function are the same basic management skills necessary to effectively manage the production function, quality function, and other related functions. Supervisors and team leaders normally find that when they have mastered basic management skills, the safety and loss prevention function can be effectively managed in the same or similar manner as the other production functions and, in fact, the supervisor or team leader will acquire more time within the workday when he or she manages rather than "putting out fires."

POLICIES

The foundation necessary to properly and effectively manage a safety and loss prevention program is the presence of safety and loss prevention policies and procedures through which the organization team members, individually or collectively, can acquire the necessary guidance regarding acceptable and unacceptable behaviors, expectations as to safety and loss prevention performance, and other basic workplace requirements. Safety and loss prevention policies and procedures should be clearly stated while removing any ambiguities or room for interpretation. Additionally, written safety and loss prevention programs outlining the essential requirements of the specific safety and loss prevention program are vital in providing continuous direction to the organization team. There is no perfect safety and loss prevention objective or goal mechanism that works for all organizations. Given substantial differences in locations, worksites, workforces, philosophy, etc., safety and loss prevention professionals should select the mechanism or method that works best for their individual situation. The key factors in safety and loss prevention program development under this management theory are that the organization team possesses a consensus safety and loss prevention goal, the objectives in attaining the goal are clearly defined and measured, the organization team is given input regarding its achievement of the safety objectives and goals, and the organization team is held accountable for the achievement of the safety and loss prevention goals.

An area that requires substantial effort in managing safety and loss prevention is achieving an open communication system with employees. All employees want to be able to work safely without injury while at work. The goal of management is the same. The confrontation normally stems from the methods used to achieve this goal. Communicating with employees, permitting employees to voice their opinions and ideas, and acquiring employee involvement in the safety and loss prevention effort are essential to the proper management of the safety and health program.

Simply complying with the OSHA standards does not guarantee a successful safety and loss prevention program. The OSHA standards are the "bare bones" and minimum conditions that the government requires all employers to meet. A safety and loss prevention program must comply with these standards, but should go far beyond these minimum standards. An efficient safety and loss prevention program should incorporate ideas and programs developed by the employees and management team that will strengthen and expand the safety and loss prevention efforts. Employees provide many of the best ideas in the safety and loss prevention area. Safety and loss prevention professionals should bear in mind that employees normally work in one area and perform one job. The employee is the expert on that particular job, and his or her ideas and input can normally provide great insight into developing safety and loss prevention programs and policies that directly affect that particular job or area.

The basic concept in managing safety and loss prevention in the workplace is to get all employees to be conscious of their own safety as well as the safety of others. Safety and loss prevention can be instilled in employees through a long-term training and education program as well as constant, consistent, and proper management of the safety and health function. Safety and loss prevention must be made an essential part of each employee's daily work habits! Employee involvement in the structure, decision making, and operation of the proactive safety and loss prevention program has been found to be successful in achieving the employee buy-in and, thus, their commitment.

Safety and loss prevention is not the sole domain of the safety director or first-line supervisor. Utilizing a team approach, the supervisor or team leader can train organization team members to take an active role in specific safety and loss prevention functions. Many organizations have found that safety and loss prevention activities that are required for the achievement of specific objectives, such as department safety inspections, personal protective equipment inspections, and other duties, can be delegated from the first-line supervisory level to the team members. In fact, the more involved the organization team members can be in the safety and loss prevention program, the more organization team members feel responsible for, and take pride in, the safety and loss prevention program. However, too much delegation of essential duties can defeat a good program. Another key area that is often overlooked in the management of a safety and loss prevention program is the accountability factor. All levels of the management team must be held accountable for their divisions, departments, or areas. The individual management team member should be involved in the development of the objectives and goals, as well as the necessary tools to enable him or her to effectively manage the safety and loss prevention function. Pertinent and timely feedback is critical.

The use of positive reinforcement has been found to be the most effective method of motivating supervisors or team leaders to achieve the specified objectives and goals. However, negative reinforcement or disciplinary action should be in place as a backup if positive reinforcement is not successful. Organizations that have embraced the proactive approach to managing safety and loss prevention have found that the benefits achieved over time far outweigh the initial costs involved. Once in place, an effectively managed safety and loss prevention program will pay dividends for years to come while minimizing potential risks and legal liabilities.

SELECTED CASE STUDY

Note: Case modified for the purposes of this text.

Luis A. Perez, et al., Plaintiffs, v. Mountaire Farms, Inc., et al., Defendants

United States District Court, D. Maryland.
Civ. No. AMD 06-121.
April 17, 2009.

Background

Current and former employees of poultry processing plant sued employer, seeking compensation under Fair Labor Standards Act (FLSA) for time required to don and doff personal protective equipment (PPE). The District Court, Andre M. Davis, J., 601 F.Supp.2d 670, certified suit as collective action, and ruled that donning/doffing constituted "work" outside Portal-to-Portal Act's exemption for wage liability.

Holdings

Following bench trial on compensability, the District Court held that:

1. donning/doffing of PPE were integral and indispensable to employees' principal activities, and thus compensable under FLSA;
2. donning/doffing were not taken outside FLSA compensable status merely because employer permitted employees to take PPE items home;
3. portions of meal breaks spent donning/doffing were compensable under FLSA;
4. any employee activity occurring between initial donning of PPE and final doffing of PPE was compensable under FLSA;
5. little weight would be accorded employer's estimate of donning/doffing time;
6. employees' videotaped study constituted sufficient evidence of donning/doffing time;

7. time spend donning/doffing was not *de minimis* and thus not excludible from compensability on that basis;
8. employer's conduct was not willful, precluding longer limitations period; and
9. award of liquidated damages was inappropriate.

Judgment for Employees

Memorandum Opinion Setting Forth Findings of Fact and Conclusions of Law Pursuant to Fed. R. Civ. P. 52

Judge

In this action arising under the Fair Labor Standards Act ("FLSA"), 29 U.S.C. § 201 et seq., as amended by the Portal-to-Portal Act, 61 Stat. 86-87, and Delaware state law, the plaintiff class, current and former employees of defendants Mountaire Farms, Inc., and Mountaire Farms of Delaware, Inc., seek compensation for the time required to don and doff personal protective equipment ("PPE"). In particular, plaintiffs assert that the time spent on (1) donning the PPE at the beginning of the shift, (2) doffing certain pieces of the PPE at the beginning of the lunch period, (3) redonning the doffed PPE at the end of the lunch period and (4) doffing the PPE at the end of the shift are compensable. Plaintiffs further assert that, under the continuous work day rule, time spent walking, sanitizing the PPE, and waiting for the principal work to commence are compensable.

Defendants present several defenses. First, defendants argue that (1) such donning and doffing is uncompensable as a matter of law and (2) lunch breaks are uncompensable per se because they primarily benefit the employee. Defendants then argue that, even if such donning and doffing is compensable under the FLSA, plaintiffs are nevertheless precluded from recovery because the time spent on donning and doffing is *de minimis*. Moreover, defendants argue that the time spent on donning and doffing items for which they have the option of taking home are properly excluded from compensable time.

I conducted a bench trial over one week, from Monday, March 23, 2009, to Friday, March 27, 2009. After careful consideration of the witness testimony, trial exhibits, and all the evidence presented, and after considering the arguments of counsel, I find that the Plaintiffs have established defendants' liability. There follows my findings of fact and conclusions of law in accordance with Fed.R.Civ.P. 52(a).

I. Findings of Fact Donning and Doffing Activities at the Millsboro, Delaware Plant

1. Defendants are Delaware corporations operating a Millsboro, Delaware, plant that slaughters, processes, and distributes chickens. Def's Stmt of Facts 1; Aristazabal Decl. ¶ 2; Pl's Ex. 8.
2. The Millsboro plant is divided into the following departments (also called lines): Receiving, Pinning, Evisceration, Rehang, Giblets, Packing, Cutup, Cone Debone, Tray Pack, Marination, WPL, Dry Cooler, MSC, PAWS, Sam's Club, Leg Debone, Thigh Debone, and Shipping. Def's Stmt of Facts 2; Aristazabal Decl. ¶ 3; Pl's Ex. 8.

3. The Millsboro plant produces 1.5 million chickens per week. Pl's Ex. 20(c).

4. Defendants' employees are paid based on "line time," which begins when the first chicken arrives at the first individual work station of each department and ends when the last chicken leaves the last individual work station of each department. *Id.*

5. Although employees are paid based on "line time," they are nevertheless required to "clock in" each day that they arrive for work. Def's Stmt of Facts 8; Aristazabal Decl. ¶ 9.

6. Supervisors use the "clock in" time to prepare a report detailing actual time worked when an employee is late for work (i.e., not present at his work station at the beginning of line time). *Id.*

7. Employees are required to wear PPE in order to comply with United States Department of Agriculture ("USDA") sanitary requirements and Occupational Safety and Health Administration ("OSHA") safety regulations. Def's Stmt of Facts 3; Aristazabal Decl. ¶ 4; Pl's Ex. 8.

8. All employees are required to wear ear plugs, bump caps, smocks (also called lab coats), hair/beard nets, and steel toed rubber boots. Def's Stmt of Facts 3; Aristazabal Decl. ¶ 4; Pl's Ex. 6.

9. Employees are required to wear a combination of other PPE (such as nitrile/latex/rubber gloves, aprons, safety glasses, mesh cut resistant gloves, chain gloves and sleeves) based on the requirements of the department in which they work. Def's Stmt of Facts 3; Aristazabal Decl. ¶ 4; Pl's Ex. 6.

10. Only the employees who work with knives or scissors (Evisceration and Debone Departments) wear cut-resistant gloves. Luisa Perez Testimony, 3/23/09; Pl's Ex. 6.

11. Employees wear bump caps in order to prevent an employee's hair from falling into the product. Pl's Ex. 26(a); Zlotorzynski Dep 17:12-15.

12. The bump caps are not of a grade or quality of a helmet that would prevent head injuries when worn. Pl's Ex. 26(c); Zlotorzynski Dep 17:6-11, 17:19-18:2, Def's Ex. 13.

13. Employees wear ear plugs to protect their ears from loud noise in the production floor. Pl's Ex. 26(d); Zlotorzynski Dep 28:10-17.

14. Different ear plugs have different OSHA ratings and employees are required to wear specific ear plugs depending on which section of the plant they work and how noisy that section is. Pl's Ex. 26(d); Zlotorzynski Dep 28:10-17.

15. Employees don and doff their PPE at various locations: by their lockers, in the bathrooms, in the production area, or in the hallways as they walk to their workstations. Def's Stmt of Facts 3; Aristazabal Decl. ¶ 4; Pl's Ex. 18.

16. The normal sequence of donning is as follows: smocks, followed by hair nets, bump caps, ear plugs, cut-resistant sleeves, apron, and safety glasses. Luisa Perez Testimony, 3/23/09; Pl's Ex. 18.

17. Before entering their department's production area, employees must wash their hands.

18. Some courts have suggested that an activity is *de minimis* if it does not exceed 10 minutes. *See Spoerle,* 527 F.Supp.2d at 868; *Reich,* 38 F.3d at 1126. The parties have presented contradictory expert testimony as to the

amount of time actually required to don, doff, and sanitize PPE and walk around the Millsboro facility. Dr. Davis has stated that the daily time workers are not compensated for donning, doffing, and walking is 10.2 minutes. Defendants *684 Ex. B at 7 and Ex. C at 5. Dr. Robert Radwin, plaintiffs' expert, measures the average time donning, doffing, walking and sanitizing to be 20 minutes. Plaintiffs Ex. 2 at 9. No court has actually explained why 10 minutes should be the benchmark for *de minimis* purposes. *See Spoerle,* 527 F.Supp.2d at 868 ("no court has explained why 10 minutes of work is worthy of compensation but 9 minutes and 59 seconds is not"). Even if this court were to accept the 10 minute rule, however, plaintiffs have presented sufficient evidence to contradict defendants. In short, this court need not engage in any fact-finding on this issue at the moment. The record is clear that plaintiffs have submitted sufficient evidence to survive summary judgment as to whether the time spent by plaintiffs on compensable activities is *de minimis.*

6 Criminal Sanctions

CHAPTER OBJECTIVES

1. Acquire an understanding of the criminal penalties under the OSH Act.
2. Acquire an understanding of the use of state criminal laws.
3. Acquire an understanding of preemption.
4. Acquire an understanding of criminal procedure.

INTRODUCTION

The OSH Act has provided the possible use of criminal sanctions since its inception in 1970. The OSH Act permitted the use of criminal penalties of up to $10,000 and up to six months in prison for willful violations causing death against the employer. However, OSHA does not have the authority to actually impose criminal penalties; thus, it must refer potential criminal cases for possible prosecution by the U.S. Department of Justice. As shown in this chapter, relatively few cases have been referred to the U.S. Department of Justice by OSHA and the U.S. Department of Justice has declined to prosecute a substantial number of these referred cases.

State plan states usually possess the authority to enforce criminal sanctions under the individual state plan program. In states with federal OSHA jurisdiction, state prosecutors can utilize the general state criminal codes and statutes to bring criminal charges against employers and individual agents of the employer or corporation. As we will discuss in greater depth in this chapter, the issue of preemption or the OSH Act preempting the use of state criminal codes in a criminal case was an issue; however, the preemption issue has been settled to a great extent, as shown in the cases in this chapter. The OSH Act criminal penalty provisions permit state prosecutors to enforce individual state criminal codes or related workplace laws for workplace injury or fatality cases.

STATE USE OF CRIMINAL SANCTIONS FOR WORKPLACE INJURIES AND FATALITIES

The use of state criminal laws for workplace fatalities and injuries can impact the scope of potential liability for many safety and loss prevention professionals. The use of state criminal laws by state and local prosecutors for injuries and fatalities that occur on the job opens a relatively new avenue of potential liability in many states, which safety and loss prevention professionals did not face in the past. The utilization of the standard state criminal laws in a workplace setting normally governed by OSHA or state plan programs continues to be controversial. However, it appears to be a viable method through which states can penalize corporate officials in situations involving fatalities or serious injury. Currently, this utilization of state criminal laws does not appear to be preempted by the OSH Act. It should be clarified that the use of criminal sanctions for workplace fatalities and injuries is not a new area of concern. In Europe, criminal sanctions for workplace fatalities are frequently used, and in the United States, as far back as 1911, criminal sanctions were used in the well-known Triangle Shirt fire in New York in which more than 100 young women were killed. In that case, the co-owners of the Triangle Shirt Company were indicted on criminal manslaughter charges (although subsequently acquitted). Safety and loss prevention professionals should, however, take note that the source of the potential criminal liability (i.e., state criminal codes in addition to the OSH Act) and the enforcement frequency (i.e., increased utilization of criminal charges under the OSH Act and state criminal codes) is a trend that began in the 1990s.

Utilizing the individual state's criminal code, state and local prosecutors are taking an active role in workplace safety and health through the enforcement of state criminal sanctions against employers for on-the-job deaths and serious injuries. This area of workplace safety and health has been exclusively within the domain of OSHA and state plan programs since the enactment of the OSH Act in 1970. State prosecutors have recently challenged OSHA's federal preemption of this area and have created an entirely new area of potential criminal liability that the safety and loss prevention professional should address on an individual and corporate basis.

FILM RECOVERY CASE

The case that propelled the issue of whether or not OSHA has jurisdiction over workplace injuries and fatalities (i.e., thus preempts state prosecution under state criminal statutes) was the first-degree murder convictions by an Illinois court of the former president, plant manager, and plant foreman of Film Recovery Systems, Inc. In this case, Stephan Golab, a 59-year-old immigrant employee from Poland, died as a direct result of his work at the Elk Grove, Illinois, factory in which he stirred tanks of sodium cyanide used in the recovery of silver from photographic films. In February 1983, Golab "walked into the plant's lunchroom, started violently shaking, collapsed and died from inhaling the cyanide fumes." After his death, both OSHA and the Cook County State's Attorney's Office investigated the accident. OSHA found 20 violations and fined the corporation $4,850. This monetary penalty was later reduced by one-half. The Cook County prosecutor's office, on the other hand, took a different view and filed charges of first-degree murder and 21 counts of reckless conduct against the corporate officers and management personnel, as well as involuntary manslaughter charges against the corporation itself.

Under Illinois law, murder charges can be brought when someone "knowingly creates a strong probability of death or great bodily harm," even if there is no specific intent to kill. The prosecutor's office initially brought charges against five officers and managers of the corporation, but these defendants successfully fought extradition from another state. The prosecutor's intent was to "criminally pierce the corporate veil" and place liability not only upon the corporation, but on the responsible individuals as well.

During the course of the trial, the Cook County prosecutor presented extensive and overwhelming testimony that the company officials were aware of the unsafe conditions, knowingly neglected the unsafe conditions, and attempted to conceal the danger from the employees. Witnesses testified to the following:

- Company officials exclusively hired foreign workers who were not likely to complain to inspectors about working conditions and would perform work in the more dangerous areas of the plant.
- Officials instructed support staff never to use the word cyanide around the workers.
- According to the testimony of two co-workers, Golab complained to plant officials shortly before his death and requested to be moved to an area where the fumes were not as strong. These pleas were ignored.
- Testimony of numerous employees recounted episodes of recurrent nausea, headaches, and other illnesses.
- Employees were not issued adequate safety equipment or warned of the potential dangers in the workplace.
- An industrial saleswoman reported trying unsuccessfully to sell safety equipment to the owners.
- Supervisors instructed employees to paint over the skull-and-crossbones on the steel containers of cyanide-tainted sludge and to hide the containers from inspectors after the employee's death.
- Workers were never told that they were working with cyanide and were never told of the hazardous nature of this substance.
- Employees were grossly overexposed on a daily basis. After the incident, the company installed emission-control devices, which dropped cyanide emissions twenty-fold.

Given the extreme circumstances in this case, the prosecution was able to obtain convictions of three corporate officials for murder and 14 counts of reckless conduct. Each was sentenced to 25 years in prison, and the corporation was convicted of manslaughter and reckless conduct and fined $24,000. The court rejected outright the company's defense of preemption of the state prosecution by the federal OSH Act.

This case opened a new era in industrial safety and, as the prosecutor appropriately stated:

> These verdicts mean that employers who knowingly expose their workers to dangerous conditions leading to injury or even death can be held criminally responsible for the results of their actions ... . Today's [criminal] verdict should send a message to employers and employees alike that the criminal justice system can and will step in to protect the rights of every worker to a safe environment and to be informed of any hazard that might exist in the work place.

The decision in _People v. O'Neil_, although appealed, opened a new era in workplace health and safety. This case marked the first time that a corporate officer had been convicted of murder in a workplace death. As expected, after the door was opened, prosecutors across the country began to initiate similar actions, such as the Imperial

Foods fire, in which the plant owner pleaded guilty to manslaughter and received a 20-year prison sentence. The Los Angeles County District Attorney instituted a "roll out" program to address workplace accidents.

The significance of the Film Recovery case, according to Professor Ronald Jay Allen of Northwestern University School of Law, "is more psychological than anything else. It will sensitize prosecutors to the possibility of bringing criminal charges against the officials of a company when an egregious accident takes place." This new attitude toward work-related deaths was summarized by Los Angeles prosecutor John Lynch when he stated, "We [prosecutors] have to raise the consciousness that it is possible to have a criminal homicide in a case where there is no gun." Prosecutors argue that this type of criminal prosecution is needed because of the lax regulatory enforcement of workplace safety by OSHA, the obligations of the local prosecutors to the employees, and the need to send a message to employers that they have a responsibility for workplace safety. On the other hand, defense lawyers argue that liability for workplace accidents and deaths should continue to be handled in the regulatory and civil arenas and should not intrude into the criminal arena.

After the decision in *People v. O'Neil*, several criminal prosecutions were initiated in other states involving work-related deaths. Prosecutors acknowledged that the preemption defense in the early cases was a major obstacle in gaining convictions against employers. As explained by Jay C. Magnuson, deputy chief of the Public Interest Bureau of the Cook County State's Attorney's office who presented the Illinois cases, "If you can't even charge someone, you're pretty much out of the ball game."

Prosecutors across the country have successfully defeated a number of defenses, including preemption, and are becoming more creative in filing charges and handling workplace injury and fatality cases. Two areas that generated a substantial amount of litigation following *People v. O'Neil* were whether preemption applied only to the general OSHA standards (including the general duty clause) or only to the specific OSHA standards, and whether a specific OSHA standard preempts a more general industry standard, even though the specific hazard "falls between the cracks" of coverage under that specific standard.

The major question to be addressed by the courts after the Film Recovery case was if the OSH Act, in total, preempts any or all state criminal actions. With *People v. O'Neil* and several other similar convictions on appeal at that time, the interest turned to the Illinois Appeal court's decision in *People v. Chicago Magnet and Wire Corp.*

CHICAGO MAGNET AND WIRE CASE

In this case, a Cook County grand jury handed down indictments against the corporation and five of its corporate officers, charging each with multiple counts of aggravated battery, reckless conduct, and conspiracy under the Illinois criminal code. The trial court dismissed all charges against the employer, finding that OSHA preempted Illinois courts from applying Illinois criminal law to conduct involved with federally regulated occupational safety and health issues within the workplace.

At the circuit court level, the state relied extensively on the U.S. Supreme Court's decision in *Silkwood v. Kerr-McGee Corp.*, which held that even though the federal government occupies the field of nuclear safety regulations, the state courts were not preempted from assessing punitive damages for work-related radiation exposure. The circuit court explicitly rejected the analogy to Silkwood on two grounds: first, that Silkwood was decided under the Atomic Energy Act rather than the OSH Act, and second, the criminal laws, unlike punitive damages, are meant to regulate conduct rather than compensate victims.

Additionally, the circuit court rejected the state's arguments of waiver of authority and inadequacy of OSHA, and held that section 18 should be interpreted narrowly, on the basis of the comprehensiveness and legislative history of the OSH Act.

On appeal, the state again contended that the indictments should be reinstated because the police power of the state was neither implied nor expressly preempted by Congress under the OSH Act. The state argued, again relying on *Silkwood v. Kerr-McGee Corporation*, that state laws are only preempted when Congress has declared its intent to occupy a certain area of law, and where preemptive intent is not expressly stated in the statute. In addition, the intent of Congress must be derived from the statutory language, the comprehensiveness of the regulatory scheme, the legislative history, and the specific conflict between the state and federal statutes in question.

The state additionally argued that the prosecution was not preempted because it was based upon the application of general criminal law rather than upon the enforcement of specific workplace health and safety regulations. To support its assertion that Congress explicitly intended to leave preexisting state criminal laws undisturbed, the state cited section 553(b)(4) of the Act, which provides:

> Nothing in this chapter shall be construed to supersede or in any manner affect any workers' compensation law or to enlarge or diminish or affect in any other manner the common law or statutory rights, duties, or liabilities of employers and employees under any law with respect to injuries, diseases, or death of employees arising out of, or in the course of employment. [29 U.S.C. Section 553(b)(4) (1982)].

The Illinois Court of Appeals rejected the Silkwood preemption approach, holding that the Atomic Energy Act, relied upon in Silkwood, was not applicable to situations governed by the OSH Act. Additionally, the court rejected the state's position on section 553(b)(4) of the OSH Act, stating, "the state would not be foreclosed from applying its criminal laws in the workplace if the prosecution charged the defendants with crimes not involving working conditions." The Court also relied on the fact that the State of Illinois had the opportunity to retain responsibility for safety and health in the workplace by initiating a state plan occupational safety and health program that would have preempted the federal OSH Act but, although a state plan was submitted, this plan was later withdrawn by the state.

In affirming the decision of the circuit court, the appeals court stated:

> The State has expressed valid and legitimate concerns about the consequences of preemption on its ability to control the activities of employers. Congress has evidenced an intent that criminal sanctions should not be imposed for activities involving workplace health and safety except in highly limited circumstances, and that health and

safety requirements should be established through standard-setting, which provides
employers with clear and detailed notice of their legal obligations. Illinois' view that
employers may be held criminally liable for workplace injuries and illnesses, regardless of their compliance with OSHA standards, would lead to piecemeal and inconsistent prosecutions of regulatory violations throughout the states, a result that Congress
sought to preclude in enacting OSHA.

On February 2, 1989, the Supreme Court of Illinois addressed the issue of "whether
the OSH Act of 1970 (OSHA) preempts the state from prosecuting the defendants,
in the absence of approval from OSHA officials, for conduct which is regulated by
OSHA occupational health and safety standards." In a landmark decision, the court
reversed and held that the OSH Act did not preempt the state from prosecuting the
defendants.

The court initially addressed the extent to which state law was preempted by federal legislation under the Supremacy Clause of the Constitution. It stated that preemption "is essentially a question of congressional intendment" and "thus, if Congress,
when acting within constitutional limits, explicitly mandates the preemption of state
law … we [court] need not proceed beyond the statutory language to determine that
state law is preempted." Even absent an express command by Congress to preempt
state law in a particular area, preemptive intent may be inferred where the scheme
of federal regulation is sufficiently comprehensive to make reasonable the inference
that Congress left no room for supplementary state regulation." The Court also noted
that preemptive intent could also be inferred "where the regulated field is one in
which the federal interest is so dominant that the federal system will be assumed to
preclude enforcement" or "where the object sought to be obtained by the federal law
and the character of obligations imposed by it … reveal the same purposes."

The defendant initially argued that Congress, in section 18(a) of the Act, explicitly
provided that the states are preempted from asserting jurisdiction unless approval
for a state plan was acquired from OSHA officials under section 18(b). (Note: section 18 provides: (a) Nothing in this chapter shall prevent any state agency or court
from asserting jurisdiction under state law over any occupational safety or health
issue with respect to which no standard is in effect under section 655 of this title;
(b) Any state which, at any time, desires to assume responsibility for development
and enforcement therein of occupational safety and health standards relating to any
occupational safety or health issue with respect to which a federal standard has been
promulgated under section 665 of this title shall submit a state plan for the development of such standards and their enforcement. 29 U.S.C. Section 667 [1982]). The
defendant argued that the narrow interpretation of section 18 by the lower courts is
consistent with the legislative history of the Act.

In spite of this argument, section 18 was interpreted by the Illinois Supreme Court
to invite state administration of its own safety and health plans and was not "intended
to preclude supplementary state regulation." Section 2 of the Act provided that states
are "to assume the fullest responsibility for the administration and enforcement of
their safety and health laws." The Court additionally examined the legislative history of the Act, noting that "it is highly unlikely that Congress considered the interaction of OSHA regulations with other common law and statutory schemes other

than worker's compensation" and "it is totally unreasonable to conclude Congress intended that OSHA's penalties would be the only sanctions available for wrongful conduct which threatens or results in serious physical injury or death to workers."

The Illinois Supreme Court further found that Congress sought to develop uniform national safety and health standards and "the purpose underlying section 18 was to ensure that OSHA would create a nationwide floor of effective safety and health standards and provide for the enforcement of those standards." Additionally, "while additional sanctions imposed through state criminal law enforcement for conduct also governed by OSHA safety standards may incidentally serve as a regulation for workplace safety, there is nothing in OSHA or its legislative history to indicate that because of its incidental regulatory effect." The court concluded that "it seems clear that the federal interest in occupational health and safety ... [is] ... not to be exclusive."

The Illinois Supreme Court, unlike the lower courts, viewed the decision in *Silkwood v. Kerr-McGee Corp.* as applicable to this preemptive issue under the Act. The lower courts had explicitly rejected the analogy to Silkwood on the grounds that Silkwood was decided under the Atomic Energy Act rather than OSHA and that criminal laws, unlike punitive damages, were meant to regulate conduct rather than compensate victims. The Illinois Supreme Court noted that the Silkwood court addressed "a question with resemblance to the one here" and "there is little if any difference in the regulatory effect of punitive damages in tort and criminal penalties under the criminal law." Additionally, the court noted, "if Congress, in OSHA, explicitly declared it was willing to accept the incidental regulation imposed by compensatory damages awards under state tort law, it cannot plausibly be argued that it also intended to preempt state criminal law because of its incidental regulatory effect on workplace safety."

The court was similarly unconvinced by the defendant's contention that "it is irrelevant that the state is invoking criminal law jurisdiction as long as the conduct charged is an indictment or information is conduct subject to regulation by OSHA." The defendant had additionally argued "that the test of preemption is whether the conduct ... is in any way regulated by OSHA ... and the conduct charged in the indictment is conduct regulated by OSHA."

The court rejected this argument, finding that "simply because the conduct sought to be regulated in a sense under state criminal law is identical to that conduct made subject to federal regulation does not result in state law being preempted. When there is no intent shown on the part of Congress to preempt the operation of state law, the inquiry is whether there exists an irreconcilable conflict between the federal and state regulatory schemes." The court noted that a conflict exists only when "compliance with both federal and state regulations is a physical impossibility" or when state law "stands as an obstacle to the accomplishment and execution of the full purposes and objectives of Congress." The court could find no existing conflict or obstacle between OSHA and state's criminal law that would prohibit state's enforcement of this criminal action.

The defendant argued that state criminal prosecution would conflict with the purposes under OSHA because Congress had intended that the federal government was to have exclusive authority to set occupational safety and health standards.

"The standards were to be set only after extensive research to assure that the standards would minimize injuries in the workplace but at the same time not be so stringent that compliance would not be economically feasible." Although the court rejected this argument, it was noted that the defendant correctly pointed out that although states are given an opportunity to enforce their own safety and health standards under a "state plan," which is "at least as effective" as the federal program, OSHA retains jurisdiction until the "state plan" is approved.

The court also rejected the defendant's closely related argument that "federal supervision over state efforts to enforce their own workplace health and safety programs would be thwarted if the state, with prior approval from OSHA officials, could enforce its criminal laws ... [and] impose(s) standards so burdensome as to exceed the bounds of feasibility or so vague as not to provide clear guidance to employers." The court, in rejecting this argument, noted there was no finding that "state prosecutions of employers for conduct which is regulated by OSHA standards would conflict with the administration of OSHA or be at odds with its standards goals or purposes." On the contrary, "prosecutions of employers who violate state criminal laws by failing to maintain safe working conditions for their employees will surely further OSHA's stated goal of assuring so far as possible every working man and woman in the Nation safe and healthful working conditions." The court went further in stating, "state criminal law can provide a valuable and forceful supplement to ensure that workers are more adequately protected and that particularly egregious conduct receives appropriate punishment."

Similarly, the defendant argued that the state did not possess the ability or the resources to enforce more stringent safety and health standards than OSHA. The court rejected this argument, noting that the defendant's interpretation would "convert the statute ... into a grant of immunity for employers responsible for serious injuries or deaths of employees." The court further noted that this would be a "consequence unforeseen by Congress" and "enforcement of state criminal law in the workplace will not stand as an obstacle to the accomplishment and execution of the full purposes and objectives of Congress."

The Illinois Supreme Court noted that the preemption issue in the instant case has been addressed by very few courts. The appellate courts of the states of Michigan and Texas held that OSHA preempts state prosecutions, whereas the appellate court of Wisconsin, in *State ex rel. Cornellier v. Black*, held to the contrary. In *Cornellier*, the officer/director of a fireworks manufacturer was charged with homicide by reckless conduct for prior knowledge of and disregard of safety violations that resulted in the death of an employee. In finding that OSHA did not bar the prosecution and that the complaint was sufficient to state probable cause, the court stated:

> There is nothing in OSHA which we believe indicates a compelling congressional direction that Wisconsin, or any other state, may not enforce its homicide laws in the workplace. Nor do we see any conflict between the Act and [the Wisconsin statute]. To the contrary, compliance with federal safety and health regulations is consistent, we believe, with the discharge of the state's duty to protect the lives of employees, and all other citizens, through penalty for violation of any safety regulations. It is only attempting to impose the sanctions of the criminal code upon one who allegedly caused the death of another person by reckless conduct. And the fact that ... conduct

may in some respects violate OSHA safety regulations does not abridge the state's historic power to prosecute crimes.

The Illinois Supreme Court gave great deference to this decision by the Wisconsin court. Subsequent to the appellate court's decision in *Chicago Magnet Wire*, the congressional committee on government operations issued a report that directly addressed the issue in the instant case. In this report, the committee concluded that "inadequate use has been made of the criminal penalty provision of the Act and recommended to Congress that OSHA should take the position that the states have clear authority under the federal OSH Act, as it is written, to prosecute employers for acts against their employees that constitute crimes under state law."

A letter supplementing this finding was issued from the Department of Justice to the committee. This letter, in part, stated that the Department of Justice shares the concern of the committee regarding the adequacy of the statutory criminal penalties provided for violations of OSHA. This letter also observed,

> As for the narrower issue as to whether the criminal penalty provisions of the OSH Act were intended to preempt criminal law enforcement in the workplace and preclude the states from enforcing against the employers the criminal laws of general application, such as murder, manslaughter, and assault, it is our view that no such general preemption was intended by Congress. As a general matter, we see nothing in the OSH Act or its legislative history which indicates that Congress intended for the relatively limited criminal penalties provided by the Act to deprive employees of the protection provided by state criminal laws of general applicability.

The defendants in Chicago Magnet and Wire argued that the congressional committee's findings and the Justice Department's letter were not binding on the Illinois Supreme Court and urged the court to provide little deference to these reports. The court agreed that these reports were not binding, but noted "it is certainly not inappropriate to note ... the view of the governmental department charged with the enforcement of OSHA."

Given the conclusions reached by the congressional committee and the Justice Department subsequent to the appellate court's review and the arguments addressed previously, the court held "that the state ... [was] not preempted from conducting prosecutions" by the OSH Act against employers who consciously exposed employees to hazards in the workplace.

As expected, this decision by the Illinois Supreme Court was appealed to the U.S. Supreme Court, however the case was not accepted. Since the *Chicago Magnet Wire* decision, two other state supreme courts have similarly rejected preemption arguments. The current decisions of the state supreme courts in rejecting the preemption issue have set a precedent for other states to test this methodology and to begin to utilize the state criminal codes in situations involving workplace accidents. A possible decision by the U.S. Supreme Court could settle the preemption issue; no cases addressing the issue are pending at the time.

The use of state criminal code enforcement in incidents involving workplace injuries and fatalities by state prosecutors has greatly expanded the potential liability to corporate officers who willfully neglect their safety and health duties and responsibilities to their employees beyond that of the OSH Act. This increased potential liability for all levels of the management hierarchy may affect the structure and methodology utilized for the future management of health and safety in the American workplace.

Safety and loss prevention professionals should be aware that individual states can create special legislation, such as Maine's workplace manslaughter law in 1989, which governs workplace injuries and fatalities within the state. This type of specialized state law is applicable only in the state in which the legislation was enacted and is normally enforceable under the powers of the individual state. This type of state-enacted law usually requires compliance with the state law in addition to the federal OSHA standards or state plan requirements. Safety and loss prevention professionals should also be aware that state legislatures may modify existing laws (such as California's SB-198, which modifies the workers' compensation laws), which can significantly change the responsibilities in the area of safety and health as well as possible civil and criminal sanctions for noncompliance.

WHAT TO EXPECT WITH STATE CRIMINAL SANCTIONS

Safety and loss prevention professionals facing a fatality, serious injury, or a multiple injury situation should be prepared for an investigation by the state or local prosecutor's office, state police, or other investigative law enforcement agencies. Although the odds are favorable that the criminal investigation will go no further than the investigation stage (unless the situation involves willful, reckless, or abnormally negligent conduct on the part of the safety and loss prevention professional and/or company), safety and loss prevention professionals should be prepared for a substantial and probing inquiry. Safety and loss prevention professionals should be aware that the prosecutor's investigation will be conducted from a criminal perspective rather than an OSHA perspective, and thus all constitutional rights should be preserved until legal counsel can be consulted.

Safety and loss prevention professionals should be aware that several jurisdictions have established programs, such as Los Angeles County's former "Roll Out" program, where workplace fatalities were specifically investigated by local law enforcement and local prosecutors as well as OSHA. Safety and loss prevention professionals should be prepared for this type of investigation and be aware of its parameters, scope, and magnitude.

In most criminal investigations, it is advisable to err toward the conservative side. As depicted on various police-related television programs, one is guaranteed specific rights under the U.S. Constitution (such as your Miranda Rights). Simply because the fatality or injury occurred at the worksite does not mean that the safety and loss prevention professional or other corporate official has waived his or her constitutional rights.

In most circumstances, the prosecutor or detective investigating the accident will not read the Miranda Rights to the company representative until an arrest is made. However, safety and loss prevention professionals should be aware that most comments, photographs, and other evidence gathered during the investigation can be used in a court of law. Additionally, after an arrest and reading of the rights, an individual possesses the ability to waive his or her rights through a verbal or written acknowledgment. Any comments made following the waiving of the Miranda Rights can be used as evidence.

In situations involving workplace fatalities or injuries, the investigation is normally concluded without immediate arrests. The prosecutor will evaluate the evidence and, if substantial, may submit the evidence to a grand jury. If the grand jury finds the evidence substantial, an indictment is rendered and arrest warrants are issued for the individuals in question.

Following arrest, there is normally a preliminary hearing or arraignment where the charges are read to the individual and he or she is asked to plead. If the individual cannot afford legal counsel, it is normally appointed at this time. A trial date is then scheduled, and bond is usually set.

Safety and loss prevention professionals should be prepared for a varying degree of isolation from the other named corporate officials and possibly from the company following arrest. Depending on the circumstances, the safety and loss prevention professional may be required to acquire his or her own legal counsel and be responsible for this cost. Abandonment or isolation of the safety and loss prevention professional by the company and its corporate officials is not uncommon.

The trial is conducted in the same manner as any other criminal trial. Normally, each individual and the corporations that are charged will have separate trials. Separate legal counsel usually is required for each of the parties. The burden is on the state to prove each element of the charge "beyond a shadow of a doubt" and that the individual committed the alleged offenses. All constitutional rights to a jury trial, cross examination, equal treatment, and self-incrimination are the same as the rights in criminal trials. If convicted, appeal rights are normally preserved and sentencing would be in accordance with the state's criminal code.

The defenses available in this type of criminal action vary, depending on the situation and the charges. The common law defenses of duress, self-defense, defense of others, defense of property, consent, and entrapment are available, in addition to the usual defenses to the charge. Because the injury or fatality occurred on the job, defenses of double jeopardy may be available if previously cited by an OSHA state plan in the same state and preemption of jurisdiction by OSHA. In addition to the factual defenses, other defenses may include, but are not limited to, lack of an applicable OSHA or state plan standard, lack of employment relationship, isolated incident defense, or even the lack of employer knowledge.

A peripheral area of concern for safety and loss prevention professionals after a workplace fatality, major workplace injury, or extensive property damage incident is the potential efficacy losses (e.g., reputation, image) that can result through widespread media distribution of information. Safety and loss prevention professionals should attempt to control the information available to the media and minimize photographs, videotape footage, and other documentation by these outside parties. Failure to control this flow of information and documentation can often result in insurmountable damage at a later time. For example, if the media acquires the name of the employee who was killed in a work-related accident, and announces this information prior to the safety and loss prevention professional (or designated person) contacting the family, there is a substantial likelihood that this impersonal method

of notification to the family may cause the family members immense pain. The
company could be perceived as uncaring, and the relationship between the family
and the safety and loss prevention professional or the company could be irreparably
damaged.

SELECTED CASE STUDY

Note: Case modified for the purpose of this text.
Seventh Circuit.

UNITED STATES OF AMERICA, PLAINTIFF-APPELLANT, V. MYR GROUP, INC., DEFENDANT-APPELLEE

No. 03-3250.

Argued Feb. 19, 2004.
Decided March 16, 2004.

Background

Federal government indicted parent corporation and its wholly owned subsidiary for
willfully violating Occupational Safety and Health Act (OSHA) in connection with
electrocution deaths of two of subsidiary's employees. The United States District
Court for the Northern District of Illinois, 274 F.Supp.2d 945, district judge, dismissed indictment as to parent corporation, and government appealed.

Holding

The Court of Appeals, Posner, Circuit Judge, held that parent could not be held criminally liable for willful violations of OSHA regulations on theory that duties created
by regulations ran to any employees, not just employees of company accused of
violating them.

Affirmed.

Before Circuit Judges.

The district judge dismissed an indictment that charged MYR Group, Inc.,
with violating section 17(e) of the Occupational Safety and Health Act, 29 U.S.C.
§ 666(e), and the government appeals. The factual record is limited to the facts
alleged in the indictment, according to which: MYR has a wholly owned subsidiary
named L.E. Myers Company (the parties call it "LEM"), which repairs high-voltage lines. MYR oversees the safety programs of its subsidiaries, provides safety
manuals and other safety instructions to the employees of the subsidiaries, and
jointly with the subsidiaries is responsible for training those employees with regard
to safety matters, including how to repair high-voltage lines without being electrocuted. Nevertheless, on two separate occasions, employees of LEM were electrocuted while repairing such lines. The indictment charges both MYR and LEM
with two counts of causing the death of an employee by willfully violating rules
promulgated under OSHA. 29 U.S.C. § 666(e). MYR is charged with violating regulations requiring, in essence, that employees be properly trained in safe working

procedures. 29 C.F.R. §§ 1910.269(a)(2)(i), (ii). LEM is charged with violations of other rules as well, and is awaiting trial in the district court.

The government's argument is a simple one. MYR is an employer, albeit not of the two workers who were electrocuted; the two workers were employees; the regulations in question state simply that "employees shall be trained in" safe working procedures. Therefore, the argument concludes, the duties created by the regulations run to anyone's employees, not merely employees of the employer accused of having violated the regulations.

In its opening brief, the government tried to make something of the fact that MYR and LEM are corporate affiliates, citing *Esmark, Inc. v. NLRB*, 887 F.2d 739 (7th Cir.1989). That, however, was a veil-piercing case, where we said that "it is solely where a parent disregards the separate legal personality of its subsidiary (and the subsidiary's own decisionmaking 'paraphernalia'), and exercises direct control over a specific transaction, that derivative liability for the subsidiary's unfair labor practices will be imposed under the theory adopted by the Board in the present case." *Id.* at 757. At argument the government made clear that it is not attempting to pierce the corporate veil and by doing so attribute the subsidiary's acts to the parent, consistent with the principles of corporate law.

Breathtaking vistas of both criminal and civil liability (the latter not dependent on proof that the violation was willful, 29 U.S.C. §§ 666(b), (c); *S.A. Healy Co. v. OSHRC*, 138 F.3d 686, 688 (7th Cir.1998)) open before our eyes. Were LEM to hire the Illinois Institute of Technology to train LEM's employees in the hazards of uninsulated high-voltage electrical cables, and IIT fell down on the job and an employee of LEM was electrocuted as a result, IIT would, if the government is right, be either criminally or civilly liable for having violated OSHA. It would be so merely by virtue of having employees, even though those were not the workers endangered by its violation. It is true that LEM and IIT are not affiliates, but the government's lawyer acknowledged that this would make no difference, for remember that it is not arguing that MYR did anything that would justify treating LEM as if it were really just a division of MYR rather than a separate corporation.

The government's argument is not limited to service providers. A firm (provided only that it had employees) that sold a defective espresso machine to a coffee shop would be subject to OSHA liability if the machine exploded and scalded a waiter. OSHA would become a products-liability statute-with criminal sanctions for its willful violation.

The government points to our decision in *United States v. Pitt-Des Moines, Inc.*, 168 F.3d 976, 984-85 (7th Cir.1999), which holds that a contractor at a construction site can be prosecuted under section 666(e) if by violating an OSHA regulation he causes the death of an employee of another contractor at the same site. However, the point of this "multi-employer" gloss (cf. *Universal Construction Co. v. OSHRC*, 182 F.3d 726, 728-30 (10th Cir.1999); *R.P. Carbone Construction Co. v. OSHRC*, 166 F.3d 815, 818 (6th Cir.1998); *Beatty Equipment Leasing, Inc. v. Secretary of Labor*, 577 F.2d 534, 536-37 (9th Cir.1978); *Marshall v. Knutson Construction Co.*, 566 F.2d 596, 599-600 (8th Cir.1977) (per curiam); *Brennan v. OSHRC*, 513 F.2d 1032, 1037-39 (2d Cir. 1975); but see *Melerine v. Avondale Shipyards, Inc.*, 659 F.2d 706, 710-11 (5th Cir.1981)) is that since the contractor is subject to OSHA's regulations of safety in construction by virtue of being engaged in the construction business, and has to comply with those regulations in order to protect his own workers at the site, it is sensible to think of him as assuming the same duty to the other workers at the site who might be injured or killed if he violated the regulations. From a safety standpoint, it is a joint-employment case. A crane operator might be killed because the contractor responsible for leveling the ground at the worksite violated a regulation requiring that the surface beneath the crane be planed smooth, and a bulldozer driver might be killed when a crane fell on him because the crane contractor had failed to comply with regulations governing the safe operation of cranes. Each employer at the worksite controls a part of the dangerous activities occurring at the site and is the logical person to be made responsible for protecting everyone at the site from the dangers that are within his power to control. See *Universal Construction Co. v. OSHRC*, *supra*, 182 F.3d at 730; *Brennan v. OSHRC*, *supra*, 513 F.2d at 1038. This case is not like that. No employee of MYR was engaged in repairing high-voltage lines, any more than a professor of electrical engineering at IIT who trained employees in the hazards of electricity would be present at the worksite.

The government's attempt to stretch the statute by filing a criminal indictment is especially questionable. Surely the proper way to proceed, if the government really thinks the statute can be stretched this far, would be to amend the regulations to bring the third-party case under them. And who by the way is "the government" in this case? No representative of the Occupational Safety and Health Administration, or for that matter anyone outside the office of the U.S. Attorney for this district, signed the government's brief. The Solicitor General of the United States had to approve the appeal, 28 C.F.R. § 0.20(b); but we have not even been told whether OSHA approves, or for that matter knows of, the extension of liability urged by the U.S. Attorney!

The dismissal of the indictment against the MYR Group is
AFFIRMED.

7 OSHA Standards and Requirements

CHAPTER OBJECTIVES

1. Acquire an understanding of the OSHA standards.
2. Understand how to develop and implement compliance programs.
3. Acquire an understanding how to develop a LOTO program.
4. Acquire an understanding how to develop a hazcom program.
5. Understand how to develop a bloodborne pathogen program.

Lock-out/Tag-out

INTRODUCTION

Depending upon the particular industry or workplace involved, some violations of OSHA standards have a greater risk of criminal and civil liability for employers than others. For example, there is a greater potential for a fatality because of a violation of the control of hazardous energy (LOTO) standard than there is from a violation of the hearing protection standard. Safety and loss prevention professionals should recognize areas of potential risk and liability for violations of particular standards. They may also wish to provide additional effort to ensure that their programs are in compliance with these riskier standards. As emphasized throughout this book, the simplest way to avoid or minimize potential liability is to ensure that the entire safety and loss prevention program is in strict compliance with OSHA standards. In addition to discussing overall efforts to achieve and maintain compliance with the OSHA requirements, this chapter highlights particular OSHA standards, and violations thereof, that have served as the foundation for civil and criminal penalties and that continue to be a source of liability in the average workplace.

Regarding monetary penalties, the OSHA violations since 2000 that have resulted in the greatest number of penalties have been in the areas of recordkeeping, ergonomic violations (cited under the general duty clause), and hazard communications. Monetary penalties proposed by OSHA for hundreds of thousands or even millions of dollars are not uncommon today. OSHA standards that provide protection against potential life-threatening hazards, such as the confined space entry and rescue standard; the control of hazardous energy (lockout and tagout) standard; and the excavation, trenching, and shoring standard, if violated, possess the greatest potential for fatalities or serious injuries in most operations. Although any violation of an OSHA standard has the potential to cause injury, illness, or death, certain standards, as applied to the individual workplace, may create a greater likelihood of serious harm or death to employees. These particular hazards must be identified by safety and loss prevention professionals and all of the necessary steps must be taken to guarantee compliance with the appropriate OSHA standard.

Of the many OSHA standards that have been promulgated since 1970, there are a few particular standards that normally cause difficulty in design, application, or management and, thus, create concern for safety and loss prevention professionals. Many of the early OSHA standards were adopted from existing standards and, in general, were related to general equipment or personal protective equipment (e.g., guardrail, machine guarding, etc.). In recent years, the standards promulgated by OSHA are far more complex, system-oriented, and performance based. These new standards offer unique challenges to the safety and loss prevention professional, who must be properly prepared to tackle these challenges.

In developing a regulatory compliance program, there is no substitute for knowledge of the OSHA standards, EPA regulations, or other applicable government regulations. Under the law, every organization covered by these regulations is required to know the law. As stated by many courts throughout history, ignorance of the law is no defense.

So how can a safety and loss prevention professional acquire a working knowledge of the OSHA regulations? The first step is to acquire a copy of the regulations. This can be accomplished by contacting the local OSHA office or purchasing the regulations online. Second, the safety and loss prevention professional must read through the regulations and become familiar with them. These OSHA standards are often used by professionals as a quick reference; one can easily reference the applicable standard through the index system located in the last section of 29 C.F.R. section 1910. Additionally, it is highly recommended that safety and loss prevention professionals attend continuing education conferences or seminars that provide current information regarding new and existing OSHA standards and regulations.

Another resource that should not be overlooked is the consulting division of OSHA, which is generally provided through the individual state's safety and health office. These consultation programs allow employers to bring OSHA or state plan consultants into the workplace to conduct audits, provide education programs, and/or assist with compliance, usually at no cost to the employer. Under the terms of most consultation programs, an OSHA consultant may not cite the employer for violations unless a situation involves imminent death. The state consultation services also usually possess a library of safety resources, which can be utilized by the management team. Another exceptional source for safety and health information is the National Institute of Occupational Safety and Health (NIOSH). This education agency is mainly focused on research and possesses publications addressing virtually every area within the safety and health realm. NIOSH does not possess the same authority to issue citations for violations as OSHA. If an OSHA standard cannot be found to specifically address a particular situation, NIOSH is an excellent resource.

When developing a regulatory compliance program, one must keep in mind the source of the standard—a government agency. Therefore, when developing a compliance program, if the program is not in writing, it is not a program! All compliance programs should be in writing and must meet the minimum requirements set forth in the applicable OSHA standard.

With most of the OSHA standards, the standard itself does not tell you how to develop the program—only what requirements must be achieved. Additionally, many of the OSHA standards reference other standards that can be found within 29 C.F.R. section 1910 or other regulatory agencies such as the American National Standards Institute (ANSI). Safety and loss prevention professionals are required to know the standard and be able to acquire further information. The OSHA standards are usually base level requirements that employers may not go below. Employers are, however, encouraged to make their programs considerably better than the OSHA requirements. OSHA normally provides only the minimum requirements to meet the standard(s). The job of the safety and loss prevention professional is to identify operational needs, identify the requirements of the standard, and develop the compliance program that best serves the operation while maintaining compliance with the applicable standard.

In addition to knowing the current OSHA standards, safety and loss prevention professionals are encouraged to review the various professional safety and health publications as well as the *Federal Register* to identify new proposed standards that may impact their operations prior to the standard becoming law. When developing a new standard, OSHA is required to publish the proposed standard in the *Federal Register*, hold open hearings, and accept comments from any source. Safety and loss prevention professionals are encouraged to voice any concerns regarding a particular proposed standard to OSHA at these hearings or in writing prior to these hearings. All final standards are published in the *Federal Register*. OSHA then gives employers a period of time to achieve compliance.

The following basic and general guideline assists the safety and health professional in developing a safety and loss prevention program for a particular standard:

- Read the OSHA standard carefully and note all requirements.
- Remember that all OSHA compliance programs must be in writing.
- Remember to visualize your program in action, add operational elements for functionality of the program, and acquire all necessary approvals.
- Develop a plan of action. Acquire management commitment and funding for the program.
- Purchase all necessary equipment. Acquire all necessary certifications, etc.
- Remember to post any required notices.
- Inform employees of the program. Acquire employee input during the development stages of the program. Inform labor organizations, if applicable.
- At this point, you may want to contact OSHA for samples of acceptable programs or check online at OSHA.gov.
- Conduct all necessary training and education. Remember to document all training!
- Conduct all of the required testing. Remember to document all testing procedures, equipment, calibrations, and so on.
- Implement the program.

- Remember to make sure that all procedures are followed. Disciplinary action given for noncompliance must be documented.
- Audit the program on a regular, periodic basis, or as required under the standard.

Again, it is important to note that most OSHA standards do not provide any guidance regarding how compliance should be achieved. Most standards require only that the employer achieve and maintain compliance. In essence, the OSHA standard tells the safety and loss prevention professional what needs to be done, but not how the compliance program is to be designed, managed, or evaluated. For example, an employer must have a facility evacuation plan in place. How the employer structures the plan and the specific details of the plan are left to the safety and loss prevention professional. Additionally, it is the responsibility of the safety and loss prevention professional to determine exactly what OSHA standards are applicable to the individual facility or worksite to ensure that the facility and worksite are in full compliance. An omission of part or all of the required safety program necessary to achieve compliance with a specific standard is a violation, just as an inadequate or mismanaged program is a violation. Errors of omission and of commission are violations of the OSH Act.

Although some specific OSHA standards will vary with the industry and type of facility, several OSHA standards, namely the hazard communication standard, the confined space entry and rescue standard, the bloodborne pathogen standard, and the control of hazardous energy (lockout/tagout) standard generally cause the most difficulty and concern within the safety and loss prevention community. These standards, as well as respiratory protection and fall protection, are discussed in the following sections.

HAZARD COMMUNICATION STANDARD

Violations of the hazard communication standard were the number one cause of OSHA citations in the last decade. The primary reason for violations of this standard is that the standard requires extensive documentation and is considered "paper intensive." Given the substantial number of different chemicals in the average facility, the hazard communication program must be managed on a daily basis to ensure continued compliance.

According to OSHA statistics, approximately 32 million U.S. workers are potentially exposed to one or more chemical hazards on a daily basis. Because of the potential health and safety hazards associated with these chemicals, and because many employers and employees know little or nothing about the chemicals that they work with, OSHA promulgated the hazard communication standard to provide not only protection from potential hazards but also education and training regarding the specific characteristics of the chemicals as well as necessary safety and health precautions. The general purpose of this standard is to ensure that all known hazards of all chemicals produced or imported are evaluated and that information relating to such hazards is transmitted to employers and employees. The transmittal of information is to be accomplished by a comprehensive hazards communication program that includes such requirements as container labeling and other forms of warning, material safety data sheets (SDSs), and employee training.

Gaining Management Commitment

A hazard communication program, like other safety and loss prevention programs, needs management commitment in order to be successful. It should be noted that management commitment is never addressed in any OSHA standard. However, without it, most safety programs will have a poor chance of success.

The hazard communication standard is mandatory in any industry possessing chemicals; this standard affects most industries and worksites. All levels of the management team need to understand the importance of this program and its required duties and responsibilities.

The most common problems encountered in maintaining this program are lack of management support, inadequate funding, and the lack of other resources necessary to develop and maintain the required elements of a hazard communication compliance program. One method that safety and loss prevention professionals have been able to use is reporting to upper management the potential OSHA penalties for noncompliance as well as the mandatory nature of the standard. It is unfortunate that this method is too often the only way to motivate management to take this standard seriously. Another method is to explain the potential harm that may occur if an employee is injured because of improper labeling or inadequate training.

Safety and loss prevention professionals should be aware of the changes to the hazardous communication standard in 2012. As identified below, OSHA adopted the Globally Harmonized System of Classification and Labeling of Chemicals, generally referred to as GHS; provided changes in the safety data sheets (formally material safety data sheets); and addressed written program and training requirements.

> The Hazard Communication Standard (HCS) is now aligned with the Globally Harmonized System of Classification and Labeling of Chemicals (GHS). This update to the Hazard Communication Standard (HCS) will provide a common and coherent approach to classifying chemicals and communicating hazard information on labels and safety data sheets. This update will also help reduce trade barriers and result in productivity improvements for American businesses that regularly handle, store, and use hazardous chemicals while providing cost savings for American businesses that periodically update safety data sheets and labels for chemicals covered under the hazard communication standard.

Hazard Communication Standard

In order to ensure chemical safety in the workplace, information about the identities and hazards of the chemicals must be available and understandable to workers. OSHA's hazard communication standard (HCS) requires the development and dissemination of such information:

- Chemical manufacturers and importers are required to evaluate the hazards of the chemicals they produce or import and prepare labels and safety data sheets to convey the hazard information to their downstream customers;
- All employers with hazardous chemicals in their workplaces must have labels and safety data sheets for their exposed workers and train them to handle the chemicals appropriately.

MAJOR CHANGES TO THE HAZARD COMMUNICATION STANDARD

- **Hazard classification**: Provides specific criteria for classification of health and physical hazards, as well as classification of mixtures.
- **Labels**: Chemical manufacturers and importers will be required to provide a label that includes a harmonized signal word, pictogram, and hazard statement for each hazard class and category. Precautionary statements must also be provided.
- **Safety Data Sheets:** Will now have a specified 16-section format.
- **Information and training:** Employers are required to train workers by December 1, 2013 on the new labels elements and safety data sheets format to facilitate recognition and understanding.
- **See** OSHA.gov.

ASSESSING THE HAZARDS

The purpose of the hazard assessment element of a hazard communication program is to determine the identity and the location of hazardous chemicals at a facility or workplace. The results of the hazard assessment are used to prepare an inventory of hazardous chemicals and their SDS as well as to define the types of hazards that need to be covered during employee training. As with most OSHA standards, if the particular worksite or facility does not have the prescribed chemicals at the location, there is no need to comply with the hazard communication standard. However, extremely few industrial workplaces, if any, can make this claim.

A major problem that can occur with this standard is omitting potential chemicals located on-site during the assessment. Therefore, it is important to locate and identify all chemicals at the worksite.

There are three primary steps in conducting a hazard assessment:

1. The first step involves understanding the definition of hazardous chemicals. As defined in the standard, a hazardous chemical is any chemical that is a physical or health hazard. This definition does not include hazardous wastes, tobacco products, wood products, drugs, food, or cosmetics used or consumed by employees at the workplace. The workplace may contain common consumer products such as household detergents and cleansers, soap, bleach, etc., that may be excluded from the hazard assessment, provided they are used in the same manner and approximate quantities as would be expected in their typical consumer applications.

2. The second step is to consider only those chemical agents that are known to be present at the workplace. The hazard assessment is not limited to only those chemicals produced or used within the facility, and must consider all potentially hazardous agents in the work environment. This includes private sector establishments as well as field survey locations. For example, if inspectors routinely go to a wide variety of industrial settings to perform their work, they should receive training designed to cover the hazards to which they may be exposed. Employers need not consider or prepare lists

of every known chemical in the world, but only those that exist within their facility or on their worksites.

3. The third step is to determine whether employees have the potential for exposure to harmful chemicals. The term potential exposure describes the ability of an employee to come into contact with a material by inhalation, ingestion, or direct contact with the skin. Exposures during normal work activities, non-routine work tasks, and foreseeable emergencies also must be considered. In many cases, the determination that a chemical is present in an employee's work area will be enough to establish the potential for exposure. However, there may be situations where it is reasonable to conclude that exposure will not occur, such as exposure to gasoline during the operation of motor vehicles.

The hazard communication standard covers chemicals in all physical forms, including such physical states as liquids, solids, gases, vapors, fumes, and mists. Potential hazards from tanks, pipes, and other containers need to be identified, in addition to normal storage tanks and containers. The possibility of hazardous dusts, vapors, or fumes that may be generated during certain operations also needs to be identified. These potential hazards may need to be included in specified documents under other standards or laws if they are regulated substances.

Finally, records of chemical purchases may help identify hazardous materials that are routinely or periodically ordered. Purchasing procedures may be developed to allow tracking of chemicals through ordering, receiving, use, and disposal.

Chemical manufacturers and importers are required to review available scientific evidence concerning the hazards of the chemicals produced or imported and report the information to their employees, distributors, and employers who use the products. Producers of and importers of chemicals are responsible for determining the hazards associated with each of the chemicals and must supply MSDS with all hazardous chemicals. The hazards associated with each of the chemicals are provided to distributors and customers in the form of MSDS and are shipped with the products. Each chemical is evaluated to determine its adverse health effects and its physical hazards (e.g., fire or explosion). In general, the chemicals listed in the following sources are considered hazardous:

- 29 C.F.R. section 1910, subpart Z, Toxic and Hazardous Substances, OSHA
- Threshold Limit Values for Chemical Substances and Physical Agents in the Work Environment, American Conference of Governmental Industrial Hygienists

Additionally, chemicals established as being or having the possibility of being a carcinogen in the following sources must be reported as such:

- National Toxicology Program (NTP), Annual Report on Carcinogens
- International Agency for Research on Cancer (IARC) Monographs (latest edition)
- 29 C.F.R. section 1910, subpart Z, Toxic and Hazardous Substances, OSHA

However, if a chemical is not listed in the previously mentioned sources, this does not mean it is not hazardous. Any chemical that poses a potential health hazard or physical hazard must be included in the hazard assessment. In general, if there is any question regarding a particular chemical, a safety and loss prevention professional should include the chemical in the hazard communication program.

KNOW THE STANDARD

It is imperative that safety and loss prevention professionals acquire and possess a firm grasp on the OSHA standard located at 29 CFR 1910.1200 and provide particular attending to the provisions found primarily in sections (e) written program; section (f) labels and warnings; section (g) SDS; and section (h) employee information and training.

IDENTIFY RESPONSIBLE PARTIES

Safety and loss prevention professionals cannot do everything themselves. Delegation is essential to ensure that all programs are in compliance at all times. Safety and loss prevention professionals should identify the specific responsibilities, such as the training function, and prepare and delegate this responsibility to appropriate and qualified staff or management team members.

DEVELOPING THE WRITTEN PROGRAM

Safety and loss prevention professionals should ensure that all required elements are addressed as well as any additional elements necessary for the efficient implementation or functioning of the written hazard communications program. Additionally, safety and loss prevention professionals should remember to write their hazard communication program, as well as all other written compliance programs, at the educational level which can be read and understood by all employees. In general, a hazard communication program contains three basic components:

1. labels and other forms of warning
2. SDS (Safety Data Sheet)
3. employee information and training

Labels and Other Forms of Warnings

Employers purchasing chemicals may rely on the labels provided by their suppliers until they transfer the chemicals into another container. At that time, it becomes the employer's responsibility to label the container, unless it is subject to the portable container exemption. The information required on a label is the identity of the material and any other warnings. This identity can be any term that appears on the label, the SDS, or the facility's list of chemicals. The identity can be a common trade name or the chemical name. The hazard warning, in general, is a brief statement of the hazardous effects of the chemical. Although there are no specific requirements for size and color at this time, labels must be in English, legible, and prominently displayed. The following should be contained in any written program for labeling and other forms of warnings:

- designation of person(s) responsible for ensuring labeling
- designation of person(s) responsible for ensuring labeling of any shipped container
- description of labeling system(s) used
- description of written alternatives to labeling of containers
- procedures to review and update label information

Maintaining Safety Data Sheets

The SDS (Safety Data Sheet) is a document that describes the physical and chemical properties of products, their physical and health hazards, and precautions for their safe handling and use. The function of an MSDS is to provide detailed information on each hazardous chemical, including its potential health effects, its physical and chemical characteristics, and recommendations for appropriate protective measures. Distributors of regulated chemicals are required to furnish their customers with a completed MSDS for each regulated chemical. Customers receiving SDSs should review them for accuracy and completeness and make sure that the latest MSDS is on file. A comparison of new and old MSDSs is useful because it may identify that there is a new hazard associated with an existing chemical or that a new ingredient is included in a currently used product.

Safety and loss prevention professionals can also use the information in an SDS in assessing and selecting the safest product for a particular job. The hazard communications standard requires that the following specified information be provided on all SDSs:

- chemical identity
- hazardous ingredients
- physical and chemical characteristics
- physical hazards
- health hazards
- special precautions, spill, leak, and cleanup procedures
- control measures
- emergency and first-aid procedures
- the responsible party

Copies of the SDS for each chemical must be readily accessible to employees during each work shift when the chemicals are in employees' work areas. An SDS booklet or binder should be placed in different locations throughout the facility so that every employee knows the locations and can easily find them. Employee representatives (such as officials) also have the right to access SDSs. The hazard communications standard permits alternative formats for the SDS of any chemical. For example, in facilities where computer terminals are readily available, SDS information could be provided to employees electronically.

The major problem regarding this element of the hazard communications program is the acquisition and maintenance of the SDS for each and every chemical at the worksite. Many organizations have instituted computerized SDS programs in conjunction with management controls to ensure that the current SDSs are available for each chemical on-site. For many organizations, the maintenance of accurate SDSs for each chemical on-site has been a problem. Some of the sources of these problems occur when purchasing agents obtain new or different chemicals and do not inform the safety and health professional, when the receiving department fails to notice unlabeled products as they enter the facility, and when updated SDSs are not incorporated into the existing MSDS booklets or binders. This type of violation is frequently cited during OSHA compliance inspections.

Employee Information and Training

The hazard communication standard requires employers to provide training programs for their employees. The standard requires that employees receive information and training on the following topics:

- all provisions of the hazard communication standard
- the types of operations in their work areas where hazardous chemicals are present
- the location and the availability of the written hazard communication program, list(s) of hazardous chemicals, and SDSs
- employee training sessions that describe methods that employees can use to detect the presence or release of toxic chemicals in the workplace (The employees receive training on the visual appearance or odor of hazardous chemicals that might be released and information on any alarm or warning systems. Employees should also be informed about the existence of any environment or medical monitoring programs.)
- the physical and health hazards associated with the chemicals in their work areas
- specific measures to protect themselves from the hazards in their work areas, such as the types of protection afforded by engineering controls, safe work practice guidelines, emergency procedures, and protective equipment
- specific components of the hazard communication program, including explanations of the facility's labeling system and the methods employees can use to obtain hazardous chemical information

AUDIT AND ASSESS YOUR PROGRAM

Although the standard does not specifically require the auditing and periodic assessment of the hazard communication program, your program must be current (especially the SDSs) at all times and changes in chemicals, equipment, or procedures must be addressed within your written program as well as other aspects of your program. Many organizations possess an annual or periodic evaluation of all compliance programs that identify necessary changes or modifications. Safety and loss prevention professionals should be aware that this is an ongoing process and the program is not complete and finished when the written program goes on the shelf. Constant assessment and improvement is required.

These additional requirements may present problems for safety and loss prevention professionals in ensuring compliance with the hazard communications standard and their ability to maintain compliance after the program is instituted. To address the continued compliance of the hazard communications program, many employers have initiated audit programs to track the performance of the program and to identify deficiencies for immediate correction. When human deficiencies are identified within the program, retraining, coaching, and progressive discipline are normally required to emphasize the importance of the hazard communication program. The safety and loss prevention professional must always remember that many employees do not have the same educational background and language skills as the professional. Continued feedback and refresher courses are critical to success.

BLOODBORNE PATHOGEN STANDARD   *Hep B liver virus*

OVERVIEW

With HIV and other diseases transmitted via bodily fluids making the headlines, the bloodborne pathogen standard has taken a priority position for many safety and loss prevention professionals. This OSHA standard was developed to protect workers in industrial, health care, and other workplaces from possible deadly exposure to HIV, hepatitis, and other bloodborne diseases. Given the nature of the protection provided under this standard, the bloodborne pathogen standard reaches beyond the normal industrial workplace, affecting such workplaces as dentist's offices, nursing homes, and hospitals. In industrial or construction workplaces, the usual areas of potential exposure include dispensaries where medical attention is provided, first-aid responders to injured employees, and other similar occurrences where an employee could be exposed to human blood. In addition to required training and personal protective equipment, the bloodborne pathogen standard requires the proper labeling and disposal of all medical waste, including a wide range of items from blood samples to used adhesive bandages.

The bloodborne pathogen standard is the first of its kind to be promulgated under the OSHA standards. Given the far-reaching requirements of this standard, many safety and loss prevention professionals have expressed concern over the potential costs of compliance, the personal rights issues involved (e.g., the hepatitis B vaccination and privacy of medical records), and the environmental issues involved in the disposal of infectious medical waste.   *drinking water?*

BLOODBORNE PATHOGEN PROGRAM GUIDELINES

The following general procedures are commonly followed by safety and health professionals in the development of a program to achieve and maintain compliance with the bloodborne pathogen standard:

- Review the standard with the management team and obtain management commitment and appropriate funding for the program.

- Develop a written program incorporating all of the required elements of the standard, including, but not limited to, universal precautions, engineering and work practice controls, personal protective equipment, housekeeping, infectious waste disposal, laundry procedures, training requirements, hepatitis B vaccinations, information to be provided to physicians, medical recordkeeping, signs and labels, and the availability of medical records (as well as limited access thereof).
- Under the area of universal precautions, employers should analyze the workplace to provide all necessary safeguards to employees who may have possible contact with human blood. Engineering controls and work practices should be examined and evaluated. Procedures should be established to avoid potential hazards, such as safe disposal of used needles and disposal or cleaning of used personal protective equipment and other equipment. Additionally, the storage of food or drink in the refrigerators, cabinets, or freezers where blood is stored should be prohibited. The standard directs that all employees be properly trained in the requirements of this standard and in safe practices, such as the prohibition from eating, drinking, smoking, applying cosmetics or lip balm, or handling contact lenses after possible exposure to body fluids.
- Where there is the potential for exposure to bodily fluids, personal protective equipment, such as surgical gloves, gowns, fluid-proof aprons, face shields, pocket masks, and ventilation devices must be provided to employees. This requirement affects first-aid responders and plant medical personnel. Supervisors and co-workers need to be trained to respond to situations where there is potential exposure to bodily fluids or to contact the appropriate person. Personal protective equipment must be appropriately located for easy accessibility. Hypoallergenic personal protective equipment should be made available to employees who may be allergic to the standard equipment.
- Employers are required to maintain a clean and sanitary worksite. A written schedule for cleaning and sanitizing (disinfecting) all applicable work areas must be implemented and included in the program. All areas and equipment that have been exposed (after an accident) must be cleaned and properly disinfected. Proper cleanup and disinfection of contaminated equipment are also required under the standard.
- All infectious waste, such as bandages, towels, or other items exposed to bodily fluids, must be placed in a closable, leak-proof container or bag with an appropriate label. This waste must be properly disposed of as medical waste according to federal, state, and local regulations.
- Contaminated uniforms, smocks, and other items of personal clothing must be laundered in accordance with the standard.
- Appropriate employees must be trained in the requirements of this standard. Employees who may be exposed are required to undergo medical examination and the hepatitis B vaccination must be provided. Training requirements include a copy of the OSHA standard and explanation of

the standard; a general explanation of the symptoms and epidemiology of bloodborne diseases; an explanation of the transmission of bloodborne pathogens; an explanation of the appropriate methods for recognizing jobs and other activities that involve exposure to blood; an explanation of the use and limitations of practices that will prevent or reduce exposure, such as engineering controls and personal protective equipment; information on the types, proper use, location, removal, handling, decontamination, and/or disposal of personal protective equipment; an explanation of the basis for the selection of personal protective equipment; information on the hepatitis B vaccine, including information on its efficacy, safety, and benefits; information on the appropriate actions to be taken and persons to contact in an emergency; an explanation of the procedure to be followed if an exposure incident occurs, including the method of reporting, medical follow-up, and medical counseling; and an explanation of the signs and labels. Additional training may be required in applicable laboratory situations and other circumstances.

- The employer is required to maintain complete and accurate medical records. These records must include the names and social security numbers of employees; copies of employee's hepatitis B vaccination records and medical evaluations; copies of the results of all physical examinations, medical tests, and follow-up procedures; written opinions of the physicians; and copies of all of the information provided to physicians. The employer is additionally responsible for maintaining the confidentiality of these records.

- The employer is required to maintain all training records. These records must include the dates of all training, names of persons conducting the training, and the name of each participant. Training records must be maintained for a minimum of five years. Employees and OSHA representatives must be able to view and copy these records upon request.

- All containers containing infectious waste, including, but not limited to, refrigerators and freezers, medical disposal containers, and all other containers, must be properly labeled. The required label must be fluorescent orange or orange-red with the appropriate symbol, and the lettering reading BIOHAZARD in a contrasting color. This label must be affixed to all containers containing infectious waste and must remain in place until the waste is properly disposed of.

CONTROL OF HAZARDOUS ENERGY (LOTO) *Lock-out Tag-out*

OVERVIEW

The lack of control of hazardous energy in the industrial workplace is responsible for approximately 10% of all serious accidents and a substantial percentage of fatalities in the workplace every year. According to the Bureau of Labor Statistics, failure to shut off power while servicing machinery and other equipment is the primary cause of injuries. To address this hazard, OSHA promulgated its control of the hazardous energy standard (commonly called lockout/tagout standard).

In general, safety and loss prevention professionals have applauded the promulgation of the lockout and tagout standard. This standard, unlike many preceding standards, is "user friendly" and provides the exact sequence of steps to be followed in order to safeguard employees. The difficulty experienced by many safety and loss prevention professionals is maintaining compliance with the standard over a period of time.

The most efficient method for safeguarding against accidental activation of machinery is the development of a lockout and tagout program that achieves and maintains compliance with the OSHA standard. By locking out and/or tagging off power sources, unauthorized use of the machine or equipment is prevented.

A lockout is simply the placement of a substantial locking mechanism upon a machine's on/off switch or electrical circuit to prevent the power supply from being activated while repairs are being made. Lockout procedures are especially effective in preventing injuries to maintenance or repair personnel, who may be placed in a hazardous situation by sudden and unexpected activation of machinery while repairs are being made. Lockouts apply not only to electrical hazards but to all other types of energy (i.e., hydraulic, pneumatic, steam, and chemical as well as to vehicles).

Locking out of a machine or equipment can be used in conjunction with or separate from the tagging procedures permitted under the standard. Tags are required to be a bright, identifiable color and marked with appropriate wording such as "Danger Do Not Operate" or similar warnings. Tags are required to be of a durable nature and must be securely affixed to all lockout locations. Tags must contain the signature(s) of the employees working on the machinery, the date and time, and the department name or number.

There are four stages in the development and management of an effective lockout and tagout program. The first stage is program development and equipment modification. In this stage, the workplace must be analyzed, and a written lockout and tagout program is developed. Identification of the potential hazards and, if possible, elimination of exposures to potential injuries by machinery or equipment should be analyzed. Additionally, appropriate equipment, including, but not limited to, padlocks, tags, T-bars, and other equipment should be purchased during this stage.

The second stage involves the education and training of appropriate personnel. Safety and health professionals must ensure that all training and education is well documented. Hands-on training is highly recommended.

The third stage involves effective monitoring and disciplinary action. The responsibility for ensuring that all employees and equipment covered under this standard are performing in a compliant manner rests solely with the employer. Disciplinary procedures and enforcement thereof are essential in ensuring continued compliance.

The fourth and final stage involves program auditing and program reassessment. The effectiveness of the compliance program can be ensured through periodic evaluation. Program deficiencies can be identified and corrective action initiated to correct these deficiencies. The lockout and tagout standard requires employers to establish a written program for locking out and tagging out machinery and equipment. The written program normally contains the following elements:

- The program must contain steps for shutting down and securing all machinery. A written program normally details the energy sources for each machine, and how it should be locked or tagged. All sources of hazardous energy must be listed and the means for releasing or blocking stored energy must be included.
- Procedural steps for applying locks and tags and their placement on the equipment or apparatus must be listed. The responsible person(s) authorized to apply locks and tags are required to be identified.
- Appropriate steps and testing of the machine/equipment is required after shutdown and lockout to verify that all energy is safely isolated.
- Procedures to be followed and steps to be taken in restarting the equipment after completion of the work must be listed.
- Employees who have been trained and authorized to lock out machinery must be identified in the written program.
- If the task requires group lockout, each employee is required to possess an individual lock. Only the person applying the lock should have a key to that lock. This ensures that, as different team members complete their tasks and remove their locks, remaining members are still fully protected from the hazardous energy.
- Shift and personnel changes must allow the continuity of the lockout/tagout protection, including provision for safe, orderly transfer of lockout/tagout devices between on and off duty personnel.
- When major replacement, repair, renovation, or modification of machines or equipment is performed, and when new machines or equipment are installed, energy-isolating devices for such machines or equipment must be designed to accept a lockout device.

Preparing for a Lockout or Tagout System (LOTO)

The following steps must be followed when preparing for a lockout and/or tagout procedure:

1. Conduct a survey to locate and identify all energy-isolating devices to make sure that switches, valves, or other energy-isolating devices apply to the equipment to be locked or tagged out.
2. Make sure that all energy sources to a specific piece of equipment have been identified. Machines usually have more than one energy source (electrical, mechanical, or others).

It is highly recommended that a list of each piece of equipment be made, with the results of the survey included, so that each time the lockout/tagout procedure needs to be performed, the mechanic or operator can consult the list.

Sequence of Lockout or Tagout System Procedure

The following steps should be taken when a machine is going to be locked and/or tagged:

1. Notify appropriate employees that a lockout or tagout system is going to be utilized. The authorized employees must know the type and magnitude of energy that the machine or equipment uses and must also understand the hazards involved.
2. If the machine or equipment is operating, shut it down by the normal stopping procedures (depress stop button, open toggle switch, etc.).
3. Operate the switch, valve, or other energy-isolating devices so that the equipment is isolated from its energy sources. Stored energy (such as that in springs, elevated machine members, rotating flywheels, hydraulic systems, and air, gas, steam, or water pressure, etc.) must be dissipated or restrained by such methods as repositioning, blocking, or bleeding down.
4. Lock out and/or tag out the energy-isolating devices with individual locks or tags.
5. After making sure that no personnel are exposed, as a check on having disconnected the energy sources, operate the push button or other normal operating controls to make certain that the equipment will not operate.
6. The equipment is now locked out or tagged out.

Restoring Machines and Equipment to Normal Production Operations

The following procedures should be done to restore machines to normal operation:

1. After the service or maintenance is complete and the equipment is ready for normal production operations, check the area around the machines or equipment to make sure that no one is exposed.
2. After all tools have been removed from the machine or equipment, the guards have been reinstalled, and employees are in the clear, remove all lockout or tagout devices. Operate the energy-isolating devices to restore energy to the machine or equipment.

Training

Effective employee training and education regarding the lockout/tagout program is a vital component in achieving compliance with the OSHA standard. Employees must be trained as follows:

- All employees must know the purpose of and use of an energy control procedure. They must also know what a tagout signifies. They must know why a machine is locked or tagged out, and what to do when they encounter a tag or lock on a switch or a device that they wish to operate. Because any person may encounter a lockout or tagout, everyone must have a general understanding of lockout/tagout safety.

- Before machinery shutdown, the authorized employee must know the type and magnitude of energy to be isolated and how to control it. Each machine or type of machine should have a written lockout procedure (preferably attached to the machine).
- Retraining to reestablish employee proficiency must take place when an employee is assigned to a different area or machine, or when written procedures change. Additionally, all new employees must be properly trained regarding the lockout/tagout program. When outside contractors are brought on-site, they should also be informed of the company's lockout procedures. The company should also be aware of a contractor's procedures and make certain that its personnel understand and comply with the outside contractor's energy control procedures (as long as they comply with the standard).
- Each employee authorized to perform maintenance should fully understand all hazardous energy as it relates to specific machinery.
- The proper sequence of locking out should be fully understood.

When a tagout system is used, employees must be trained in the following limitations of tags:

- Tags are essentially warning devices affixed to energy-isolating devices; they do not provide the physical restraint on those devices that is provided by locks.
- When a tag is attached to an energy-isolating means, it is not to be removed without authorization of the person responsible for it, and a tag is never to be bypassed, ignored, or otherwise defeated.
- To be effective, tags must be legible and understandable by all authorized employees, all affected employees, and all other employees whose work operations may be located in the area.
- Tags and their means of attachment must be made of materials that will withstand the environmental conditions encountered in the workplace.
- Tags may evoke a false sense of security; their meaning must be understood as part of the overall energy control program.
- Tags must be securely attached to energy-isolating devices so that they cannot be inadvertently detached during use.

A violation of the previously listed OSHA standards poses a substantial risk of great bodily harm or death. Exposure of employees to hazardous chemicals, infected bodily fluids, or harmful energy, without the protection afforded under the OSHA standards, places the safety and loss prevention professional and other company officers at risk for potential civil or criminal sanctions under the OSH Act or state laws in the event of an accident. When developing and managing compliance programs, safety and health professionals are advised to ensure compliance with each and every element prescribed in the OSHA standard, to completely document compliance with each element of the standard, and to properly discipline trained employees who are not complying with the prescribed compliance procedures. Safety and loss prevention professionals who are unsure of requirements of the listed standards, or any other applicable standard, should obtain assistance in order to ensure compliance.

RESPIRATORY PROTECTION

The OSHA standard on respiratory protection, 29 C.F.R. section 1910.134, states that any facility or operation that contains hazards to the respiratory system must institute a written respiratory protection program. These hazards include, but are not limited to, breathing air contaminated with harmful dusts, fogs, fumes, mists, gases, smokes, sprays, and vapors.

REQUIREMENTS FOR A MINIMAL ACCEPTABLE RESPIRATORY PROTECTION PROGRAM

Requirements for a minimal acceptable respiratory protection program are as follows:

- The respiratory protection program must be in writing.
- Written guidelines must be made regarding the selection and use of the proper respiratory protection (i.e., respirators and self-contained breathing apparatus).
- Respiratory protection will be selected on the basis of the hazards encountered by the worker.
- The worker must be instructed and trained in the proper use and limitation of the respiratory protection program.
- If possible, the respiratory protection equipment should be assigned to individuals for their exclusive use.
- Respiratory protection must be regularly cleaned and disinfected. Protection issued for exclusive use should be cleaned and disinfected at least daily. Protection used by more than one individual must be cleaned and disinfected after each use.
- Respiratory protection must be stored in a convenient, clean, and sanitary location.
- Respiratory protection used routinely shall be inspected during cleaning. Worn or deteriorated parts must be replaced.
- Respiratory protection for emergency use shall be inspected monthly and after each use.
- Appropriate surveillance of work area conditions and the degree of employee exposure or stress must be maintained.
- There must be regular inspections and evaluations to determine the continued effectiveness of the program.
- Workers should not be assigned to tasks requiring use of respirators unless it has been determined that they are physically able to perform the task and are able to use the equipment. The local physician may determine what health and physical conditions are pertinent. The respirator user's medical status should be reviewed periodically.

- Approved or accepted respirators shall be used when they are available. The furnished respiratory protection must provide adequate protection against the particular hazard for which it is designed in accordance with standards established by competent authorities. The U.S. Department of the Interior, Bureau of Mines, and the U.S. Department of Agriculture are recognized as such authorities. Although respirators listed by the U.S. Department of Agriculture continue to be acceptable for protection against specified pesticides, the U.S. Department of Labor, Mine Safety and Health Administration is the agency that is currently responsible for testing and approving pesticide respirators.

RESPIRATORY PROTECTION PROGRAM GUIDELINE

The respiratory protection program must be in writing. This includes all standard operating procedures governing the selection and use of respirators and self-contained breathing apparatus, a complete training format, respirator and self-contained breathing apparatus storage locations, documented cleaning and inspection procedures, documented parts replacement on respirators and self-contained breathing apparatus, annual health and physical condition examinations for appropriate personnel, and surveillance/air sampling procedures. The selection, type, manufacturer, and location of all respirators and self-contained breathing apparatus should also be contained in the written program.

SELECTION OF RESPIRATORS AND SELF-CONTAINED BREATHING APPARATUS

All respirators and self-contained breathing apparatus should be selected on the basis of the hazards to which the employees may be exposed. Only approved or accepted respirators and self-contained breathing apparatus should be purchased. The respirators and self-contained breathing apparatus should provide adequate respiratory protection against the particular hazard for which it is designed, in accordance with the established standards. All respirators and self-contained breathing apparatus must be approved by the U.S. Department of Labor, Mine Safety and Health Administration, and the U.S. Department of Agriculture.

Safety and loss prevention professionals should be aware that OSHA recently published a new guidance document focused on mandatory respiratory selection provisions, which was added to the existing respiratory protection standard. This important new publication provides safety and loss prevention professionals with guidance with regards to the selection of respirators for their employees who may be exposed to airborne contaminates.

In 2006, OSHA revised the existing respiratory protection standard to add APF and MUC requirements to the standard. "APF means the workplace level of respiratory protection that a respirator or class of respirators is able to provide to workers. The higher the APF number (5 to 10,000), the greater the level of protection provided to the user." The MUC (maximum use concentration) identifies the limits at which the class of respirators is expected to provide protection to employees. "MUC means the maximum atmospheric concentration of a hazardous substance for which the worker can be expected to be protected when wearing a respirator." In essence, when the exposure limits of the airborne contaminate exceed the MUC, safety and loss prevention professionals should select a respirator with a higher level of protection, as identified by the APF.

APF and MUC are mandatory respirator selection requirements that can only be used after respirators are properly selected and are used in compliance with the entire standard. The respiratory protection standard requires fit testing, medical evaluations, specific training, and proper respirator use. The standard applies to general industry, construction, longshoring, and shipyard and marine terminal workplaces.

TRAINING

Training provided to employees should include hands-on sessions and instruction regarding respiratory equipment limitations. The employees shall be trained about the hazards that they may be confronted with and the signs and symptoms of health effects associated with exposure. Training must be provided to appropriate personnel at least annually by a competent, certified instructor.

CLEANING AND MAINTENANCE

The cleaning and disinfecting of respirators and self-contained breathing apparatus should be conducted monthly and be properly documented. Respiratory protection issued for exclusive use shall be cleaned daily or when necessary. All respiratory equipment must be thoroughly cleaned and disinfected after each use. All respirators and self-contained breathing apparatus should be properly marked and stored in a convenient, clean, and sanitary location. All respirators and self-contained breathing apparatus should be inspected on a weekly basis and after each use. All worn, broken, missing, or deteriorated parts should be replaced immediately. Remember to document all cleaning, disinfecting, and maintenance activities performed on all respirators and self-contained breathing apparatus.

MONITORING

Monitoring and inspecting work area conditions and employee exposure should be performed daily by appropriate and competent personnel. Monitoring should be emphasized regarding the types of exposures that may be present in a particular facility. It is recommended that if the facility has a need for a respiratory protection program, it also has the need to purchase equipment to assist personnel in monitoring. Such equipment can include the following items:

- drag tube type air monitoring kit
- oxygen deficiency monitor
- hydrogen sulfide monitor
- radiation monitor
- other appropriate testing equipment for respiratory hazards in a prospective facility or operation

Auditing the Program

The respiratory protection program should be audited annually. When new respiratory hazards exist, appropriate personnel should investigate the characteristics of the material and purchase proper respiratory protection. For assistance in this task, a checklist is helpful to determine the strengths and weakness of the respiratory protection program. After the audit is completed, modifications should be instituted in the program as necessary.

Enforcement of the Program

Enforcement of the respiratory program is of the utmost importance. A safety program is only as effective as its enforcement mechanisms. Employees must know that upper management will not tolerate employees violating the procedures established under the respiratory protection program. Employees who violate the respiratory protection program must be disciplined under the establishment's policies.

Respiratory Protection Checklist

- Has the facility been analyzed for potential respiratory hazards?
- Has the respiratory protection policy been posted in strategic areas of the facility?
- Have employees been trained about the types and proper use of respiratory equipment, including the location of hazard zones where respiratory protection is required?
- Have employees been trained about the proper use, limitations, and locations of respiratory protection equipment?
- Have employees been trained to clean and disinfect respiratory equipment?
- Is respiratory equipment stored in a convenient, clean, and sanitary location?
- Have employees been trained about the hazards that they may be confronted with, including signs and symptoms of the possible health effects associated with such hazards?
- Is training provided to employees at least annually by a competent certified instructor?
- Have appropriate employees had a physical examination to determine whether they can physically perform a task while wearing respiratory protection?
- Are all worn, broken, missing, or deteriorated parts replaced immediately after their detection?
- Are work area conditions monitored and inspected on a daily basis?
- Does respiratory equipment provide adequate protection against the hazards for which they were designed?
- Is the respiratory protection program audited annually?
- Have employees been instructed about the facility's disciplinary policy for noncompliance?
- Are respiratory hazard areas posted with warning signs stating that respiratory protection must be worn?

Safety and loss prevention professionals should be aware that OSHA provides several e-tools on their website, located at OSHA.gov, to assist professionals in developing respiratory protection training programs.

PERSONAL PROTECTIVE EQUIPMENT

Safety and loss prevention professionals should be aware of the relatively recent final rule regarding the payment for personal protective equipment for employees. In summary, this final rule requires employers to provide their employees with personal protective equipment (PPE) when the equipment is necessary to protect employees from work-related injuries and illnesses and to provide this safety equipment to employees at no cost. The final rule specifies which PPE, such as hard hats, safety glasses, fall protection equipment, and other PPE the employer is required to provide to employees at no cost. However, safety and loss prevention professionals should be aware that there are exceptions to this general rule that employers must pay for employees' PPE, which may be applicable to certain circumstances and situations. The exceptions specifically addressed in the final rule include safety-toe protective footwear, non-specialty prescription safety eye-wear, long pants, and ordinary rain gear in a list of PPE that may be exempt from this rule. Safety and loss prevention professionals should thoroughly research this final rule when addressing the issue as to whether or not the employer is responsible for incurring the cost of the specified personal protective equipment.

FALL PROTECTION PROGRAM (PREVENTING SLIPS AND FALLS IN YOUR FACILITY)

Overview

Slips and falls are a potential hazard in most businesses and are risks that are often hidden until after an accident occurs. The usual types of injuries sustained in slip and fall accidents involve the back area and appendages (i.e., arms, legs, knees, etc.), which can be extremely difficult to rehabilitate, in comparison to other injuries. The two major types of slip and fall accidents are employees who fall because of slippery floors and employees who fall from elevated areas.

The following guidelines help prevent slips and falls in a facility:

- Identify areas that are potentially slippery. Review the injury records to identify areas in which slip and fall accidents have occurred. Inspect all areas of the facility and initiate action to correct areas that are potentially slippery.
- Keep work areas free from debris and clean up spills immediately. Clean work areas frequently during the work shift if necessary.
- Wear appropriate footwear to provide proper traction. Inspect boots on a periodic basis. Make sure that soles and treads are in good shape.
- Provide nonslip work surfaces in areas that are exposed to water, blood, fat, and other products on a daily basis.

- Develop work rules prohibiting running and horseplay in the facility. Insist upon strict enforcement of these rules.
- Instruct employees to avoid sudden movements and quick changes of direction in areas that are potentially slippery.
- Educate employees to walk flat-footed in smaller, measured steps.
- When employees are lifting or reaching, the employee's balance is affected, causing a higher probability of slipping and falling. Employees should be educated and trained in proper lifting and reaching techniques.
- Use nonslip surfaces on all rungs of ladders, walkways, stairways, and other frequently used areas.

DEVELOPMENT OF A FALL PROTECTION PROGRAM

When employees are working above the floor surface, protection must be provided to prevent the employee from falling. The usual method of preventing employees from falling is the use of standard guardrails and toe-boards. In some circumstances, guardrails are not feasible because of the nature of the work or the structure of the facility. In these circumstances, the employer is required to protect employees through other measures, such as safety harnesses and safety nets. The following guide will assist in the development of a program:

- Evaluate the facility and identify the jobs or areas that require fall protection procedures. Such jobs may include employees climbing a material handling storage rack, maintenance employees climbing to lubricate equipment, and other related tasks.
- A written fall protection program must be developed in order to document all efforts made. A written program should include, but is not limited to, jobs, types of equipment, training, equipment testing, inspection procedures, and responsibilities for this program.
- Evaluate and select the proper fall protection equipment for each situation. Remember to consider fall arresting devices and climb protection equipment. (Note: See OSHA standard regarding the use of harnesses.)
- Install and test the secure points to which the fall protection equipment will be anchored prior to possible use.
- Properly test all fall protection equipment prior to use.
- Develop a documented inspection program for all fall protection equipment.
- Equipment should be inspected before and after each use, and on a periodic basis as necessary. All broken, worn, or malfunctioning equipment should be replaced immediately.
- Include the fall protection program in the safety audit assessment.

In summary, the previous guidelines are but a few of the many OSHA standards that may directly or indirectly affect your work place. Preparation is the key to achieving and maintaining compliance. Read the applicable standard carefully to make sure that you fully and completely understand every aspect of the standard. OSHA does not tell you how to achieve compliance, only the level of compliance that must be achieved. Remember, prepare your programs in a defensible manner so that they can withstand careful scrutiny. If an incident occurs that places your programs at issue, a court and other appropriate party will evaluate your program utilizing the benefit of hindsight—and hindsight is 20/20!

SELECTED CASE STUDY

Note: Case modified for the purpose of this text.
United States Court of Appeals
District of Columbia Circuit.

STALEY MANUFACTURING CO., Petitioner,
v. SECRETARY OF LABOR, Respondent

No. 00-1530.

Argued Jan. 17, 2002.
Decided July 23, 2002.

Corn refiner petitioned for review of final order of Occupational Safety and Health Review Commission (OSHRC), finding that refiner committed 89 willful violations of hazardous locations standard and hazard communication standard under the Occupational Safety and Health Act (OSHA). The Court of Appeals, Garland, Circuit Judge, held that refiner was plainly indifferent to its OSHA violations, rendering them willful.

Petition denied.

Opinion for the Court filed by Circuit Judge.

Circuit Judge

A.E. Staley Manufacturing Company petitions for review of a final order of the Occupational Safety and Health Review Commission (OSHRC). The Commission found that Staley committed 89 willful violations of 29 C.F.R. § 1910.307(b) (the "hazardous locations standard"), which mandates that electrical equipment in hazardous locations be approved for use in such locations. OSHRC also concluded that Staley committed two willful violations of 29 C.F.R. § 1910.1200(h) (the "hazard communication standard"), which requires employers to provide employees with effective information and training regarding hazardous chemicals in their work areas. Staley does not dispute that it committed the violations, but contends that the Commission erred in deeming them willful. Finding no error, we deny the petition for review.

I.

Staley is a corn refiner that produces corn starch, corn oil, fructose, and dextrose at a number of facilities. This case concerns Staley's Decatur, Illinois plant. In 1990, the plant included over 130 buildings and had approximately 833 hourly employees. In July of that year, the Occupational Safety and Health Administration (OSHA) began an inspection of the plant, prompted by a May 1990 accident in which an employee was fatally asphyxiated. As a result of the inspection, the Secretary of Labor, acting through OSHA, issued two sets of citations alleging hundreds of violations of the Occupational Safety and Health Act ("OSH Act"), 29 U.S.C. §§ 651-678. One set of citations alleged violations of safety standards, including the hazardous locations standard, 29 C.F.R. § 1910.307(b). The other set charged violations of health standards, including the hazard communication standard, *Id.* § 1910.1200(h), and the asbestos standard, *Id.* § 1910.1001(j)(2), (k)(1).

Partial settlements led to the withdrawal of all citations for non-willful violations. Staley contested the remaining 177 safety and 4 health citations, and the Commission assigned an Administrative Law Judge (ALJ) to hear the case. The ALJ upheld 171 of the safety citations and all 4 of the health citations. However, he concluded that only 87 of the safety violations (all for hazardous locations) and 2 of the health violations (both for asbestos) were willful. He found the remaining violations, including the 2 violations of the hazard communication standard, to be serious but not willful. Both parties appealed the ALJ's decision to OSHRC.

The Commission affirmed all of the ALJ's findings of violations, as well as all of his findings of willfulness. In addition, it upgraded several violations from serious to willful, including two further hazardous locations violations (making a total of 89) and the two hazard communication violations (which it grouped as one for penalty purposes). In finding willfulness, the Commission relied on evidence that previous dust explosions, internal audits, and a survey by the National Institute for Occupational Safety and Health had put Staley on notice of serious safety and health problems, including the location of non-approved electrical equipment in the vicinity of combustible dust and a lack of employee training concerning dangerous chemicals. The Commission concluded that Staley's continued failure to take corrective action in the face of these widespread problems supported a determination of willfulness. *1345 **78 A.E. Staley Mfg. Co., 19 O.S.H. Cas. (BNA) 1199, 1221-22 (O.S.H.R.C.2000). Staley then filed a petition for review in this court pursuant to section 10(c) of the OSH Act, 29 U.S.C. § 659(c).

II.

[1] A reviewing court must uphold the factual findings of the Commission if they are "supported by substantial evidence on the record considered as a whole," 29 U.S.C. § 660(a), and must uphold its other conclusions as long as they are not arbitrary, capricious, an abuse of discretion, or otherwise contrary to law, 5 U.S.C. § 706(2)(A). See *Anthony Crane Rental, Inc. v. Reich*, 70 F.3d 1298, 1302 (D.C.Cir.1995). Moreover, "[w]e defer to the Secretary's interpretation of the Act and regulations, upholding such interpretations so long as they are consistent with the statutory language and otherwise reasonable." *Id.* (citing *Martin v. OSHRC*, 499 U.S. 144, 150-51, 111 S.Ct. 1171, 1175-76, 113 L.Ed.2d 117 (1991)).

[2] Staley does not contest the Commission's findings that it committed serious violations of the OSH Act. It disputes only the findings that the 89 hazardous locations and 2 hazard communication violations were willful. In the OSH Act context, a willful violation is "an act done voluntarily with either an intentional disregard of, or plain indifference to, the Act's

requirements." *Kaspar Wire Works, Inc. v. Secretary of Labor,* 268 F.3d 1123, 1127 (D.C.Cir.2001) (quoting *Conie Constr., Inc. v. Reich,* 73 F.3d 382, 384 (D.C.Cir.1995)). The Commission based its findings of willfulness on its determination that Staley was plainly indifferent to the requirements of the Act.

Staley argues that the Commission committed two errors in deeming its violations willful. First, Staley contends that substantial evidence does not support the Commission's determination that the company was plainly indifferent to its violations of OSHA standards. Second, with respect to the hazardous locations standard, Staley maintains that even if there were evidence of its plain indifference to that standard, the Commission legally erred in finding willfulness without finding that the company knew of each specific violation cited by OSHA. We consider these two arguments in Parts III and IV below.

III.

[3] The Commission found Staley willful because it demonstrated plain indifference to its violations of the standards requiring it: (1) to use only approved electrical equipment in hazardous locations, and (2) to train and inform its employees regarding hazardous chemicals in their work areas. Staley maintains that there is an absence of substantial evidence to support findings of plain indifference with respect to either standard. We disagree.

A.

The 89 violations of the hazardous locations standard, affirmed by the Commission and undisputed by Staley, all involved the presence of non-approved electrical equipment in Class II, Division 2 locations. The equipment at issue included exposed wiring, bulbs without protective globes, and improperly sealed junction boxes—all potential ignition sources for combustible dust. The evidence of Staley's plain indifference to those violations is as follows.

In May 1987, Staley conducted a mock OSHA inspection and found unsafe electrical equipment in places that Staley considered Class II, Division 2 locations, including the elevators and Buildings 9, 44, and 75—all places in which OSHA inspectors subsequently found violations at issue in this case. Joint Appendix (J.A.) at 325, 327–28, 339. Two years later, in April 1989, an internal audit conducted by Staley safety engineer Ken Page turned up more instances of unprotected electrical equipment in hazardous locations. Page's handwritten report warned that Building 44 had "literally gone to hell in a handbasket" and contained "hundreds of safety type violations." J.A. at 361. Page testified that those violations included uncovered electrical boxes and exposed wires. Supplemental Appendix (S.A.) at 47–48. He recommended that a wall-to-wall audit of the plant be conducted as soon as possible, and noted that Staley's potential liability for OSHA penalties was very high. J.A. at 361–62.

Page's recommendation for a wall-to-wall audit was not approved. Page testified that his supervisor, Lynn Elder, director of Staley's department of environmental sciences and safety, told Page that he had advised Bob Jansen, corporate vice president of operations, of Page's findings. Jansen reportedly replied that he was aware of the problems in the plant, but that another Staley project had priority. Page also testified that Elder told him that he should not distribute his report because "the legal department would crucify us." Instead, Elder suggested that Page's report be either destroyed, or stamped "privileged and confidential" and sent to the legal department. Page, however, did not destroy the report. He kept a copy for himself and gave copies to three others, including Bob Trent, Decatur's chief of plant protection, and Jim Brinkmeyer, the corporate industrial hygienist. Elder did permit Page to make an oral presentation to plant staff during which he explained his findings in detail. But Elder assigned another auditor, J.B. Webb, the supervisor of the Decatur safety department, to revise Page's written report. The revised report substantially toned down Page's language and omitted references to hundreds of the specific electrical and safety violations that Page had observed. J.A. at 363–65; S.A. at 38–43, 51–52.

In May 1989, a major dust explosion and fire occurred at the Decatur plant. It was not the first: other explosions and fires had occurred in several buildings over the years. In the same month, an insurance loss-control report for Building 44 noted open electrical junction boxes, and warned that "[m]oisture, lint, dust and combustible materials can easily come in contact with the exposed wiring and create a potential ignition." S.A. at 10–11, 56–58. The following year, after the fatal May 1990 accident, Staley's president instructed Page to conduct another audit of the plant. Page submitted a report that identified exposed wires, conduits, breaker boxes, junction boxes, and bulbs in hazardous locations. The locations again included the elevators and Buildings 9 and 75. J.A. at 122, 373–74, 390, 417–19; S.A. at 59–67.

Two months later, OSHA conducted the inspection that resulted in the findings of 89 violations of the hazardous locations standard. That inspection also revealed that many of Staley's supervisory personnel were not properly trained regarding the hazards presented by Class II, Division 2 areas. Some were not even aware that the areas they supervised were classified locations. Others received no training on the classification of areas or the requirements for such areas. For example, Michael Slimbarski, the plant operations manager, had not been trained concerning the hazardous locations standard and, with one exception, did not know which areas of the plant were classified. J.A. at 269–70. Shift coordinators Gordon Green and Ron Young were also untrained on the standard and unaware that faulty electrical equipment could produce dust explosions. J.A. at 283–89, 304–05. Even electricians lacked training regarding the existence of classified hazardous locations in the plant. S.A. at 183.

We agree with the Commission that the evidence just recited constitutes substantial evidence that the 89 hazardous locations violations were part of a pattern or practice of plain indifference to violations of that standard. A.E. Staley Mfg. Co., 19 O.S.H. Cas. (BNA) at 1222. The series of internal reports between 1987 and 1990 put Staley on notice of unsafe electrical equipment in hazardous locations, and of the persistence of that problem over the years covered by those reports. A series of dust explosions and fires, although not themselves caused by faulty electrical equipment, also put the company on heightened notice of the dangers of combustible dust. Yet despite this notice, the company failed to train its employees about such hazards, attempted to suppress Page's internal audit report, and ignored his recommendations for correction. Within months of Page's last audit, OSHA inspectors found the same kinds of unsafe equipment in many of the same locations that Page did. This evidence is more than sufficient to sustain the Commission's determination of willfulness. See, e.g., *Caterpillar, Inc. v. OSHRC*, 122 F.3d 437, 441 (7th Cir.1997) (holding that the fact that an employer "rejected or ignored the recommendations of the very person" it had asked to make safety recommendations showed plain indifference to employee safety).

[4] [5] Staley's principal attacks on the sufficiency of the evidence require only brief mention. First, the company contends that the 89 violations were too few to demonstrate plain indifference, as the equipment involved represented only a small percentage of all of the company's electrical equipment. That is not an adequate defense. Even a single violation of the OSH Act may be found willful, regardless of whether the workplace is otherwise safe. See *Kaspar Wire Works*, 268 F.3d at 1128 (holding that an employer cannot "contend that it was entitled to rely on its lack of prior violations to undermine a finding of willfulness," because then "an employer with no prior citations could choose to violate a regulatory obligation without risking a finding of willfulness"); *Valdak Corp. v. OSHRC*, 73 F.3d 1466, 1469 (8th Cir.1996).

[6] Staley also argues that Page's report did not heighten its awareness of safety problems at Decatur because his handwritten notes were never given to plant staff. As described above, there is substantial evidence to the contrary: Page gave his notes to both Bob Trent and Jim Brinkmeyer; they were read by his supervisor, Lynn Elder; they were orally reported by Elder to vice president Jansen; and Page recounted them in a detailed oral presentation to the staff. That knowledge is properly imputed to the company. See *Caterpillar, Inc. v. Herman*, 154 F.3d 400, 402 (7th Cir.1998). Moreover, to the extent that Page's report was not more widely disseminated, it was only because Elder directed that it not be distributed, and that a revised report—omitting references to hundreds of specific violations—be distributed instead. S.A. at 38–43, 51–52. Such willful blindness is no defense at all. *See United States v. Schnabel*, 939 F.2d 197, 203 (4th Cir.1993).

[7] Finally, Staley argues that the Commission ignored evidence of its good faith efforts to comply with the hazardous locations standard, pointing specifically to its plant-wide (joint) safety committee and to the 25 separate departmental safety committees that were scheduled to meet monthly to address safety issues. The joint committee, however, lacked authority to initiate or direct corrective action, while the departmental safety committees held their meetings only half the time. S.A. at 115–16. Moreover, the record shows that many members of the departmental committees were dismayed at Staley's safety program: many resigned because safety problems were not corrected, meetings were canceled, and management either did not attend meetings or sent different managers each month. S.A. at 103–04, 120–21. Far from being evidence of good faith, then, the record of Staley's safety committees offers only further evidence of the company's plain indifference to its violations of safety standards.

B.

[8] Staley also contends that substantial evidence does not support the Commission's determination that the company exhibited plain indifference to the two hazard communication violations. Those violations were: (1) failing to provide hazard communication training for twelve substances (including silica sand, filteraid, asbestos, and feed dust) for which Staley did not have material safety data sheets (MSDSs); and (2) failing to train employees regarding the meaning of the hazard communication symbols used on the company's ethylene oxide, propylene oxide, caustic, and sulfuric acid storage tanks. The evidence of Staley's plain indifference is as follows.

In November 1988, the Decatur plant's joint health and safety committee warned of serious deficiencies in Staley's hazard communication program. In a memorandum, the committee noted that bags of filteraid were strewn around and that the material was tracked all over Building 11—one of the buildings specifically named in OSHA's 1990 hazard communication citation. The committee further noted that "[b]uilding personnel need to be trained on the danger of this product." A.E. Staley Mfg. Co., 19 O.S.H. Cas. (BNA) at 1204; S.A. at 22.

In March 1989, the National Institute for Occupational Safety and Health (NIOSH) conducted an evaluation of health hazards at the Decatur plant. NIOSH industrial hygienists observed that Staley employees were improperly trained with respect to the hazards of toxic chemicals. NIOSH also found that employees disregarded an alarm that sounded when ethylene oxide and propylene oxide leaked into the air, that they did not have immediately available respirators, and that they engaged in work practices that increased their exposure to chemicals. S.A. at 3–5. Following its evaluation, NIOSH sent J.B. Webb, the Decatur safety supervisor, a summary of its findings:

> [W]e feel there is significant potential for chemical overexposures in the starch reaction area, and possibly throughout the starch stream. This appears to be due to improper work practices, poor management oversight, and emphasis of production over worker safety. Included among the chemicals used in this area are ethylene oxide, propylene oxide, and vinyl acetate. Ethylene oxide is currently regulated as a cancer hazard by [OSHA]. Propylene oxide is very similar in chemical structure to ethylene oxide and is currently being evaluated by NIOSH with regard to potential carcinogenicity ... . In addition, lack of training and demand for product seems to have circumvented measures which were specifically implemented to reduce potential exposure.
> S.A. at 5.

Ken Page's April 1989 audit disclosed further serious deficiencies in employee training regarding hazardous substances. Page noted that hazardous chemicals were not properly labeled, annual training was not being conducted in several buildings, and employees lacked access to MSDSs. He also observed large quantities of filteraid on the floor of two buildings, including Building 11, and continuing problems with the material in a third. A.E. Staley Mfg. Co., 19 O.S.H. Cas. (BNA) at 1205; J.A. at 341–45, 347–48, 350–53, 359–60. A second auditor, Robert Moore, notified his supervisor about the results of the audit and the presentation Page made to plant staff:

> The staff was told that the Hazard Communications Compliance, Respiratory Protection Compliance, and Hazardous Material Control had deteriorated since we last conducted such a survey (1986). In our opinion, an OSHA inspection prompted by the NIOSH visit could potentially result in the assessment of major penalties.
> S.A. at 6 (emphasis added).

Notwithstanding these warnings, Staley management told Page that completing another project had priority over correcting the problems he had identified. A year later, when Page undertook another audit following the fatal May 1990 accident, he found that containers of hazardous materials were improperly labeled in two buildings, updated MSDSs were absent in two buildings, and hazard communication training had not been conducted for years. Page received no feedback from Staley regarding his May 1990 report. J.A. at 400, 406, 415; S.A. at 43, 87.

As noted above, when OSHA inspected the plant in July 1990, it cited Staley for failing to train employees regarding twelve hazardous substances for which Staley did not have MSDSs, including silica sand, filteraid, asbestos, and feed dust; and for failing to train employees concerning the symbols used to label the ethylene oxide, propylene oxide, caustic, and sulfuric acid storage tanks. The inspection found that many of Staley's managers had received little or no training about OSHA compliance, and that both managers and hourly employees were untrained regarding the cited hazardous chemicals. MSDSs were mostly either unavailable or kept in locked offices. Of particular concern, employees working in the vicinity of hazardous chemicals like ethylene oxide and propylene oxide were untrained in the labeling system and had no idea what the colors and numbers meant. As a consequence, they were unaware of the hazards posed by the chemicals and of how to protect themselves. J.A. at 186–207, 234–40, 277, 298; S.A. at 105–14, 169–86.

We again agree with the Commission that this record evidence is more than sufficient to sustain its findings that Staley was plainly indifferent to its violations of the hazard communication standard. The NIOSH evaluation and internal surveys and audits gave Staley a heightened awareness of its hazard communication problems, including problems with the specific chemicals for which Staley was later cited. Yet, the company responded unappreciatively to those warnings and substantially failed to ensure the required training of its managers and employees. In the Commission's words, "Staley's HazCom program remained grossly deficient." A.E. Staley Mfg. Co., 19 O.S.H. Cas. (BNA) at 1205.

Staley seeks to minimize the scope of its failure by noting that the training violations concerned "only" 15 of 120 buildings and 12 of 1200 hazardous materials used at the Decatur plant. As we have already noted, however, an otherwise safe workplace does not prevent findings of willfulness with respect to those violations that do occur. Moreover, five of the chemicals cited in the two health violations— ethylene oxide, propylene oxide, sulfuric acid, asbestos, and silica—are among Staley's "mean fifteen" chemicals, the most dangerous in the plant. All five may produce acute or chronic health effects in exposed employees. S.A. at 22–23, 88–89;

see 29 C.F.R. § 1910.1200(c). Accordingly, Staley's effort to minimize its violations of OSHA's health standards is unavailing, and we conclude that the Commission was justified in describing Staley's attitude toward its violations as "plain indifference to [] the Act's requirements." Kaspar Wire Works, 268 F.3d at 1127.

IV.

[9] Staley's second major argument is that, even if it was plainly indifferent to the requirements of the OSH Act, its violations of the hazardous locations standard were not willful because it was unaware of the specific conditions for which it was cited. There is no evidence in the record, Staley notes, that its management knew of the precise uncovered electrical boxes and exposed wires discovered by the OSHA inspectors. Although these conditions were of the same kind and in the same locations as problems found in earlier internal audits, Staley stresses that OSHA cannot prove that they were the same pieces of noncompliant equipment. In the company's view, "implicit to a finding of willfulness is employer knowledge of the existence of a condition, an awareness that the condition does not meet the Act's requirements ..., and a conscious decision not to correct the condition... ." Staley Br. at 14.

[10] Staley offers little support for this position, and we reject it. The OSH Act authorizes its most severe civil penalties for any employer who "willfully" violates a health or safety standard. 29 U.S.C. § 666(a); *see supra* note 2. The Act does not itself define "willfully." In its decision below, the Commission, citing its own precedents, defined a willful violation as one "committed with intentional, knowing or voluntary disregard for the requirements of the Act, or with plain indifference to employee safety." A.E. Staley Mfg. Co., 19 O.S.H. Cas. (BNA) at 1202 (quoting Falcon Steel Co., 16 O.S.H. Cas. (BNA) 1179, 1181 (OSHRC 1993), and A.P. O'Horo Co., 14 O.S.H. Cas. (BNA) 2004, 2012 (OSHRC 1991)) (emphasis added).

8 Legislation and Trends

CHAPTER OBJECTIVES

1. Acquire an understanding of the safety profession.
2. Acquire a knowledge of the current legislation.
3. Acquire a knowledge of the current trends in safety and loss prevention.
4. Acquire a knowledge of new risks and potential liabilities.

SAFETY AND HEALTH PROFESSION

A pivotal question for those of us working in the safety, health and loss prevention arena is whether or not the safety, health and loss prevention function is truly a "profession" or is it simply an operational position within a corporation or a function required to manage regulatory requirements? In comparing safety, health and loss prevention to other identifiable "professions" such as medicine or law, arguable safety, health and loss prevention appears to be deficient in many of the requirements for a "profession". According to the Cambridge English dictionary, a "profession" is defined as "any type of work, especially one that needs a high level of education or a particular skill." In analyzing the safety and health profession, there are no specific educational requirements proscribed by law to be a safety and health professional. There are no specific qualifications necessary to be a safety, health and loss prevention professional. And there is no competency testing required to manage the safety and health function.

In analyzing the medical and legal professions, each of these professions required substantial and high levels of education, both possess licensure requirements established by the state, possess competency testing, require continuing education and have a mandatory code of professional conduct which is enforceable. When analyzing the safety, health and loss prevention profession, there is no educational requirement, no licensure requirements, no competency testing, no required continuing education requirements, and no mandatory and enforceable code of professional conduct. The safety and health profession does have several safety and health focused organizations, such as the American Society of Safety Professionals and the National Safety Council, which offer voluntary competency education and specified levels and do possess a code of professional conduct, however this code is largely unenforceable outside of the membership.

When comparing the requirements of the legal or medical professions with that of the safety, health and loss prevention profession, the safety, health and loss prevention profession appears structurally deficient. Arguably, the safety, health and loss prevention profession, a function which safeguards the lives and welfare of large number of workers throughout the United States and beyond, does not meet the criteria to be called a "profession." In its current status, individual companies or organizations can determine the criteria from which they select the individual(s) who will be responsible for the safety and health of their employees. The individual company or organization established the educational requirements and enforces the individual's work performance and conduct through internal company policy. However, unlike law or medicine, a company or organization can establish the educational level at a high school level and hire any individual a high school diploma and no safety, health or loss prevention education or experience and simply designate this individual as their safety, health and loss prevention professional.

For the individuals reading this text who are or plan to work within the safety, health and loss prevention function, please ask yourself the question whether or not safety and health is truly a profession? For OSHA and our governmental entities, please ask what qualifications and level of education should the individual(s) responsible for the safety and health of the American workforce possess to be effective in reducing or eliminating risks and thus injuries and illnesses in the workplace? Has the risk of injury or illness in the workplace reduced or increased through permitting unqualified individuals to serve in the important safety, health and loss prevention function in the workplace?

Arguably, for the safety, health and loss prevention function to truly be consider a profession along the lines of the medical and legal professions, federal and/or state regulatory bodies should consider the establishment of a licensure requirement for individuals working within the safety and health function with minimum continuing education requirements. Furthermore, the federal and/or state regulatory bodies should establish a mandatory code of professional conduct that is enforceable through suspension or removal of the licensure as well as civil and criminal penalties. To enter the safety, health and loss prevention profession, the federal and/or state regulatory bodies should establish minimum educational levels, and appropriate competency testing should be established to ensure the skill and ability levels of those entering the safety, health and loss prevention profession.

OSHA STANDARDS

To achieve the ultimate goal of safeguarding employees from on-the-job injuries and illnesses, the field of safety and loss prevention is constantly evolving and changing. Safety and loss prevention professionals of today must keep abreast of new OSHA standards, legislation, trends, and technology that affect their workplace and their employees. Simply complying with mandatory regulations and standards minimizes potential liability for individual safety and loss prevention professionals and their organizations.

How can a safety and loss prevention professional keep abreast of proposed legislation in Congress and new standards promulgated by OSHA? Although there are numerous service organizations that can provide varying levels of information on a periodic basis, the easiest and most cost-effective method for keeping current is to review professional publications for information on new standards and to review the *Federal Register* for notices of proposed rulemaking, hearings, and final notices.

Safety and loss prevention professionals should realize that the employer is responsible for identifying the OSHA standards that apply to the facility or operation and for ensuring compliance with those standards. Most OSHA standards do not provide specific instructions on how to comply, but they normally spell out the required end result. Several of the newer standards, like control of hazardous energy and confined space entry and rescue standards, do provide specific management or system-based requirements for implementing and managing compliance with the standard. In comparing the early OSHA standards promulgated from 1970 to approximately 1980, the newer standards address significantly more complicated situations and issues (e.g., chemical process safety standards and ergonomics). These newer standards place emphasis on the structure and management of compliance efforts and so require upper management commitment and support for these programs.

Under the current administration there is emphasis to reduce the number of regulations and reduce the amount of paperwork. However, OSHA maintains a robust agenda including such areas as workplace violence, LOTO, and emergency response and preparedness. Safety and loss preventions should maintain a recognition of the possible new or revised standards which are in the prerule stage, proposed rule stage, and especially those standards in the final rule stage. Safety and loss prevention professionals should be cognizant that although most standards are vertical in nature and impact a specific industry or segment of a particular industry, some new standards are horizontal in nature and encompass virtually every industry and workplace.

Unified Agenda
(Modified for the purpose of this text)

Fall 2018 Unified Agenda of Federal Regulatory and Deregulatory Actions

Agency	Agenda Stage of Rulemaking	Title	RIN
DOL/OSHA	Prerule Stage	Communication Tower Safety	1218-AC90
DOL/OSHA	Prerule Stage	Emergency Response and Preparedness	1218-AC91
DOL/OSHA	Prerule Stage	Mechanical Power Presses Update	1218-AC98
DOL/OSHA	Prerule Stage	Powered Industrial Trucks	1218-AC99
DOL/OSHA	Prerule Stage	Lockout/Tagout Update	1218-AD00
DOL/OSHA	Prerule Stage	Tree Care Standard	1218-AD04
DOL/OSHA	Prerule Stage	Prevention of Workplace Violence in Health Care	1218-AD08
DOL/OSHA	Prerule Stage	Blood Lead Level for Medical Removal	1218-AD10
DOL/OSHA	Prerule Stage	Occupational Exposure to Crystalline Silica; Revisions	1218-AD18
DOL/OSHA	Proposed Rule Stage	Amendments to the Cranes and Derricks in Construction	1218-AC81

Fall 2018 Unified Agenda of Federal Regulatory and Deregulatory Actions

Agency	Agenda Stage of Rulemaking	Title	RIN
DOL/OSHA	Proposed Rule Stage	Update to the Hazard Communication Standard	1218-AC93
DOL/OSHA	Proposed Rule Stage	Cranes and Derricks in Construction: Exemption Expansions	1218-AD07
DOL/OSHA	Proposed Rule Stage	Puerto Rico State Plan	1218-AD13
DOL/OSHA	Proposed Rule Stage	Exposure to Beryllium NPRM to Review General Industry	1218-AD20
DOL/OSHA	Final Rule Stage	Standards Improvement Project IV	1218-AC67
DOL/OSHA	Final Rule Stage	Quantitative Fit-Testing Protocol: Amendment to the Final Rule on Respiratory Protection	1218-AC94
DOL/OSHA	Final Rule Stage	Rules of Agency Practice and Procedure Concerning OSHA	1218-AC95
DOL/OSHA	Final Rule Stage	Crane Operator Qualification in Construction	1218-AC96
DOL/OSHA	Final Rule Stage	Technical Corrections to 35 OSHA Standards and Regulations	1218-AD12
DOL/OSHA	Final Rule Stage	Tracking of Workplace Injuries and Illnesses	1218-AD17
DOL/OSHA	Final Rule Stage	Occupational Exposure to Beryllium and Beryllium Compounds in Construction and Shipyard Sectors	1218-AD21
DOL/OSHA	Long-term Actions	Occupational Injury and Illness Recording and Reporting	1218-AC45
DOL/OSHA	Long-term Actions	Infectious Diseases	1218-AC46
DOL/OSHA	Long-term Actions	Process Safety Management and Prevention of Major Chemical Accidents	1218-AC82
DOL/OSHA	Long-term Actions	Shipyard Fall Protection—Scaffolds, Ladders, and Other	1218-AC85

(Taken from OSHA website located at www.osha.gov)

OSHA STANDARDS DEVELOPMENT

Although OSHA promulgates hundreds of new compliance standards each year, some of the standards are broadly applied and quite difficult to implement and maintain on an ongoing basis. The newer standards tend to be much more technical in nature and cover a broader spectrum of potential workplace hazards than the older standards. Many of the new standards, however, provide substantial guidance for the implementation stage and in the management systems to be utilized. A commitment by upper management to the safety and loss prevention compliance effort is essential with these new standards in order to minimize budgetary constraints, support considerations, and other common impediments to a successful program.

OSHA can begin standards-setting procedures on its own initiative or in response to petitions from other parties, including the secretary of Health and Human Services (HHS), the National Institute for Occupational Safety and Health (NIOSH), state and local governments, any nationally recognized standards-producing organization, employer or labor representatives, or any other interested person.

ADVISORY COMMITTEES

If OSHA determines that a specific standard is needed, any of several advisory committees may be called upon to develop specific recommendations. There are two standing committees, and ad hoc committees may be appointed to examine special areas of concern to OSHA. All advisory committees, standing or ad hoc, must have members representing management, labor, and state agencies as well as one or more designees of the secretary of HHS. The two standing advisory committees are:

- National Advisory Committee on Occupational Safety and Health (NACOSH), which advises, consults with, and makes recommendations to the secretary of HHS and to the secretary of labor on matters regarding administration of the Act.
- Advisory Committee on Construction Safety and Health, which advises the secretary of labor on formulation of construction safety and health standards and other regulations.

NIOSH RECOMMENDATIONS

Recommendations for standards also may come from NIOSH, established by the Act as an agency of the Department of HHS.

NIOSH conducts research on various safety and health problems, provides technical assistance to OSHA, and recommends standards for OSHA's adoption. While conducting its research, NIOSH may make workplace investigations, gather testimony from employers and employees, and require that employers measure and report employee exposure to potentially hazardous materials. NIOSH may also require employers to provide medical examinations and tests to determine the incidence of occupational illness among employees. When such examinations and tests are required by NIOSH for research purposes, they may be paid for by NIOSH rather than the employer.

STANDARDS ADOPTION

Once OSHA has developed plans to propose, amend, or revoke a standard, it publishes these intentions in the *Federal Register* as a "Notice of Proposed Rulemaking," or often as an earlier "Advance Notice of Proposed Rulemaking." Prior to publication of proposed and final major rules OSHA consults with OMB under procedures established by executive order. OSHA consults with small business on proposed rules that significantly affect them through a panel with participation by the Small Business Administration and OMB, as required by the Small Business Regulatory Enforcement and Fairness Act (SBREFA.)

An advance notice is used, when necessary, to solicit information that can be used in drafting a proposal. The notice of proposed rulemaking will include the terms of the new rule and provide a specific time (at least 30 days from the date of publication, usually 60 days or more) for the public to respond.

Interested parties who submit written arguments and pertinent evidence may request a public hearing on the proposal when none has been announced in the notice. When such a hearing is requested, OSHA will schedule one, and will publish, in advance, the time and place for it in the *Federal Register.*

After the close of the comment period and public hearing, if one is held, OSHA must publish in the *Federal Register* the full, final text of any standard amended or adopted and the date it becomes effective, along with an explanation of the standard and the reasons for implementing it. OSHA may also publish a determination that no standard or amendment needs to be issued.

EMERGENCY TEMPORARY STANDARDS

Under certain limited conditions, OSHA is authorized to set emergency temporary standards that take effect immediately and are in effect until superseded by a permanent standard. OSHA must determine that an emergency standard is needed to protect workers who are in grave danger due to exposure to toxic substances, agents determined to be toxic or physically harmful, or to new hazards. Then, OSHA publishes the emergency temporary standard in the *Federal Register,* where it also serves as a proposed permanent standard. It is then subject to the usual procedure for adopting a permanent standard except that a final ruling should be made within six months. The validity of an emergency temporary standard may be challenged in an appropriate U.S. Court of Appeals.

APPEALING A STANDARD

No decision on a permanent standard is ever reached without due consideration of the arguments and data received from the public in written submissions and at hearings. Any person who may be adversely affected by a final or emergency standard, however, may file a petition (no later than the 59th day after the rule's promulgation) for judicial review of the standard with the U.S. Court of Appeals for the circuit in which the objector lives or has his or her principal place of business. Filing an appeals petition, however, will not delay the enforcement of a standard unless the Court of Appeals specifically orders it.

VARIANCES

Employers may ask OSHA for a variance from a standard or regulation if they cannot fully comply by the effective date due to shortages of materials, equipment, or professional or technical personnel (i.e., temporary variances), or can prove their facilities or methods of operation provide employee protection "at least as effective" as that required by OSHA (permanent variances).

Employers located in states with their own occupational safety and health programs should apply to the state for a variance. If, however, an employer operates facilities in states under federal OSHA jurisdiction and also in state plan states, the employer may apply directly to federal OSHA for a single variance applicable to all the establishments in question. OSHA will then work with the state plan states involved to determine if a variance can be granted that will satisfy state as well as federal OSHA requirements.

TEMPORARY VARIANCE

A temporary variance may be granted to an employer who cannot comply with a standard or regulation by its effective date due to unavailability of professional or technical personnel, materials or equipment, or because the necessary construction or alternation of facilities cannot be completed in time.

Employers must demonstrate to OSHA that they are taking all available steps to safeguard employees in the meantime, and that the employer has put in force an effective program for coming into compliance with the standard or regulation as quickly as possible.

A temporary variance may be granted for the period needed to achieve compliance or for one year, whichever is shorter. It is renewable twice, each time for six months. An application for a temporary variance must identify the standard or portion of a standard from which the variance is requested and the reasons why the employer cannot comply with the standard. The employer must document those measures already taken and to be taken (including dates) to comply with the standard and establish that all available steps to safeguard employees against the hazards covered by the standard are being taken.

The employer must certify that workers have been informed of the variance application, that a copy has been given to the employees' authorized representative, and that a summary of the application has been posted wherever notices are normally posted. Employees also must be informed that they have the right to request a hearing on the application.

The temporary variance will not be granted to an employer who simply cannot afford to pay for the necessary alterations, equipment, or personnel.

PERMANENT VARIANCE

A permanent variance (an alternative to a particular requirement or standard) may be granted to employers who prove their conditions, practices, means, methods, operations, or processes provide a safe and healthful workplace as effectively as would compliance with the standard.

In making a determination, OSHA weighs the employer's evidence and arranges a variance inspection and hearing where appropriate. If OSHA finds the request valid, it prescribes a permanent variance detailing the employer's specific exceptions and responsibilities under the ruling.

When applying for a permanent variance, the employer must inform employees of the application and of their right to request a hearing. Anytime after six months from the issuance of a permanent variance, the employer or employees may petition OSHA to modify or revoke it. OSHA also may do this of its own accord.

INTERIM ORDER

So that employers may continue to operate under existing conditions until a variance decision is made, they may apply to OSHA for an interim order. Application for an interim order may be made either at the same time as, or after, application for a variance. Reasons why the order should be granted may be included in the interim order application.

If OSHA denies the request, the employer is notified of the reason for denial.

If the interim order is granted, the employer and other concerned parties are informed of the order and the terms of the order are published in the *Federal Register*. The employer must inform employees of the order by giving a copy to the authorized employee representative add by posting a copy wherever notices are normally posted.

EXPERIMENTAL VARIANCE

If an employer is participating in an experiment to demonstrate or validate new job safety and health techniques and that experiment has been approved by either the secretary of labor or the secretary of HHS, a variance may be granted to permit the experiment.

OTHER

In addition to temporary, permanent, and experimental variances, the secretary of labor may also find certain variances justified when the national defense is impaired. For further information and assistance in applying for a variance, contact the nearest OSHA office.

Variances are not retroactive. An employer who has been cited for a standards violation may not seek relief from the citation by applying for a variance. The fact that a citation is outstanding, however, does not prevent an employer from filing a variance application.

PUBLIC PETITIONS

OSHA continually reviews its standards to keep pace with developing and changing industrial technology. Therefore, employers and employees should be aware that, just as they may petition OSHA for the development of standards, they also may petition OSHA for modification or revocation of standards.

COMBUSTIBLE DUST HAZARDS

An issue which has been on the front burner in recent years has been issues related to combustible dust. "Combustible dusts are solids finely ground into fine particles, fibers, chips, chunks or flakes that can cause a fire or explosion when suspended in air under certain conditions. Types of dusts include metals (aluminum and magnesium), wood, plastic or rubber, coal, flout, sugar and paper, among others."

"Since 1980, more than 130 workers have been killed and more than 780 injured in combustible dust explosions. These include 14 people who were killed in a dust explosion February 7, 2008, at the Imperial Sugar Co. plant in Georgia and three workers who were burned in April 2009 in an Illinois pet food plant dust explosion." Secretary of Labor Hilda L Solis stated, "Over the years, combustible dust explosions have caused many deaths and devastating injuries that could have been prevented … . OSHA is reinvigorating the regulatory process to ensure workers receive the protections they need while also ensuring that employers have the tools needed to make their workplaces safer."

Safety and loss prevention professionals possessing operations generating or possessing dusts with the potential of combustion should carefully evaluate these operations. Additional information can be found at http://www.osha.gov/dsg/combustibledust/index.html.

FOOD FLAVORINGS CONTAINING DIACETYL

In April 2009, the secretary of labor announced that OSHA will convene a Small Business Regulatory Enforcement Fairness Act (SBREFA) panel to draft a proposed rule on occupational exposure to diacetyl and food flavorings containing diacetyl. As stated by Secretary of Labor Solis, "I am alarmed that workers exposed to food flavorings containing diacetyl may continue to be at risk of developing a potentially fatal lung disease. Exposure to this harmful chemical already has been linked to the deaths of at least three workers. These deaths are preventable, and it is imperative that the Labor Department move quickly to address these hazards."

VOLUNTARY PROTECTION PROGRAM

Throughout most of this text we discuss the potential liabilities that OSHA can create for employers and individuals who do not comply with its standards and regulations. Conversely, OSHA and many of the state plan programs offer a very positive program for the "elite" companies who possess exceptional safety and loss prevention programs called the Voluntary Protection Program (VPP). This program is a voluntary program in which the company asks OSHA to review its safety and loss prevention programs, conduct a voluntary inspection, talk with its employees, and conduct a complete and total evaluation of its safety and loss prevention efforts. If the company's programs achieve the requisite levels, it can be rewarded by OSHA in terms of recognition and relief from future inspection.

At this time, there are approximately 280 companies who have qualified for the VPP in the United States. The vast majority of these companies are large corporations. OSHA has, however, initiated a very successful pilot program for small employers to participate in VPP and safety and loss prevention. Professionals are encouraged to consider participating.

Employers considering participation in the VPP should be prepared for a rigorous application and evaluation process. Information regarding the application and evaluation process can be obtained from your regional OSHA office. This process usually includes specific qualifications (such as your program being in place for at least one year), an extensive written application (including numerous documents such as your OSHA 200 log, written compliance programs, safety committee minutes, etc.), and one or more on-site evaluations.

When an employer achieves VPP, the employer would then qualify for participation in the Voluntary Protection Programs Participants' Association (VPPPA). This is a very active, nonprofit, charitable organization that is dedicated to ensuring the best practices in workplace safety, health, and environmental protections.

Safety and loss prevention professionals who have transported their programs to the highest level may want to consider the VPP. This type of joint partnership with OSHA has proven to be extremely beneficial in creating successful safety and loss prevention programs.

ERGONOMIC HAZARDS: NO STANDARD BUT CITED UNDER GENERAL DUTY CLAUSE

INTRODUCTION

For several years OSHA has been citing cumulative trauma illnesses and ergonomic hazards under the general duty clause because of the fact that a specific standard had not been promulgated. For many companies, the direct and indirect costs being incurred through these occupational illnesses have been the driving force to establish ergonomic programs to minimize or eliminate the potential risk factors that lead to cumulative trauma illnesses. Consider the following synopsis when establishing an ergonomics program.

DEFINITIONS

Ergonomics itself is an inexact and emerging science. For terminology purposes, "ergo" basically means "the act of work" and "nomic" loosely means "law." Therefore, the general definition of ergonomics is "the law of work or the workplace." Several specific terms are associated with ergonomics and cumulative trauma illnesses, including:

- Cumulative trauma disorders (CTDs): CTDs are health disorders arising from repeated or cumulative stress placed on the human body or parts of the body. These disorders result from chronic exposure of a particular body part to repeated stress, e.g., a meat cutter using the same arm and hand motions while boning a particular product repeatedly over a period of time. CTDs are generally categorized as occupational illnesses rather than occupational injuries.
- Carpal Tunnel Syndrome: One of the most common CTDs, this is a compression of the medial nerve in the carpal tunnel, a passage in the wrist through which the finger tendons and a major nerve passes to the hand from the forearm. Its symptoms include tingling, pain, and/or numbness in the thumb and the first three fingers.

- Tendinitis: The muscle–tendon junction and adjacent muscle tissues become inflamed, resulting from repeated stress of a body member.
- Tennis elbow: This CTD is characterized by inflammation of tissue in the elbow.
- Trigger finger: With this condition, the finger frequently flexes against resistance.

ANTICIPATED REQUIREMENTS

Given the higher frequency of cumulative trauma illnesses in specific industries, such as meatpacking, OSHA has published guidelines for this industry. The structure utilized in these guidelines will serve as the basic framework of the proposed general industry standard.

Management Commitment and Employee Involvement

These guidelines require management commitment to the ergonomic compliance program. The management group would be required to provide necessary resources as well as be a motivating force in the ergonomic program. Employee involvement and feedback to identify existing and potential hazards will be instrumental in developing and implementing this program.

The Ergonomic Team

The proposed standard would require that ergonomic teams be developed to identify and correct ergonomic hazards in the workplace. These teams should consist of a wide range of personnel, including employees, managers, ergonomic specialists, and other related personnel.

WORKSITE ANALYSIS

Worksite analysis is a vital part of any ergonomic evaluation process. This analysis should be performed by the ergonomic team and should involve examining and identifying existing hazards or conditions and operations that could create a hazard. The worksite analysis should include identifying work positions that need an ergonomic hazard analysis, such as the following situations:

- using an ergonomic checklist that includes components such as posture, force, repetition, vibration, and various upper extremity factors
- identifying work positions that place employees at risk for developing cumulative trauma disorders
- verifying low-risk factors for light-duty jobs, restricting awkward work positions
- verifying risk factors for work positions that have already been evaluated and corrected
- providing results of worksite analysis for use in assigning light-duty jobs
- reevaluating all planned, new, and modified facilities, processes, materials, and equipment to ensure that workplace alterations contribute to reducing or eliminating ergonomic hazards

The proposed standard could also require that surveys be conducted on an annual or periodic basis, when operations change or the need arises.

HAZARD PREVENTION AND CONTROL

The ergonomic standard could require that hazards be identified through the work-site analysis and that specific design measures be used to prevent or control these hazards. Ergonomic hazards are primarily prevented by the effective design of the workstation, tools, and the job.

ENGINEERING CONTROLS

Engineering techniques are the preferred method of correcting ergonomic hazards. The purpose of engineering controls is to make the job fit the person, not make the person fit the job. This can be accomplished by designing or modifying the workstation, work methods, and tools in order to eliminate excessive exertion and awkwardness. This methodology might be included in the proposed standard.

Workstation Design

The proposed ergonomic standard could require that workstations be designed to accommodate the person who actually works on a given job; designing for the "average" or "typical" worker may not be adequate. Workstations should be easily adjustable and either designed or selected to fit a specific task so that they are comfortable for the workers who are using them.

Design of Work Methods

The proposed standard could require that work methods be designed to reduce static, extreme and awkward postures, repetitive motions, and the use of excessive force. The standard could also require that the production system be analyzed and that the tasks be designed or modified to eliminate any stressors.

Tool and Handle Design

The proposed standard could require that tool and handle designs be evaluated and modified to eliminate or minimize the following situations:

- chronic muscle contraction or steady force
- extreme or awkward finger, hand, or arm positions
- repetitive, forceful motions
- tool vibration
- excessive gripping, pinching, and pressing with the hand and fingers

WORK PRACTICE CONTROLS

The proposed ergonomic standard may include practices such as the training of proper work habits, proper work techniques, employee conditioning, regular monitoring, feedback, maintenance, adjustments, modifications, and enforcement. Appropriate training and practice time for employees in proper work techniques could include the following activities:

- proper tool techniques, including work methods that improve posture and reduce stress and strain on extremities
- correct lifting techniques, including proper body mechanics, such as using the legs while lifting, not the back
- proper use and maintenance of pneumatic and other types of power tools
- correct use of ergonomically designed workstations and fixtures
- gradual integration of new and returning employees into a full workload
- regular workplace monitoring to ensure that employees continue to use proper work practices, including the periodic review of techniques to ensure that the procedures being used are the proper ones
- adjustments and modifications when changes occur at the workplace, including line speeds, staffing of positions, and the type, size, weight, or temperature of the product handled.

ADMINISTRATIVE CONTROLS

Administrative controls should reduce the duration, frequency, and severity of ergonomic stressors. These might include the following methods:

- Reduce the total number of repetitions per employee by decreasing production rates and limiting overtime work.
- Provide rest pauses to relieve fatigued muscle–tendon groups.
- Increase the number of employees assigned to a task, thus alleviating severe conditions, especially while lifting heavy objects.
- Rotate jobs as a preventive measure. The principle of job rotation is to alleviate the physical fatigue and stress of a particular set of muscles and tendons by rotating employees among other jobs that use different motions. The ergonomic team should analyze each operation to ensure that the same muscle–tendon groups are not used.
- Provide sufficient standby and relief personnel for a foreseeable condition on production lines.
- Enlarge jobs.
- Perform preventive maintenance for mechanical and power tools and equipment, including power saws and knives.

PERSONAL PROTECTIVE EQUIPMENT

The proposed standard could require that personal protective equipment (PPE) be selected with ergonomic stressors in mind, be provided in a variety of sizes, and accommodate the physical requirements of workers. Safety and loss prevention professionals should realize that there is the possibility that PPE may increase some ergonomic risk factors. PPE may include the following items:

- gloves that facilitate grasping tools that are needed for a particular job while protecting the worker from injury
- clothing for protection against extreme temperatures
- braces, splints, and back belts for support

MEDICAL MANAGEMENT

A medical management system could be a major component of the proposed ergonomic program and might include the following items:

- injury and illness recordkeeping
- early recognition and reporting
- systematic evaluation and referral
- conservative treatment
- conservative return to work
- systematic monitoring
- adequate staffing and facilities11

PERIODIC WORKPLACE WALK-THROUGH

The proposed standard might require that medical management teams conduct workplace walk-throughs to keep abreast of operations and work practices, to identify potential light-duty jobs, and to maintain close contact with employees.

SYMPTOMS SURVEY AND SYMPTOMS SURVEY CHECKLIST

Under the proposed standard, an anonymous survey of employees could be required to measure their awareness of work-related disorders and to report the location, frequency, and duration of possible discomfort. An industry-specific checklist may also be required.

TRAINING AND EDUCATION

Training and education might be required under the proposed standard to educate employees about the potential risks of illnesses and injuries, their causes and early symptoms, the means of prevention, and treatment. A training program should include employees, engineers and maintenance personnel, supervisors, managers, and medical personnel.

General Training

Under the proposed standard, employers could be required to give employees who might be exposed to ergonomic hazards formal instruction on the potential hazards associated with their jobs and with their equipment. This may include information defining the varieties of cumulative trauma disorders, identifying their risk factors, recognizing and reporting symptoms, and preventing the disorders.

Job-Specific Training

Under the proposed guidelines, employers could be required to give new employees and reassigned workers initial orientation and hands-on training before they are placed in a full-time production job. The initial training program could include the following skills:

- care, use, and handling techniques of tools
- use of special tools and devices associated with individual workstations
- use of appropriate lifting techniques and devices
- use of appropriate guards and safety equipment, including PPE

Supervisor Training

Supervisors might be required to undergo training comparable to that of employees, as well as additional training that will help them to recognize early signs and symptoms of cumulative trauma disorders to recognize hazardous work practices and to correct such practices.

Manager Training

Managers might be required to receive training in ergonomic issues at each workstation and in the production process so that they could effectively carry out their responsibilities.

Maintenance Training

Maintenance personnel might be required to undergo training in preventing and correcting ergonomic hazards through job and workstation design and proper maintenance.

AUDITING

The ergonomic team might be required to review the ergonomic program to ensure that new and existing ergonomic hazards are identified and corrected. If the need arises, corrective measures should be taken to eliminate or minimize potential hazards.

DISCIPLINARY ACTION

In the event that employees refuse to comply with any safety program, disciplinary action may be required.

CONSTRUCTION FOCUS

In recent years, there has been an emphasis on construction industry related hazards. Issues ranging from scaffolding to trenching have received attention. Safety and loss prevention professionals with construction operations should be aware of the separate category of standards for the construction industry and ensure compliance with all applicable standards.

Top 10 Most Frequently Cited Construction Violations - 2017 (as of Sept. 30, 2017)

1. Fall Protection - General Requirements (1926.501)
2. Scaffolding (1926.451)
3. Ladders (1926.1053)
4. Fall Protection — Training (1926.503)
5. Eye and Face Protection (1926.102)
6. Hazard Communication (1910.1200)
7. Head Protection (1926.100)
8. Aerial Lifts (1926.453)
9. General Safety and Health Provisions (1926.20)
10. Fall Protection - Systems Criteria and Practices (1926.502)

Taken from the OSHA website located at www.osha.gov

SELECTED CASE STUDY

SECRETARY OF LABOR, Complainant, v. SUMMIT CONTRACTORS, INC., Respondent

OSHRC Docket No. 03-1622

Decision

BY THE COMMISSION:

Statement of the Case

This case is before the Commission on remand from the United States Court of Appeals for the Eighth Circuit. Solis v. Summit Contractors, Inc., 558 F.3d 815 (8th Cir. 2009). In its initial decision, a Commission majority held that 29 C.F.R. § 1910.12(a)—a regulation that describes the reach of the Occupational Safety and Health Administration ("OSHA") construction standards—precluded the Secretary from citing a "controlling employer" under her multi-employer citation policy for a violation it did not create and to which none of its own employees were exposed. The Commission, therefore, vacated a citation alleging that general contractor Summit Contractors, Inc. ("Summit") violated 29 C.F.R. § 1926.451(g)(1)(vii) because the cited conditions were created by a subcontractor whose employees were the only ones exposed. Summit Contractors, Inc., 21 BNA OSHC 2020, 2025, 2007 CCH OSHD ¶ 32,888, p. 53,264 (No. 03-1622, 2007).

On appeal by the Secretary, the court vacated the Commission's decision and remanded the case, holding that the plain language of § 1910.12(a) "is unambiguous in that it does not preclude OSHA from issuing citations to employers for violations when their own employees are not exposed to any hazards related to the violations." Summit, 558 F.3d at 825. For the following reasons, we affirm the citation.

Issues

The primary issue before the Commission on remand is whether Summit exercised sufficient control over the worksite to prevent or detect and abate a hazardous condition created by its subcontractor, All Phase Construction, Inc. ("All Phase"), to which none of its own employees were exposed. If so, Summit can be properly cited as a "controlling employer" under the Secretary's multi-employer citation policy for the violation in question.

As a threshold matter, we also address Summit's contention before the Commission that the Secretary could not lawfully apply the multi-employer citation policy "without first adopting it through the informal rulemaking process of" the Administrative Procedure Act (the "APA"). Summit, 558 F.3d at 826 n.6 (internal citations omitted).

Findings of Fact

In June 2003, OSHA conducted an inspection of a college dormitory construction site in Little Rock, Arkansas, for which Summit was the general contractor. On June 18 and 19, an OSHA compliance officer ("CO") observed and photographed employees of subcontractor All Phase working on scaffolds from elevations over ten feet above a lower level without fall protection. At approximately ten o'clock in the morning of June 18, the CO took photographs of the cited conditions from the street, but did not enter the worksite until the next day, when he returned around nine o'clock in the morning. On both days, the CO observed that All Phase employees were working on the same scaffold without fall protection. On the second day, the CO also observed All Phase employees working on a second scaffold, again without fall protection. At Summit's request, the CO agreed to hold an opening conference several days later to discuss the violative conditions he observed, but by then the scaffolds were no longer standing.

At issue before the Commission is a decision of Administrative Law Judge Peggy S. Ball denying Respondent's motion for relief from a final order under Federal Rule of Civil Procedure 60(b)(1) after Respondent filed an untimely notice of contest. Upon review, for the reasons that follow, we set aside the judge's decision and remand this case to the judge for further proceedings. On June 1, 2017, the Occupational Safety and Health Administration issued Respondent a citation that was sent via certified mail and signed for on June 5, 2017, by an employee at Respondent's Carrollton, Texas branch office. Respondent's notice of contest was due on June 26, 2017, but it was not filed until September 18, 2017. Because Respondent's notice of contest SECRETARY OF LABOR, TEXAS MANAGEMENT DIVISION, INC. Complainant, Respondent. v. OSHRC Docket No. 17-1861 2 was untimely pursuant to section 10(a) of the Occupational Safety and Health Act, the citation became a final order of the Commission. See 29 U.S.C. § 659(a) (failure to contest citation within fifteen working days results in citation becoming final order of Commission).

On October 24, 2017, Respondent filed a Rule 60(b)(1) motion seeking relief from the final order, which was denied by the judge. In its Rule 60(b)(1) motion, Respondent argued that it was entitled to relief from a final order based on "mistake, inadvertence[,] o[r] excusable neglect." See Fed. R. Civ. P. 60(b) ("On motion and just terms, the court may relieve a party ... from a final judgment, order, or proceeding for the following reasons: (1) mistake, inadvertence, surprise, or excusable neglect ... ."). According to Respondent, management at its headquarters in Houston, Texas did not become aware of the citation until August 8, 2017; the company asserts that it subsequently filed its notice of contest after conducting an internal investigation, communicating with the OSHA area office, and retaining counsel. Respondent also submitted a declaration from its corporate controller in support of its motion and requested an evidentiary hearing with an opportunity for limited discovery.

The secretary filed an opposition to Respondent's motion, asserting that "Respondent has not established any basis for relief" under Rule 60(b)(1). In support of his opposition, the secretary submitted: (1) a declaration from the OSHA Assistant Area Director; (2) a handwritten statement that OSHA obtained during the inspection from Respondent's former branch manager; and (3) copies of correspondence sent from OSHA to Respondent during the period between the inspection and the issuance of the citation. On January 17, 2018, the judge issued the parties an order to appear for a "Telephonic Motions Hearing" on February 21, 2018, stating that "[t]he parties should be prepared to address the current status of the case, and any pending issues before this Court." That same day, the judge issued a "Notice of Trial," scheduling trial dates of July 18 and 19, 2018. After apparently communicating with Respondent's counsel, the secretary's counsel sent an email on February 12, 2018, to the judge's legal assistant, in which she copied Respondent's counsel and sought Respondent claims that following receipt of the citation at Respondent's branch office, it was given to Respondent's branch manager, who thereafter resigned on or about June 20, 2017. Respondent asserts that this branch manager did not forward the citation to its corporate headquarters as he was required to do pursuant to company policy; thus, Respondent was not aware of the citation until August 8, 2017, when a successor employed in the position of branch manager found in a pile of papers left in the predecessor's office a notice of delinquency regarding the citation clarification about the judge's telephonic hearing order:

"The parties do not interpret the Order as requiring the parties to put on a telephonic evidentiary hearing involving witnesses, exhibits and a court reporter. Please let the parties know if our understanding is incorrect." There is nothing in the record to indicate that a response to this email was ever sent, nor is there any documentation of what transpired during the telephonic hearing. In denying Respondent's Rule 60(b)(1) motion, the judge found that even if its initial handling of the citation constituted excusable neglect because the former branch manager did not follow the company's established procedure of transmitting the citation to corporate headquarters, Respondent's "continued failure to respond in a timely manner" after receiving a second copy of the citation at its headquarters was unjustified. See *A. W. Ross Inc.*, 19 BNA OSHC 1147, 1148 (No. 99-0945, 2000) ("A key factor in evaluating whether a party's delay in filing was due to excusable neglect is 'the reason for the delay' including whether it was within the reasonable control of the movant.") (citations omitted). In her decision, the judge relied primarily on the parties' declarations to make her factual findings. There is no mention—nor any record—of the judge having ruled on Respondent's requests for limited discovery and an evidentiary hearing.

On review, Respondent raises several arguments that center largely on the judge's failure to explicitly rule on these specific requests. Respondent contends that it was not made aware of the "purpose" of the telephonic hearing and therefore, had "no opportunity to put on evidence and testimony" regarding its Rule 60(b)(1) motion. In addition, Respondent challenges the judge's reliance on what it asserts are hearsay statements in the declaration submitted by the secretary. We agree that the absence of express rulings from the judge on Respondent's requests is problematic, particularly where an opportunity for limited discovery may have addressed the hearsay concerns evident in the declarations submitted by both parties. It also appears that the parties may have been confused by the judge's simultaneous issuance of orders scheduling a telephonic "hearing" and a subsequent "trial" in the case. Indeed, based on the record before us, the judge also rejected Respondent's alternative claim that it was unfairly surprised by the citation because, as a staffing agency, it did not believe it was the subject of OSHA's inspection or that it could be the recipient of a citation. In rejecting Respondent's argument, the judge found that employees of Respondent participated in various stages of the inspection, and in any event, Respondent's allegation that "it did not expect to be issued a citation … does not justify its failure to file a timely notice of contest" once a citation was received. The parties were never informed that the telephonic hearing would constitute the dispositive hearing on the Rule 60(b)(1) issue.

Under these circumstances, we remand this case to the judge to: (1) address Respondent's request for limited discovery and provide a written explanation of the basis for her ruling; and (2) conduct an evidentiary hearing to afford the parties an opportunity to develop the record. See *Elan Lawn and Landscape Serv.*, 22 BNA OSHC 1337, 1339-40 (No. 08-0700, 2008) (remanding case for evidentiary hearing where Respondent was confused about opportunity to establish potential basis for Rule 60(b)(1) relief); see, e.g., *Rheem Mfg. Co., Inc.*, 25 BNA OSHC 1838, 1839 (No. 15-1248, 2016) (remanding case for evidentiary hearing where judge's order. Our dissenting colleague, while noting that "Respondent provided the judge with an explanation for its failure to file its notice of contest before the June 26 statutory deadline …", places a great deal of emphasis on Respondent's "delay" in filing its notice of contest after August 8, when it became aware of the citation.

Other than hearsay evidence upon which the judge relied, there is no record evidence as to the substance of Respondent's communications with OSHA, either during the inspection, after the closing conference, or up until September 18. In addition, there is no legal basis to say that Respondent was obliged to file a notice of contest within fifteen days after receiving the second copy of the citation, as requesting a copy did not restart the clock on the 15-day contest period. The issue in determining whether relief is warranted under Rule 60(b)(1) is whether there was excusable neglect in failing to timely respond before the June 26 statutory deadline. That is not to say that the reason(s) for the delay and the conduct of Respondent should not be weighed, but this consideration should not artificially mandate a new deadline; rather, the Respondent's untimeliness should be considered together with other such circumstances as might be relevant. *Pioneer Inv. Servs. Co. v. Brunswick Assocs. Ltd. P'ship*, 507 U.S. 380, 395 (1993) (these circumstances include "the danger of prejudice to the debtor, the length of the delay and its potential impact on judicial proceedings, the reason for the delay, including whether it was within the reasonable control of the movant, and whether the movant acted in good faith").

Unlike our colleague, we do not create an artificial deadline that a delay of six weeks might pass muster but, without a full consideration of the Pioneer factors, a delay of almost three months tips the scale to too many "passing day[s]" and merits a finding of no excusable neglect. Further, our decision today does not, despite our colleague's concern, create an "unfettered right to have discovery and an evidentiary hearing on [Respondent's] motion." Our dissenting colleague poses the loaded question of why we believe that discovery and an evidentiary hearing is necessary, given her assumption that Respondent already possesses the information it seeks. However, this unfounded assertion fails to acknowledge that discovery was not allowed, the judge relied upon hearsay information, and Respondent was unable to cross-examine the proffered hearsay evidence. Rather, we simply find that the circumstances here warrant directing the judge to consider Respondent's request to engage in limited discovery and then conduct a hearing on its motion for Rule 60(b)(1) relief. 5 denying Rule 60(b)(1) relief based on excusable neglect was "premature based on the limited record").

Accordingly, we set aside the judge's decision and remand this case to the judge for further proceedings. SO ORDERED. /s/ Heather L. MacDougall Chairman /s/ James J. Sullivan, Jr. Dated: July 31, 3018

Commissioner ATTWOOD, Commissioner, dissenting: Because I find no error in the judge's decision to deny Respondent's Rule 60(b)(1) motion without first conducting an evidentiary hearing, I dissent. In my view, Respondent has not only failed to assert facts sufficient to support a finding that the untimely filing of its notice of contest was due to excusable neglect; it has also failed to even hint at how an evidentiary hearing on the issue would remedy that failure of proof. After filing its notice of contest almost three months past the statutory deadline, Respondent filed a Rule 60(b)(1) motion for relief with the judge, as well as a reply to the Secretary's opposition to its motion. See 29 U.S.C. § 659(a) (failure to contest citation within fifteen working days results in citation becoming final order of Commission). Both parties submitted sworn declarations and exhibits with their filings. In her decision, the judge considered the parties' arguments and the record before her, then sufficiently explained the basis for her findings. She noted the applicability of *Pioneer Investment Services Co. v. Brunswick Associates Limited Partnership*, 507 U.S. 380, 395 (1993), and following the Commission's decision in *A. W. Ross Inc.*, 19 BNA OSHC 1147, 1148 (No. 99-0945, 2000), focused on the reasons for Respondent's delay. Specifically, she concluded that even if Respondent's claim that its corporate management did not learn of the citation until August 8, 2017 was sufficient to excuse its failure to contest the citation prior to August 8, 2017, Respondent failed to provide a sufficient explanation for why it "still did not file its notice of contest until September 18, 2017." The judge found that this continued delay cuts "against a finding that Respondent was maintaining orderly procedures for the handling of important documents such as the [c]itation[,] and making a good faith effort to comply with the Act" and therefore denied the motion. I agree with the judge's ruling that Respondent failed to establish that its almost six-week delay in filing its notice of contest after OSHA provided it with a second copy of the citation was due to excusable neglect. Under *Pioneer*, the factors to consider in determining whether a party has established excusable neglect such that it is entitled to relief under Rule 60(b)(1) "include ... the danger of prejudice to the [opposing party], the length of the delay and its potential impact on judicial proceedings, the reason for the delay, including whether it was within the reasonable control of the movant, and whether the movant acted in good faith." 507 U.S. at 395. Although Respondent provided the judge with an explanation for its failure to file its notice of contest prior to August 8, it made no attempt what-soever to justify its delay after that date. Indeed, in its 7 filings here and below, Respondent simply stated that "[a]fter conducting an internal investigation (which is ongoing to date) and communicating with the OSHA Area Office, TMD retained the undersigned counsel to represent it in this matter." Respondent made no claim either before the judge or in its petition to us that these activities prevented it from promptly filing a notice of contest.

Compounding this failure, Respondent has not asserted any other reason for its lengthy delay, which was fully within Respondent's control. Cf. *Nw. Conduit Corp.*, 18 BNA OSHC 1948, 1950-51 (No. 97-851, 1999) (granting Rule 60(b)(1) relief where employer's president and attorney showed "diligence in pursuing their reme-dies" and attorney personally delivered notice of contest to OSHA the same day he discovered it had not been timely filed). Absent any attempt at an explanation, I agree with the judge that this further delay in filing demonstrates Respondent's failure to maintain orderly procedures for handling important documents and to proceed in good faith and does not constitute excusable neglect. *La.-Pac. Corp.*, 13 BNA OSHC 2020, 2021 (No. 86-1266, 1989) ("Even during a management transition, a business must maintain orderly procedures for handling important documents.")

The majority asserts that "there is no legal basis to say that Respondent was obliged to file a notice of contest within fifteen days after receiving the second copy of the citation." Neither the judge nor I posit anything of the sort, what I am asserting is that with every passing day after Respondent received the second copy of the citation, its argument that its delay was due to excusable neglect became more difficult to sustain. Moreover, the majority erroneously asserts that I have created "an artificial deadline that a delay of six weeks might pass muster but, ... a delay of almost three months tips the scale to too many 'passing day[s].'" But like the judge, I simply rely upon the delay from August 8 to September 18 because, whatever happened between June 26 and August 8, Respondent failed to put forth any facts justifying its further six-week delay following receipt of the second copy of the citation. Thus, the judge committed no error in finding that Respondent's delay after August 8 was not excusable. My colleagues find significance in the fact that "other than hearsay evidence upon which the judge relied, there is no record evidence as to the Respondent's communications with OSHA, either during the inspection, after the closing conference, or up until September 18." And they label as an "unfounded assertion" my statement that all of that evidence was known to Respondent. However, since Respondent presumably participated in all of its "communications with OSHA," evidence of those communications must, by application of elementary logic, have been known to Respondent. Given that it was Respondent's burden to demonstrate that it was entitled to relief, its failure to do so before the judge is fatal to its motion. One can only wonder why the majority believes that Respondent's failure to put forth evidence unquestionably in its possession entitles it to discovery and an evidentiary hearing. 8 (citation omitted). Respondent argues, and my colleagues agree, that the judge erred in denying Respondent's Rule 60(b)(1) motion without ruling on its request to have limited discovery and an evidentiary hearing on the motion. Although the judge did not explicitly deny this request (which was only included in the conclusion of Respondent's motion and reply and was not supported by any assertions, facts, or argument), she effectively denied it in rejecting Rule 60(b)(1) relief. Thus, the real issue is whether in the circumstances of this case, the judge erred in failing to afford Respondent an opportunity for discovery and/or an evidentiary hearing on its motion. In contrast to my colleagues, I find nothing in the record before us that would support a conclusion that the judge erred in this respect. Thus, at most the judge's failure to expressly deny Respondent's request is harmless error. Nothing in the Occupational Safety and Health Act, the Commission's Rules of Procedure, the Federal Rules of Civil Procedure, or the Administrative Procedure Act grants Respondent the unfettered right to have discovery and an evidentiary hearing on its motion. And Respondent has made no showing that it is legally entitled to either, nor has it shown how it was harmed by this alleged error. As any attorney who routinely engages in a motions practice is well aware, in the absence of some showing by Respondent that discovery and an evidentiary hearing were required here, the judge was under no compulsion to grant that request prior to ruling on Respondent's motion.

Moreover, Respondent has not put forth a single fact that it claims it could have proven in support of its motion for relief but for the judge's denial of discovery and an evidentiary hearing. See *Williams Enters. Inc.*, 13 BNA OSHC 1249, 1250-51 (No. 85-355, 1987) (party that "seeks to have a judgment set aside because of an erroneous ruling carries the burden of showing that prejudice resulted") (citation omitted). Since all of the facts that would be relevant to any justification for Respondent's delay in filing its notice of contest from August 8 until September 18 are exclusively within Respondent's control, discovery and an evidentiary hearing could add nothing beyond what it already knows in this regard. Respondent, with benefit of counsel, chose at its peril not to assert before the judge any facts related to its almost six-week delay in filing after receipt of the second copy of the citation. I would also reject Respondent's argument that it was unfairly surprised by the citation for the reasons stated in the judge's decision beyond its unsupported claim that it was conducting an internal investigation.

Its failure to do so should not now provide a justification for a needless remand. Thus, I find not even a shred of support for my colleagues' conclusion that Respondent was deprived of a sufficient opportunity to show that the untimely filing of its notice of contest was due to excusable neglect.

Accordingly, I would affirm the judge's decision. In any event, Respondent did not need to obtain any facts before immediately filing its notice of contest upon obtaining the second copy of the citation—completion of an "internal investigation" to gather information regarding either the merits of the citation or to support its motion for relief under Rule 60(b)(1) was not a necessary predicate for taking such action. Respondent and my colleagues assert that the judge's simultaneous issuance of an order to appear for a "Telephonic Motions Hearing" and a "Notice of Trial" was potentially confusing to the parties. On the contrary, it was perfectly reasonable for the judge to schedule a hearing on the merits at the same time she scheduled a "Telephonic Motions Hearing." Presumably, prior to that hearing the judge had not determined that the motion should be denied. And early scheduling of a hearing on the merits benefits the parties, works to expedite the adjudicative process, and can be easily undone at a later date should a hearing no longer be required. Nor do I sign on to my colleague's apparent view that the judge overstepped her authority by relying on the parties' declarations instead of holding an evidentiary hearing. Indeed, the Commission has routinely relied on declarations and related documents in ruling on Rule 60(b) motions. See, e.g., *Burrows Paper Corp.*, 23 BNA OSHC 1131, 1132 (No. 08-1559, 2010); *Acrom Constr. Serv., Inc.*, 15 BNA OSHC 1123, 1126 (No. 88-2291, 1991); *J.F. Shea Co.*, 15 BNA OSHC 1092, 1093-94 (No. 89-976, 1991); *La.-Pac. Corp.*, 13 BNA OSHC at 2021.

As discussed above, it is Respondent's failure to even assert facts essential to its position that renders its motion unsupportable, and as the judge found, "[n]o essential dispute of facts material to this [d]ecision and [o]rder has been established." See, e.g., *Evergreen Envtl. Serv.*, 26 BNA OSHC 1982, 1984 (No. 16-1295, 2017). Finally, my colleagues rely on *Rheem Manufacturing Co., Inc.*, to support their decision to remand for an evidentiary hearing, but in that case the Commission was not faced with a lengthy filing delay following the employer's receipt of the citation after the contest period in that case; on the contrary, the employer filed its late notice of contest only two days after discovering the citation. 25 BNA OSHC 1838, 1839 (No. 15-1248, 2016). *Elan Lawn and Landscape Services* is also inapposite—the employer in that case, initially appearing pro se, filed its notice of contest only one day late and was not fully aware of its procedural position "[a]bsent any communication or contact ... from either the judge or the Secretary."

9 Safety and Loss Prevention and the Americans with Disabilities Act

CHAPTER OBJECTIVES

1. Acquire an understanding as to qualifications for a "Qualified Individual with a Disability".
2. Acquire an understanding of the five titles within the ADA.
3. Acquire a level of knowledge of how the ADA impacts the safety and loss prevention function.
4. Acquire a knowledge as to achieving and maintaining compliance with the ADA.

OVERVIEW AND IMPACT

The Americans with Disabilities Act of 1990 (ADA) has opened a huge new area of regulatory compliance that will directly or indirectly affect most safety and loss prevention professionals. In a nutshell, the ADA prohibits discriminating against qualified individuals with physical or mental disabilities in all employment settings. Given the impact of the ADA on the job functions of a safety and loss prevention professional, especially in the areas of workers' compensation, restricted duty programs, facility modifications, and other areas, it is critical for safety and loss prevention professionals to possess a firm grasp of the scope and requirements of this law.

From most estimates, the ADA has afforded protection to approximately 43 to 45 million individuals or, in other terms, approximately one in five Americans. In terms of the effect on the American workplace, the estimates of protected individuals compared to the number of individuals currently employed in the American workplace (approximately 200 million), employers can expect that approximately one in four currently employed individuals (or potential employees) could be afforded protection under the ADA.

Structurally, the ADA is divided into five titles, and all titles possess the potential of substantially impacting the safety and loss prevention function in covered public or private sector organizations. Title I contains the employment provisions that protect all individuals with disabilities who are in the United States, regardless of their national origin or immigration status. Title II prohibits discriminating against qualified individuals with disabilities or excluding them from the services, programs, or activities provided by public entities. Title II contains the transportation provisions of the Act. Title III, entitled "Public Accommodations," requires that goods, services, privileges, advantages, and facilities of any public place be offered "in the most integrated setting appropriate to the needs of the individual."

Title IV also covers transportation offered by private entities and addresses telecommunications. Title IV requires that telephone companies provide telecommunication relay services and that public service television announcements that are produced or funded with federal money include closed-captioning. Title V includes the miscellaneous provisions. This title notes that the ADA does not limit or invalidate other federal and state laws providing equal or greater protection for the rights of individuals with disabilities and addresses related insurance, alternate dispute, and congressional coverage issues.

Title I of the ADA went into effect for all employers and industries engaged in interstate commerce with 25 or more employees on July 26, 1992. On July 26, 1994, the ADA became effective for all employers with 15 or more employees. Title II, which applies to public services such as fire departments, and Title III, requiring public accommodations and services operated by private entities, became effective on January 26, 1992, except for specific subsections of Title II, which went into effect on July 26, 1990. A telecommunication relay service required by Title IV became effective on July 26, 1993.

Title I prohibits covered employers from discriminating against a "qualified individual with a disability" with regard to job applications, hiring, advancement, discharge, compensation, training, and other terms, conditions, and privileges of employment.

Section 101(8) defines a "qualified individual with a disability" as any person who, with or without reasonable accommodation, can perform the essential functions of the employment position that such individual holds or desires ... consideration shall be given to the employer's judgment as to what functions of a job are essential, and if an employer has prepared a written description before advertising or interviewing applicants for the job, this description shall be considered evidence of the essential function of the job.

The Equal Employment Opportunity Commission (EEOC) provides additional clarification of this definition by stating, "an individual with a disability who satisfies the requisite skill, experience and educational requirements of the employment position such individual holds or desires, and who, with or without reasonable accommodation, can perform the essential functions of such position."

Congress did not provide a specific list of disabilities covered under the ADA because "of the difficulty of ensuring the comprehensiveness of such a list." Under the ADA, an individual has a disability if he or she possesses:

- a physical or mental impairment that substantially limits one or more of the major life activities of such individual,
- a record of such an impairment, or
- is regarded as having such an impairment.

For an individual to be considered "disabled" under the ADA, the physical or mental impairment must limit one or more "major life activities." Under the U.S. Justice Department's regulation issued for section 504 of the Rehabilitation Act, "major life activities" are defined as, "functions such as caring for one's self, performing manual tasks, walking, seeing, hearing, speaking, breathing, learning and working." Congress clearly intended to have the term "disability" broadly construed. However, this definition does not include simple physical characteristics nor limitations based on environmental, cultural, or economic disadvantages.

The second prong of this definition is "a record of such an impairment disability." The Senate Report and the House Judiciary Committee Report each stated:

> This provision is included in the definition in part to protect individuals who have recovered from a physical or mental impairment which previously limited them in a major life activity. Discrimination on the basis of such a past impairment would be prohibited under this legislation. Frequently occurring examples of the first group (i.e., those who have a history of an impairment) are people with histories of mental or emotional illness, heart disease or cancer; examples of the second group (i.e., those who have been misclassified as having an impairment) are people who have been misclassified as mentally retarded.

The third prong of the statutory definition of a disability extends coverage to individuals who are "being regarded as having a disability." The ADA has adopted the same "regarded as" test that is used in section 504 of the Rehabilitation Act:

> "Is regarded as having an impairment" means (A) has a physical or mental impairment that does not substantially limit major life activities but is treated ... as constituting such a limitation; (B) has a physical or mental impairment that substantially limits major life activities only as a result of the attitudes of others toward such impairment; (C) has none of the impairments defined (in the impairment paragraph of the Department of Justice regulations) but is treated ... as having such an impairment.

Under the EEOC's regulations, this third prong covers three classes of individuals:

1. The first class includes persons who have physical or mental impairments that do not limit a major life activity, but who are nevertheless perceived by covered entities (employers and places of public accommodation) as having such limitations. (For example, an employee with controlled high blood pressure that is not, in fact, substantially limited, is reassigned to less strenuous work because of his employer's unsubstantiated fear that the individual will suffer a heart attack if he continues to perform strenuous work. Such a person would be "regarded" as disabled.)
2. The second class includes persons who have physical or mental impairments that substantially limit a major life activity only because of a perception that the impairment causes such a limitation. (For example, an employee has a condition that periodically causes an involuntary jerk of the head, but no limitations on his major life activities. If his employer discriminates against him because of the negative reaction of customers, the employer would be regarding him as disabled and acting on the basis of that perceived disability.)

3. The third class includes persons who do not have a physical or mental impairment, but are treated as having a substantially limiting impairment. (For example, a company discharges an employee based on a rumor that the employee is HIV-positive. Even though the rumor is totally false and the employee has no impairment, the company would nevertheless be in violation of the ADA.)

Thus, a "qualified individual with a disability" under the ADA is any individual who can perform the essential or vital functions of a particular job with or without the employer accommodating the particular disability. The employer is provided the opportunity to determine the essential functions of the particular job before offering the position through the development of a written job description. This written job description will be considered evidence about which functions of the particular job are essential and which are peripheral. In deciding the "essential functions" of a particular position, the EEOC will consider the employer's judgment, whether the written job description was developed prior to advertising or beginning the interview process, the amount of time spent performing the job, the past and current experience of the individual to be hired, relevant collective bargaining agreements, and other factors.

The EEOC defines the term essential function of a job as meaning "primary job duties that are intrinsic to the employment position the individual holds or desires" and precludes any marginal or peripheral functions that may be incidental to the primary job function. The factors provided by the EEOC in evaluating the essential functions of a particular job include the reason that the position exists, the number of employees available, and the degree of specialization required to perform the job. This determination is especially important to safety and loss prevention professionals who may be required to develop the written job descriptions or to determine the essential functions of a given position.

Of particular concern to safety and loss prevention professionals is the treatment of the disabled individual, who, as a matter of fact or due to prejudice, is believed to be a direct threat to the safety and health of others in the workplace. To address this issue, the ADA provides that any individual who poses a direct threat to the health and safety of others that cannot be eliminated by reasonable accommodation may be disqualified from the particular job. The term direct threat to others is defined by the EEOC as creating "a significant risk of substantial harm to the health and safety of the individual or others that cannot be eliminated by reasonable accommodation." The determining factors that safety and health professionals should consider in making this determination include the duration of the risk, the nature and severity of the potential harm, and the likelihood that the potential harm will occur.

Additionally, safety and health professionals should consider the EEOC's Interpretive Guidelines, which state:

[If] an individual poses a direct threat as a result of a disability, the employer must determine whether a reasonable accommodation would either eliminate the risk or reduce it to an acceptable level. If no accommodation exists that would either eliminate the risk or reduce the risk, the employer may refuse to hire an applicant or may discharge an employee who poses a direct threat.

Safety and loss prevention professionals should note that Title I additionally provides that if an employer does not make reasonable accommodations for the known limitations of a qualified individual with disabilities, it is considered to be discrimination. Only if the employer can prove that providing the accommodation would place an undue hardship on the operation of the employer's business can discrimination be disproved. Section 101(9) defines a "reasonable accommodation":

a. making existing facilities used by employees readily accessible to and usable by the qualified individual with a disability and includes:
b. job restriction, part-time or modified work schedules, reassignment to a vacant position, acquisition or modification of equipment or devices, appropriate adjustments or modification of examinations, training materials, or policies, the provisions of qualified readers or interpreters and other similar accommodations for ... the QID (qualified individual with a disability).

The EEOC further defines "reasonable accommodation" as:

1. any modification or adjustment to a job application process that enables a qualified individual with a disability to be considered for the position such qualified individual with a disability desires, and which will not impose an undue hardship on the ... business; or
2. any modification or adjustment to the work environment, or to the manner or circumstances which the position held or desired is customarily performed, that enables the qualified individual with a disability to perform the essential functions of that position and which will not impose an undue hardship on the ... business; or
3. any modification or adjustment that enables the qualified individual with a disability to enjoy the same benefits and privileges of employment that other employees enjoy and does not impose an undue hardship on the ... business.

In essence, the covered employer is required to make "reasonable accommodations" for any/all known physical or mental limitations of the qualified individual with a disability, unless the employer can demonstrate that the accommodations would impose an "undue hardship" on the business or that the particular disability directly affects the safety and health of that individual or others. Included under this section is the prohibition against the use of qualification standards, employment tests, and other selection criteria that can be used to screen out individuals with disabilities, unless the employer can demonstrate that the procedure is directly related to the job function. In addition to the modifications to facilities, work schedules, equipment, and training programs, employers must initiate an "informal interactive (communication) process" with the qualified individual to promote voluntary disclosure of his or her specific limitations and restrictions to enable the employer to make appropriate accommodations that will compensate for the limitation.

Job restructuring, according to section 101(9)(B), means modifying a job so that a disabled individual can perform its essential functions. This does not mean, however, that the essential functions themselves must be modified. Examples of job restricting might include:

- eliminating nonessential elements of the job
- re-delegating assignments
- exchanging assignments with another employee
- redesigning procedures for task accomplishment
- modifying the means of communication that are used on the job

Section 101(10)(a) defines "undue hardship" as "an action requiring significant difficulty or expense," when considered in light of the following factors:

- the nature and cost of the accommodation
- the overall financial resources and workforce of the facility involved
- the overall financial resources, number of employees, and structure of the parent entity
- the type of operation, including the composition and function of the workforce, the administration, and the fiscal relationship between the entity and the parent

Section 102(c)(1) of the ADA prohibits discrimination through medical screening, employment inquiries, and similar scrutiny. Safety and loss prevention professionals should be aware that underlying this section was Congress's conclusion that information obtained from employment applications and interviews "was often used to exclude individuals with disabilities—particularly those with so-called hidden disabilities such as epilepsy, diabetes, emotional illness, heart disease and cancer—before their ability to perform the job was even evaluated."

Under section 102(c)(2), safety and loss prevention professionals should be aware that conducting preemployment physical examinations of applicants and asking prospective employees whether they are qualified individuals with disabilities are prohibited. Employers are further prohibited from inquiring as to the nature or severity of the disability, even if the disability is visible or obvious. Safety and loss prevention professionals should be aware that individuals may ask whether any candidates for transfer or promotion who have a known disability can perform the required tasks of the new position if the tasks are job related and consistent with business necessity. An employer is also permitted to inquire about the applicant's ability to perform the essential job functions prior to employment. The employer should use the written job description as evidence of the essential functions of the position.

Safety and loss prevention professionals may require medical examinations of employees only if the medical examination is specifically job related and is consistent with business necessity. Medical examinations are permitted only after the applicant with a disability has been offered the job position. The medical examination may be given before the applicant starts the particular job, and the job offer may be contingent upon the results of the medical examination if all employees are subject to the medical examinations and information obtained from the medical examination is maintained in separate, confidential medical files. Employers are permitted to conduct voluntary medical examinations for current employees as part of an ongoing medical health program, but again, the medical files must be maintained separately and in a confidential manner.

The ADA does not prohibit safety and loss prevention professionals from making inquiries or requiring medical or "fit for duty" examinations when there is a need to determine whether or not an employee is still able to perform the essential functions of the job, or where periodic physical examinations are required by medical standards or federal, state, or local law.

Of particular importance to safety and loss prevention professionals is the area of controlled substance testing. Under the ADA, the employer is permitted to test job applicants for alcohol and controlled substances prior to an offer of employment under section 104(d). The testing procedure for alcohol and illegal drug use is not considered a medical examination as defined under the ADA. Employers may additionally prohibit the use of alcohol and illegal drugs in the workplace and may require that employees not be under the influence while on the job. Employers are permitted to test current employees for alcohol and controlled substance use in the workplace to the limits permitted by current federal and state law. The ADA requires all employers to conform to the requirements of the Drug-Free Workplace Act of 1988. Thus, safety and loss prevention professionals should be aware that most existing preemployment and postemployment alcohol and controlled substance programs that are not part of the preemployment medical examination or ongoing medical screening program will be permitted in their current form.

Individual employees who choose to use alcohol and illegal drugs are afforded no protection under the ADA. However, employees who have successfully completed a supervised rehabilitation program and are no longer using or addicted are offered the protection of a qualified individual with a disability under the ADA.

Of importance to safety and loss prevention professionals with responsibilities in food-processing facilities, meatpacking plants, jail facilities, and other food-related functions is section 103(e)(1). This section was designed specifically for food-handling employees. The secretary of health and human services is required to develop and publish a list of infectious or communicable diseases that can be transmitted through the handling of food. If an employee possesses one or more of the listed diseases, and if the risk cannot be eliminated through reasonable accommodation by the employer, the employer may refuse to assign the employee to or remove the employee from a job involving food handling.

Title II of the ADA is designed to prohibit discrimination against disabled individuals by public entities. This title covers the provision of services, programs, activities, and employment by public entities. A public entity under Title II includes:

- a state or local government
- any department, agency, special purpose district, or other instrumentality of a state or local government
- the National Railroad Passenger Corporation (Amtrak), and any commuter authority as this term is defined in section 103(8) of the Rail Passenger Service Act

Title II of the ADA prohibits discrimination in the area of ground transportation, including buses, taxis, trains, and limousines. Air transportation is excluded from the ADA, but is covered under the Air Carrier Access Act. Covered organizations may be affected in the purchasing or leasing of new vehicles and in other areas, such as the transfer of disabled individuals to the hospital or other facilities. Title II requires covered public entities to make sure that new vehicles are accessible to and usable by qualified individuals, including individuals in wheelchairs. Thus, vehicles must be equipped with lifts, ramps, wheelchair space, and other modifications unless the covered public entity can justify that such equipment is unavailable despite a good faith effort to purchase or acquire this equipment. Covered organizations may want to consider alternative methods to accommodate the qualified individual, such as use of ambulance services or other alternatives.

Title III of the ADA builds upon the foundation establishing by the Architectural Barriers Act and the Rehabilitation Act. This title basically extended the discrimination prohibition to apply to all privately operated public accommodations. Title III focuses on the accommodations in public facilities, including such covered entities as retail stores, law offices, medical facilities, and other public areas. This section requires that goods, services, and facilities of any public place provide "the most integrated setting appropriate to the needs of the (qualified individual with a disability)" except where that individual may pose a direct threat to the safety and health of others that cannot be eliminated through modification of company procedures, practices, or policies. Prohibited discrimination under this section includes prejudice or bias against the individual with a disability in the "full and equal enjoyment" of these services and facilities.

The ADA makes it unlawful for public accommodations not to remove architectural and communication barriers from existing facilities or transportation barriers from vehicles "where such removal is readily achievable." This statutory language is defined as "easily accomplished and able to be carried out without much difficulty or expense," for example, moving shelves to widen an aisle, lowering shelves to permit access, etc. The ADA also requires that when a commercial facility or other public accommodation is undergoing a modification that affects the access to a primary function area, specific alterations must be made to afford accessibility to the qualified individual with a disability.

Title III also requires that "auxiliary aids and services" be provided for the qualified individual with a disability including, but not limited to, interpreters, readers, amplifiers, and other devices (not limited or specified under the ADA) to provide that individual with an equal opportunity for employment, promotion, etc. Congress did, however, provide that auxiliary aids and services do not need to be offered to customers, clients, and other members of the public if the auxiliary aid or service creates an undue hardship on the business. Safety and loss prevention professionals may want to consider alternative methods of accommodating the qualified individual with a disability. This section also addresses the modification of existing facilities to provide access to the individual and requires that all new facilities be readily accessible and usable by the individual.

Title IV requires all telephone companies to provide "telecommunications relay service" to aid hearing and speech impaired individuals. The Federal Communications Commission issued a regulation requiring the implementation of this requirement by

July 26, 1992, and established guidelines for compliance. This section also requires that all public service programs and announcements funded with federal monies be equipped with closed-captioning for the hearing impaired.

Title V assures that the ADA does not limit or invalidate other federal or state laws that provide equal or greater protection for the rights of individuals with disabilities. Some unique features of Title V are the miscellaneous provision and the requirement of compliance to the ADA by all members of Congress and all federal agencies. Additionally, Congress required that all state and local governments comply with the ADA and permitted the same remedies against the state and local governments as any other organizations.

Congress expressed its concern that sexual preferences could be perceived as a protected characteristic under the ADA or that the courts could expand the ADA's coverage beyond Congress's intent. Accordingly, Congress included section 511(b), which contains an expansive list of conditions that are not to be considered within the ADA's definition of disability. This list includes individuals such as transvestites, homosexuals, and bisexuals. Additionally, the conditions of transsexualism; pedophilia; exhibitionism; voyeurism; gender identity disorders not resulting from physical impairment; and other sexual behavior disorders are not considered as a qualified disability under the ADA. Compulsive gambling, kleptomania, pyromania, and psychoactive substance use disorders (from current illegal drug use) are also not afforded protection under the ADA.

Safety and loss prevention professionals should be aware that all individuals associated with or having a relationship to the qualified individual with a disability are extended protection under this section of the ADA. This inclusion is unlimited in nature, including family members, individuals living together, and an unspecified number of others. The ADA extends coverage to all "individuals," legal or illegal, documented or undocumented, living within the boundaries of the United States, regardless of their status. Under section 102(b)(4), unlawful discrimination includes "excluding or otherwise denying equal jobs or benefits to a qualified individual because of the known disability of the individual with whom the qualified individual is known to have a relationship or association." Therefore, the protections afforded under this section are not limited to only familial relationships. There appears to be no limits regarding the kinds of relationships or associations that are afforded protection. Of particular note is the inclusion of unmarried partners of persons with AIDS or other qualified disabilities.

As with the OSH Act, the ADA requires that employers post notices of the pertinent provisions of the ADA in an accessible format in a conspicuous location within the employer's facilities. A prudent safety and loss prevention professional may wish to provide additional notification on job applications and other pertinent documents.

Under the ADA, it is unlawful for an employer to "discriminate on the basis of disability against a qualified individual with a disability" in all areas, including the following examples:

- recruitment, advertising, and job application procedures
- hiring, upgrading, promoting, awarding tenure, demotion, transfer, layoff, termination, the right to return from layoff, and rehiring

- rate of pay or other forms of compensation and changes in compensation
- job assignments, job classifications, organization structures, position descriptions, lines of progression, and seniority lists
- leaves of absence, sick leave, or other leaves
- fringe benefits available by virtue of employment, whether or not administered by the employer
- selection and financial support for training, including apprenticeships, professional meetings, conferences and other related activities, and selection for leave of absence to pursue training
- activities sponsored by the employer, including social and recreational programs
- any other term, condition, or privilege of employment

The EEOC has also noted that it is "unlawful ... to participate in a contractual or other arrangement or relationship that has the effect of subjecting the covered entity's own qualified applicant or employee with a disability to discrimination." This prohibition includes referral agencies, labor unions (including collective bargaining agreements), insurance companies and others providing fringe benefits, and organizations providing training and apprenticeships.

Safety and loss prevention professionals should note that the ADA possesses no recordkeeping requirements, has no affirmative action requirements, and does not preclude or restrict antismoking policies. Additionally, the ADA possesses no retroactivity provisions.

The ADA has the same enforcement and remedy scheme as Title VII of the Civil Rights Act of 1964, as amended by the Civil Rights Act of 1991. Compensatory and punitive damages (with upper limits) have been added as remedies in cases of intentional discrimination, and there is also a correlative right to a jury trial. Unlike Title VII, there is an exception when there is a good faith effort at reasonable accommodation.

For now, the enforcement procedures adopted by the ADA mirror those of Title VII of the Civil Rights Act. A claimant under the ADA must file a claim with the EEOC within 180 days from the alleged discriminatory event or within 300 days in states with approved enforcement agencies such as the Human Rights Commission. These are commonly called dual agency states or Section 706 agencies. The EEOC has 180 days to investigate the allegation and sue the employer or to issue a right-to-sue notice to the employee. The employee will have 90 days to file a civil action from the date of this notice.

The original remedies provided under the ADA included reinstatement (with or without back pay) and reasonable attorney fees and costs. The ADA also provided protection from retaliation against the employee for filing the complaint or against others who might assist the employee in the investigation of the complaint. The ADA remedies are designed, as with the Civil Rights Act, to make the employee "whole" and to prevent future discrimination by the employer. All rights, remedies, and procedures of section 505 of the Rehabilitation Act of 1973 are also incorporated into the ADA. Enforcement of the ADA is also permitted by the attorney general or by private lawsuit. Remedies under these titles included the ordered modification of a facility and civil penalties of up to $50,000 for the first violation and $100,000 for any subsequent violations. Section 505 permits reasonable attorney fees and litigation costs for the prevailing party in an ADA action but, under section 513, Congress encourages the use of arbitration to resolve disputes arising under the ADA.

With the passage of the Civil Rights Act of 1991, the remedies provided under the ADA were modified. Employment discrimination (whether intentional or by practice) that has a discriminatory effect on qualified individuals may include hiring, reinstatement, promotion, back pay, front pay, reasonable accommodation, or other actions that will make an individual "whole." Payment of attorney fees, expert witness fees, and court fees are still permitted, and jury trials also allowed.

Compensatory and punitive damages were also made available if intentional discrimination is found. Damages may be available to compensate for actual monetary losses, future monetary losses, mental anguish, and inconvenience. Punitive damages are also available if an employer acted with malice or reckless indifference. The total amount of punitive and compensatory damages for future monetary loss and emotional injury for each individual is limited and is based upon the size of the employer. Punitive damages are not available against state or local governments.

In situations involving reasonable accommodation, compensatory or punitive damages may not be awarded if the employer can demonstrate that "good faith" efforts were made to accommodate the individual with a disability.

Safety and loss prevention professionals should be aware that the Internal Revenue Code may provide tax credits and/or tax deductions for expenditures incurred while achieving compliance with the ADA. Programs such as the Small Business Tax Credit and Targeted Jobs Tax Credit may be available upon request by the qualified employers. Additionally, expenses incurred while achieving compliance might be considered a deductible expense or capital expenditure (permitting depreciation over a number of years under the Internal Revenue Code).

TITLE I—EMPLOYMENT PROVISIONS

The two most common questions asked by safety and loss prevention professionals are whether or not their organizations must comply with the ADA and who is a protected individual under the ADA. These are vitally important questions that must be addressed by safety and loss prevention professionals in order to ascertain whether compliance is mandated and, if so, whether current employees, job applicants, and others who may directly affect the operation are within the protective scope of the ADA.

QUESTION 1: WHO MUST COMPLY WITH TITLE I OF THE ADA?

All private sector employers that affect commerce; state, local, and territorial governments; employment agencies; labor unions; and joint labor-management committees fall within the scope of a "covered entity" under the ADA. Additionally, Congress and its agencies are covered, but they are permitted to enforce the ADA through internal administrative procedures. The federal government, government-owned corporations, Indian tribes, and tax-exempt private membership clubs (other than labor organizations that are exempt under section 501(c) of the Internal Revenue Code) are excluded from coverage under the ADA.

Covered employers cannot discriminate against qualified applicants and employees on the basis of disability. Congress did provide a time period to enable employers to achieve compliance with Title I. Coverage for Title I was phased in by two steps, in order to allow additional time to smaller employers.

State and local governments, regardless of size, are covered by employment non-discrimination requirements under Title II of the ADA and must have complied by January 26, 1992. Certain individuals who were appointed by elected officials of state and local governments are covered by the same special enforcement procedures established for Congress.

Similar to the coverage requirements under Title VII of the Civil Rights Act of 1964, an "employer" is defined to include persons who are agents of the employer, such as safety and health managers, supervisors, personnel managers, and others who act on behalf of the employer. Therefore, the corporation or legal entity is responsible for the acts and omissions of their managerial employees and other agents who might violate the provisions of the ADA.

In calculating the number of employees for compliance purposes, employers should include part-time employees who have worked for them for 20 or more calendar weeks in the current or preceding calendar year. The definition of "employees" also includes U.S. citizens working outside of the U.S. for U.S.-based corporations. However, the ADA provides an exemption from coverage for any compliance action that would violate the law of a foreign country in which the actual workplace is located.

Employers should be aware that the ADA is structured to afford protection against discrimination to "individuals" rather than "citizens" or "Americans." There is no distinction made under the ADA between individuals with disabilities who are illegal or undocumented versus U.S. citizens. ADA protection does not require an individual to possess a permanent resident alien card (known as a "green card"). In addressing this issue, the judiciary committee stated, "[as] in other civil rights laws ... the ADA should not be interpreted to mean that only American citizens are entitled to the protection afforded by the Act."

It should be noted that religious organizations are covered by the ADA, but some religious organizations may demonstrate employment preference by employing individuals of their own religion or religious organizations.

After an employer has ascertained that his or her organization or company is a "covered" entity, the employer will ask which individuals are protected under Title I and how these protected individuals can be identified. These questions can be answered by asking the following questions:

- Who is protected by Title I?
- What constitutes a disability?
- Is the individual specifically excluded from protection under the ADA?

QUESTION 2: WHO IS PROTECTED BY TITLE I?

The ADA prohibits discrimination against "qualified individuals with disabilities" in such areas as job applications, hiring, testing, job assignments, evaluations, disciplinary actions, medical examinations, layoff/recall, discharge, compensation, leave, promotion, advancement, compensation, benefits, training, social activities, and other terms, conditions, and privileges of employment. A qualified individual with a disability is defined as

> an individual with a disability who meets the skill, experience, education, and other job-related requirements of a position held or desired, and who, with or without reasonable accommodation, can perform the essential functions of a job.

Additionally, unlawful discrimination under the ADA includes excluding or otherwise denying equal jobs or benefits to a qualified individual because of the known disability or an individual with whom the qualified individual is known to have a relationship or association.

This clause is designed to protect individuals who possess no disability themselves, but who may be discriminated against because of their association or relationship to a disabled person. The protection afforded under this clause is not limited to family members or relatives, but extends in an (apparently) unlimited fashion to all associations and relationships. However, in an employment setting, if an employee is hired and then violates the employer's attendance policy, the ADA will not protect the individual from appropriate disciplinary action. The employer owes no accommodation duty to an individual who is not disabled.

QUESTION 3: WHAT CONSTITUTES A DISABILITY?

Section 3(1) of the ADA provides a three-part definition to ascertain who is and is not afforded protection. A person with a disability is an individual who meets the following requirements:

- Test 1: has a physical or mental impairment that substantially limits one or more of his or her major life activities
- Test 2: has a record of such an impairment
- Test 3: is regarded as having such an impairment

This definition is comparable to the definition of "handicap" under the Rehabilitation Act of 1973. Congress adopted this terminology in an attempt to use the most current acceptable terminology, but also intended that the relevant case law developed under the Rehabilitation Act be applicable to the definition of "disability" under the ADA. It should be noted, however, that the definition and regulations applying to "disability" under the ADA are more favorable to the disabled individual than the "handicap" regulations under the Rehabilitation Act.

The first prong of this definition includes three major subdivisions that further define who is a protected individual under the ADA. These are: (1) a physical or mental impairment (2) that substantially limits (3) one or more of his or her major life activities. These provide additional clarification regarding the definition of a "disability" under the ADA.

A Physical or Mental Impairment

The ADA does not specifically list all covered entities. Congress noted that

> it is not possible to include in the legislation a list of all the specific conditions, diseases or infections that would constitute physical or mental impairments because of the difficulty in ensuring the comprehensiveness of such a list, particularly in light of the fact that new disorders may develop in the future.

A "physical impairment" is defined by the ADA as:

> any physiological disorder, or condition, cosmetic disfigurement, or anatomical loss affecting one or more of the following body systems: neurological, musculoskeletal, special sense organs, respiratory (including speech organs), cardiovascular, reproductive, digestive, genital-urinary, hemic and lymphatic, skin, and endocrine.

A "mental impairment" is defined by the ADA as:

> any mental or psychological disorder, such as mental retardation, organic brain syndrome, emotional or mental illness, and specific learning disabilities.

A person's impairment is determined without regard to any medication or assisting devices that the individual may use. For example, an individual with epilepsy who uses medication to control the seizures, or a person with an artificial leg would be considered to have an impairment even if the medicine or prosthesis reduced the impact of the impairment.

The legislative history is clear that an individual with AIDS or HIV is protected by the ADA. A contagious disease, such as tuberculosis, would also constitute an impairment; however, the employer would not have to hire or retain a person with a contagious disease that poses a direct threat to the health and safety of others. This is discussed in detail later in this section.

The physiological or mental impairment must be permanent in nature. Pregnancy is considered temporary and thus is not afforded protection under the ADA, but it is protected under other federal laws. Simple physical characteristics, such as hair color, left handedness, height, or weight (within the normal range), are not considered impairments. Predisposition to a certain disease is also not an impairment within this definition. Environmental, cultural, or economic disadvantages, such as lack of education or possession of prison records, are not impairments. Similarly, personality traits, such as poor judgment, quick temper, or irresponsible behavior, are not impairments. Conditions, such as stress and depression, may or may not be considered an impairment, depending on whether or not the condition results from a documented physiological or mental disorder.

Case law under the Rehabilitation Act, applying similar language as the ADA, has identified the following as some of the protected conditions: blindness, diabetes, cerebral palsy, learning disabilities, epilepsy, deafness, cancer, multiple sclerosis, allergies, heart conditions, high blood pressure, loss of leg, cystic fibrosis, hepatitis B, osteoarthritis, and numerous other conditions.

Substantially Limits

Congress clearly intended that the term disability be construed broadly. An impairment is only a "disability" under the ADA if it "substantially limits" one or more major life functions. An individual must be unable to perform or be significantly limited in performance of a basic activity that can be performed by an average person in America.

To assist in this evaluation, a three-factor test was provided to determine whether or not an individual's impairment substantially limits a major life activity:

- the nature and severity of the impairment
- how long the impairment will last or is expected to last
- the permanent and long-term impact, or expected impact of the impairment

The determination of whether or not an individual is substantially limited in a major life activity must be made on a case-by-case basis. The three-factor test is helpful because it is not the name of the impairment or condition that determines whether or not an individual is protected, but rather the effect of the impairment or condition on the life of that person. While some impairments, such as blindness, AIDS, and deafness, are substantially limiting by nature, other impairments may be disabling for some individuals and not for others, depending on the type of the impairment and the particular activity in question.

Individuals with two or more impairments, neither of which by itself substantially limits a major life activity, may be combined together to impair one or more major life activities. Temporary conditions, such as a broken leg, common cold, and sprains/strains, are generally not protected because of the extent, duration, and impact of the impairment. However, such temporary conditions may evolve into a permanent condition that substantially limits a major life function if complications arise.

In general, it is not necessary to determine whether an individual is substantially limited in a work activity if the individual is limited in one or more major life activities. An individual is not considered to be substantially limited in working if he or she is substantially limited in performing only a particular job or unable to perform a specialized job in a particular area. An individual may be considered substantially limited in working if the individual is restricted in his or her ability to perform either a class of jobs or a broad range of jobs in various classes when compared to an average person of similar training, skills, and abilities. The following factors should considered:

- the type of job from which the individual has been disqualified because of his or her impairment
- the geographic area in which the person may reasonably expect to find a job
- the number and types of jobs using similar training, knowledge, skill, or abilities from which the individual is disqualified within the geographic area

- the number and types of other jobs in the area that do not involve similar training, knowledge, skill, or abilities from which the individual is also disqualified because of his or her impairment.

In evaluating the number of jobs from which an individual might be excluded, the EEOC regulations note that it is only necessary to show the approximate number of jobs from which the individual would be excluded.

Major Life Activities

An impairment must substantially limit one or more major life activities to be considered a "disability" under the ADA. A major life activity is an activity that an average person can perform with little or no difficulty. Examples include the following basic activities:

- walking
- speaking
- breathing
- performing manual tasks
- standing
- lifting
- seeing
- hearing
- learning
- caring for oneself
- working
- reading
- sitting

The above list of examples is not all-inclusive. All situations should be evaluated on a case-by-case basis.

The second test of this definition of disability requires that an individual possess a record of having an impairment as specified in Test 1. Under this test, the ADA protects individuals who possess a history of, or who have been misclassified as possessing, a mental or physical impairment that substantially limits one or more major life functions. A record of impairment would include such documented items as educational, medical, or employment records. Safety and loss prevention professionals should note that merely possessing a record of being a "disabled veteran" or record of disability under another federal or state program does not automatically qualify the individual for protection under the ADA. The individual must meet the definition of "disability" under Test 1 and possess a record of such disability under Test 2.

The third test of the definition of "disability" includes an individual who is regarded or treated as having a covered disability even though the individual does not possess a disability as defined under Tests 1 and 2. This part of the definition protects individuals who do not possess a disability that substantially limits a major life activity from the discriminatory actions of others because of their perceived disability. This protection is necessary because "society's myths and fears about disability and disease are as handicapping as are the physical limitations that flow from actual impairment."

Three circumstances in which protection would be provided to the individual include the following situations:

1. when the individual possesses an impairment that is not substantially limiting, but the individual is treated by the employer as having such an impairment;
2. when an individual has an impairment that is substantially limiting because of the attitude of others toward the condition; and
3. when the individual possesses no impairment but is regarded by the employer as having a substantially limiting impairment.

To acquire the protection afforded under the ADA, an individual must not only be an individual with a disability, but also must qualify under the previously noted tests. A "qualified individual with a disability" is defined as a person with a disability who:

> satisfies the requisite skills, experience, education and other job-related requirements of the employment position [that] such individual holds or desires, and who, with or without reasonable accommodation, can perform the essential functions of such position.

Safety and loss prevention professionals should be aware that the employer is not required to hire or retain an individual who is not qualified to perform a particular job.

QUESTION 4: IS THE INDIVIDUAL SPECIFICALLY EXCLUDED FROM PROTECTION UNDER THE ADA?

The ADA has a specific provision that excludes certain individuals from its protection. As set forth under sections 511 and 512(a), (b), the following individuals are not protected:

- Individuals who are currently engaged in the use of illegal drugs are not protected when an employer takes action due directly to their continued use of illegal drugs. This includes the use of illegal prescription drugs as well as illegal drugs. (However, individuals who have undergone a qualified rehabilitation program and are not currently using drugs illegally are afforded protection under the ADA.)
- Homosexuality and bisexuality are not impairments and are therefore not considered disabilities under the ADA.
- The ADA does not consider transvestism, transsexualism, pedophilia, exhibitionism, voyeurism, gender identity disorders (not resulting from physical impairment), and other sexual behavior disorders as disabilities and thus are not afforded protection.

- Other areas not afforded protection include compulsive gambling, klepto-mania, pyromania, and psychoactive substance use disorders resulting from illegal drug use.

A major component of Title I is the "reasonable accommodation" mandate, which requires employers to provide a disabled employee or applicant with the necessary "reasonable accommodations" that would allow the disabled individual to perform the essential functions of a particular job. Safety and loss prevention professionals should note that "reasonable accommodation" is a key nondiscrimination requirement in order to permit individuals with disabilities to overcome unnecessary barriers that could prevent or restrict employment opportunities.

The EEOC regulations define "reasonable accommodation" as meaning

1. any modification or adjustment to a job application process that enables a qualified individual with a disability to be considered for the position such qualified individual with a disability desires, and which will not impose an undue hardship on the … business, or
2. any modification or adjustment to the work environment, or to the manner or circumstances which the position held or desired is customarily performed, that enables the qualified individual with a disability to perform the essential functions of that position and which will not impose an undue hardship on the … business, or
3. any modification or adjustment that enables the qualified individual with a disability to enjoy the same benefits and privileges of employment that other employees enjoy and does not impose an undue hardship on the … business.

Section 101(9) of the ADA states that reasonable accommodation includes two components. First, there is the accessibility component. This sets forth an affirmative duty for the employer to make physical changes in the workplace so that the facility is readily accessible and usable by individuals with disabilities. This component "includes both those areas that must be accessible for the employee to perform the essential job functions, as well as non-work areas used by the employer's employees for other purposes."

The second component is the modification of other related areas. The EEOC regulations set forth a number of examples of modification that an employer must consider

- job restructuring;
- part-time or modified work schedules;
- reassignment to vacant position;
- appropriate adjustment or modification of examinations and training materials;
- acquisition or modification of equipment or devices;
- the provision of qualified readers or interpreters.

Safety and loss prevention professionals should note that the employer possesses no duty to make an accommodation for an individual who is not otherwise qualified for a position. In most circumstances, it is the obligation of the individual with a disability to request a reasonable accommodation from the employer. The individual with a disability possesses the right to refuse an accommodation, but if the individual with a disability cannot perform the essential functions of the job without the accommodation, that individual may not be qualified for the job.

An employer is not required to make a reasonable accommodation that would impose an undue hardship on the business. An undue hardship is defined as an action that would require "significant difficulty or expense" in relation to the size of the employer, the employer's resources, and the nature of its operations. Although the undue hardship limitations will be analyzed on a case-by-case basis, several factors have been set forth to determine whether or not an accommodation would impose an undue hardship. First, the undue hardship limitation states that the accommodation should not be unduly costly, be extensive or substantial in nature, be disruptive to the operation, or fundamentally alter the nature or operation of the business. Additionally, the ADA provides four factors to be considered in determining whether or not an accommodation would impose an undue hardship on a particular operation:

1. the nature and the cost of the accommodation needed;
2. the overall financial resources of the facility (or facilities) making the accommodation, the number of employees in the facility, and the effect on expenses and resources of the facility;
3. the overall financial resources, size, number of employees, and type and location of facilities of the entity covered by the ADA; and
4. the type of operation of the covered entity, including the structure and functions of the workforce, the geographic separateness, and the administrative or fiscal relationship of the facility involved in making the accommodation to the larger entity.

Other factors, such as the availability of tax credits and tax deductions, the type of enterprise, etc., can also be considered when evaluating an accommodation situation for the undue hardship limitation. Safety and loss prevention professionals should note that the requirements to prove undue hardship are substantial in nature and cannot easily be utilized to circumvent the purposes of the ADA.

The ADA prohibits the use of preemployment medical examinations, medical inquiries, and requests for information regarding workers' compensation claims prior to an offer of employment. An employer, however, may condition a job offer (i.e., conditional or contingent job offer) upon the satisfactory results of a post-offer medical examination if the medical examination is required of all applicants or employees in the same job classification. Questions regarding other injuries and workers' compensation claims may also be asked following the offer of employment. A post-offer medical examination cannot be used to disqualify an individual with a disability who is currently able to perform the essential functions of a particular job because of speculation that the disability may cause future injury or workers' compensation claims.

Safety and loss prevention professionals should note that if an individual is not employed because his or her medical examination revealed a disability, the reason for not hiring the qualified individual with a disability must be business-related and necessary for the particular business. The burden of proving that a reasonable accommodation would not have enabled the individual with a disability to perform the essential functions of the particular job or that the accommodation was unduly burdensome falls squarely on the employer.

As often revealed in the post-offer medical examination, the physician should be informed that the employer possesses the burden of providing that a qualified individual with a disability should be excluded because of the risk to the health and safety of other employees or individuals. To address this issue, Congress specifically noted that the employer may possess a job requirement that specifies that "an individual not impose a direct threat to the health and safety of other individuals in the workplace." A "direct threat" has been defined as "a significant risk to the health and safety of others that cannot be eliminated or reduced by reasonable accommodation."

The burden of proving this requirement falls upon the employer (and, thus, often the safety and loss prevention professional). The safety and loss prevention professional must prove that a significant risk to the safety of others or property exists, not a speculative or remote risk, and that no reasonable accommodation is available that can remove that risk. The safety and loss prevention professional should identify the specific aspect of the disability that is causing the direct threat and if an individual poses a direct threat as a result of a disability, the safety and loss prevention professional should determine whether a reasonable accommodation would either eliminate the risk or reduce it to an acceptable level. If no accommodation exists that would either eliminate the risk or reduce the risk, the safety and loss prevention professional may recommend that the employer refuse to hire an applicant or may discharge the employee who poses a direct threat.

In most circumstances, the safety and loss prevention professional will work closely with the attending physician to ascertain the medical condition of the disabled individual. The physician's evaluation of any future risk must be supported by valid medical analysis and be based on the most current medical knowledge and/or best available objective evidence about the individual. The safety and loss prevention professional should not rely only on the attending physician's opinion, but on the best available objective evidence to support any decision in this area. Other areas of expertise that may be called upon include physicians who specialize in a particular disability, advice of rehabilitation counselors, experience of an individual with a disability in previous jobs or other activities, or the various disability organizations. The safety and loss prevention professional is encouraged to discuss possible accommodations with the individual with a disability.

Safety and loss prevention professionals should be aware that the direct threat evaluation is vitally important in evaluating disabilities that involve contagious diseases. The leading case in this area is *School Board of Nassau County v. Arline*. This case sets forth the test to be used in evaluating a direct threat to others:

- the nature of the risk
- the duration of the risk

- the severity of the risk
- the probability that the disease will be transmitted and will cause varying degrees of harm

For safety and loss prevention professionals working in the food industry, Congress specifically addressed the issue of infectious or communicable diseases in the ADA. If an individual has a disease that is on the list published by the secretary of health and human services, the employer may refuse to assign the individual to a job involving food handling if no other reasonable accommodation exists that would avoid the risks of contamination.

The ADA imposes a very strict limitation on the use of information acquired through post-offer medical examinations or inquiries. All medical-related information must be collected and maintained on separate forms and be kept in separate files. These files must be maintained in a confidential manner in which only designated individuals are provided access. Medical-related information may be shared with appropriate first-aid and safety personnel when applicable (in an emergency situation). Supervisors and other managerial personnel may be informed about necessary job restrictions or job accommodations. Appropriate insurance organizations may acquire access to medical records when they are required for health or life insurance. State and federal officials may acquire access to medical records for compliance and other purposes.

In the area of insurance, the ADA specifies that nothing within the Act should be construed to prohibit or restrict "an insurer, hospital, or medical service company, health maintenance organization, or any agent, or entity that administers benefit plans, or similar organization from underwriting risks, or administering such risks that are based on or not inconsistent with State laws." However, an employer may not classify or segregate an individual with a disability in a manner that adversely affects not only the individual's employment but also any provisions or administration of health insurance, life insurance, pension plans, or other benefits. In essence, this means that if an employer provides insurance or benefits to all employees, the employer must also provide the same coverage to the individual with a disability. An employer cannot deny insurance or subject the individual with a disability to different terms or conditions based upon the disability alone, if the disability does not pose an increased insurance risk. An employer cannot terminate or refuse to hire an individual with a disability because the individual's disability or a family member or dependent's disability is not covered under its current policy or because the individual poses a future risk of increased health costs. The ADA does not, however, prohibit the use of preexisting condition clauses in insurance policies.

An employer is prohibited from shifting away the responsibilities and potential liabilities under the ADA through contractual or other arrangements. An employer may not do anything through a contractual relationship that it cannot do directly. This provision applies to all contractual relationships that include insurance companies, employment and referral agencies, training organizations, agencies used for background checks, and labor unions.

Labor unions are covered by the ADA and have the same responsibilities as any other covered employer. Employers are prohibited from taking any action through a collective bargaining agreement (i.e., union contract) that it may not take directly by itself. A collective bargaining agreement may be used as evidence in a decision regarding undue hardship or in identifying the essential elements in a job description.

Although not required under the ADA, a written job description describing the essential elements of a particular job is the first line of defense for most ADA-related claims. A written job description that is prepared before advertising or interviewing applicants for a job will be considered as evidence of the essential elements of the job along with other relevant factors.

In order to identify the essential elements of a particular job and whether or not an individual with a disability is qualified to perform the job, the EEOC regulations have set forth three key factors, among others, which must be considered:

- the reason the position exists
- the limited number of employees available to perform the function, or among whom the function can be distributed
- if the task function is highly specialized, and the person in the position is hired for special expertise or ability to perform the job

A substantial number of safety and loss prevention professionals have developed job safety analysis (JSA) or job hazard analysis (JHA) procedures. In essence, a job analysis for ADA purposes is usually performed in the development of a written job description. Job analysis is not required by the ADA, but is highly recommended in order to appropriately evaluate each job. A job analysis can be a formal process through which the essential job functions are identified and possible reasonable accommodations explored. The focus of a job analysis should be on the purpose of the job and the importance of the actual job function in achieving that purpose.

TITLE II—PUBLIC SERVICES

Title II is designed to prohibit discrimination against disabled individuals by public entities. Title II covers all services, programs, activities, and employment by government or government units. Title II adopted all of the rights, remedies, and procedures provided under section 505 of the Rehabilitation Act of 1973, and the undue financial burden exception is also applicable. The effective date for Title II was January 26, 1992.

The public entities that Title II applies to include state or local government, any department, agency, special purpose district or other instrumentality of a state or local government, the National Railroad Passenger Corporation (Amtrak), and any commuter authority as defined in the Rail Passenger Service Act.

Title II possesses two basic purposes: to extend the prohibition against discrimination under the Rehabilitation Act of 1973 to state and local governments and to clarify section 504 of the Rehabilitation Act for public transportation entities that receive federal assistance. Given these purposes, the main emphasis of Title II is directed at public sector organizations and possesses minimal impact on private sector organizations.

The vast majority of Title II's provisions cover the transportation that is provided by public entities to the general public, such as buses and trains. The major requirement under Title II mandates that public entities who purchase or lease new buses, rail cars, taxis, or other vehicles must make certain that these vehicles are accessible to and usable by qualified individuals with disabilities. This accessibility requirement includes disabled individuals who may use wheelchairs and requires that all vehicles be equipped with lifts, ramps, wheelchair spaces, or other special accommodations unless the public entity can prove that such equipment is unavailable despite a good faith effort to locate it.

Many public entities purchase used vehicles or lease vehicles because of the substantial cost of such vehicles. The public entity must make a good faith effort to obtain vehicles that are readily accessible and usable by individuals with disabilities. As provided under the ADA, it is considered discrimination to remanufacture vehicles to extend their useful life for five years or more without making the vehicle accessible and usable by individuals with disabilities. Historic vehicles, such as the trolley cars, may be excluded if the modifications alter the historical character of the vehicle.

Of particular importance to police, fire, and other emergency organizations is Title II's impact on 911 systems. Congress observed that many 911 telephone numbering systems were not directly accessible to hearing impaired and speech impaired individuals. Congress cited an example of a deaf woman who died of a heart attack because the police organization did not respond when her husband tried to use his telephone communication device for the deaf (TDD) to call 911. In response to such examples, Congress stated, "As part of its prohibition against discrimination in local and state programs and services, Title II will require local governments to ensure that these telephone emergency number systems (911) are equipped with technology that will give hearing impaired and speech impaired individuals a direct line to these emergency services." Thus, public safety organizations must ensure compliance with this requirement.

Of importance for state governments is the fact that section 502 eliminates immunity of a state in state or federal court under the Eleventh Amendment for violations of the ADA. A state can be found liable in the same manner and is subject to the same remedies (including attorney fees) as private sector covered organizations.

Additionally, the claims procedures for instituting a complaint against a state or local government are significantly different from instituting a complaint against a private covered entity. The ADA provides that a claim can be filed with any of seven federal government agencies, including the EEOC and the Justice Department, or the EEOC may assist in such litigation. A procedure for instituting complaints against a public organization without going to court is provided in the Justice Department's regulations. The statute of limitations on filing such a claim with the designated federal agency is 180 days from the date of the act of discrimination, unless the agency extends the time limitation for good cause. If the responsible agency finds a violation, the violation will be corrected through voluntary compliance, negotiations, or intervention by the attorney general.

This procedure is totally voluntary. An individual may file suit in court without filing an administrative complaint, or an individual may file suit at any time while an administrative complaint is pending. No exhaustion of remedies is required.

Under the Department of Justice's regulations, public entities with 50 or more employees are required to designate at least one employee to coordinate efforts to comply with Title II. The public entity must also adopt grievance procedures and designate at least one employee who will be responsible for investigating any complaint filed under this grievance procedure.

TITLE III—PUBLIC ACCOMMODATIONS

Title III builds upon the foundation established by Congress under the Architectural Barriers Act and the Rehabilitation Act. Title III basically extends the prohibition of discrimination that has existed for facilities constructed by or financed by the federal government to all private sector public facilities. Title III requires that all goods, services, privileges, advantages, or facilities of any public place be offered "in the most integrated setting appropriate to the needs of the [disabled] individual," except when the individual poses a direct threat to the safety or health of others. Title III additionally prohibits discrimination against individuals with disabilities in the "full and equal enjoyment" of all goods, services, facilities, etc.

Title III covers public transportation offered by private sector entities, in addition to all places of public accommodation, without regard to size. Congress wanted small businesses to have ample time to comply with this mandatory change without fear of civil action. To achieve this, Congress provided that no civil action could be brought against businesses that employ 25 or fewer employees and have annual gross receipts of $1 million or less between January 26, 1992, and July 26, 1992. Additionally, businesses with fewer than 10 employees and having gross annual receipts of $500,000 or less were provided a grace period from January 26, 1992 to January 26, 1993 to achieve compliance. Residential accommodations, religious organizations, and private clubs were made exempt from these requirements.

Title III provides categories and examples of places requiring public accommodations:

- places of lodging such as inns, hotels, and motels, except for those establishments located in the proprietor's residence and not more than five rooms are for rent
- restaurants, bars, or other establishments serving food or drink
- motion picture houses, theaters, concert halls, stadiums, or other places of exhibition or entertainment
- bakeries, grocery stores, clothing stores, hardware stores, shopping centers, or other sales or retail establishments
- laundromats, dry cleaners, banks, barber shops, beauty shops, travel services, funeral parlors, gas stations, offices of accountants or lawyers, pharmacies, insurance offices, professional offices of healthcare providers, hospitals, or other service establishments
- terminal depots, or other stations used for specified public transportation
- parks, zoos, amusement parks, or other places of entertainment
- nursery, elementary, secondary, undergraduate, or post-graduate private schools, or other places of education

- day care centers, senior citizen centers, homeless shelters, food banks, adoption agencies, or other social service center establishments
- gymnasiums, health spas, bowling alleys, golf courses, or other places of exercise or recreation

Safety and loss prevention professionals should note that it is considered discriminatory under Title III for a covered entity to fail to remove structural, architectural, and communication barriers from existing facilities when the removal is "readily achievable," easily accomplished, and can be performed with little difficulty or expense. Some factors to be considered include the nature and cost of the modification, the size and type of the business, and the financial resources of the business. If the removal of a barrier is not "readily achievable," the covered entity may make goods and services readily available and achievable to individuals with disabilities through alternative methods.

Safety and loss prevention professionals should be aware that employers may not use application or other eligibility criteria that can be used to screen out individuals with disabilities, unless they can prove that doing so is necessary to providing goods or services to the public. Title III additionally makes discriminatory the failure to make reasonable accommodations in policies, business practices, and other procedures that afford access to and the use of public accommodations to individuals with disabilities. Employers who deny access to goods and services because of the absence of "auxiliary aids" (unless the providing of such auxiliary aids would fundamentally alter the nature of the goods or services or would impose an undue hardship) are also discriminatory. The ADA defines "auxiliary aids and services" as

a. qualified interpreters or other effective methods of making aurally delivered materials available to individuals with hearing impairments
b. qualified readers, taped texts, or other effective methods of making visually delivered materials available to individuals with visual impairments
c. acquisition or modification of equipment or devices
d. other similar services or actions

Title III does not specify the type of auxiliary aids that must be provided, but requires that individuals with disabilities be provided equal opportunity to obtain the same result as individuals without disabilities.

Title III provides an obligation to provide equal access, requires the modification of policies and procedures to remove discriminatory effects, and provides an obligation to provide auxiliary aids, in addition to other requirements. The safety and health exception and undue burden exception are available under Title III, in addition to the "structurally impracticable" and possibly the "disproportionate cost" defenses for covered organizations.

TITLE IV—TELECOMMUNICATIONS

Title IV amends Title II of the Communication Act of 1934 mandate that telephone companies provide "telecommunication relay services in their service areas by July 26, 1993." Telecommunication relay services provide individuals with speech related disabilities the ability to communicate with hearing individuals through the use of tele-communication devices like the TDD systems or other nonvoice transmission devices.

The purpose of Title IV is in large measure "to establish a seamless interstate and intrastate relay system for the use of TDDs (telecommunication devices for the deaf) that will allow a communication-impaired caller to communicate with anyone who has a telephone, anywhere in the country." Title IV contains provisions affording the disabled access to telephone and telecommunication services equal to those that the nondisabled community enjoys. In actuality, Title IV is not a new regulation, but simply an effort to ensure that the general mandates of the Communication Act of 1934 are made effective. Title IV consists of two sections. Section 401 adds a new section (section 225) to the Communication Act of 1934, and section 402 amends section 711 of the Communications Act.

Regulations governing the implementation of Title IV were specifically issued by the Federal Communication Commission (FCC). These regulations establish the minimum standards, guidelines, and other requirements mandated under Title IV, in addition to establishing regulations requiring around-the-clock relay service oper-ations, operator-maintained confidentiality of all messages, and rates for the use of the telecommunication relay systems that are equivalent to current voice communi-cation services. Title IV additionally prohibits the use of relay systems under certain circumstances, encourages the use of state-of-the-art technology where feasible, and requires that public service announcements and other television programs that are partially or fully funded by the federal government contain closed-captioning.

TITLE V—MISCELLANEOUS PROVISIONS

Title V contains many provisions that address a wide assortment of related coverages under the ADA. First, Title V permits insurance providers to continue to underwrite insurance, to continue to use the preexisting condition clauses, and to classify risks, as long as they are consistent with state-enacted laws. Title V also permits insurance carriers to provide bona fide benefit plans based upon risk classifications, but prohib-its denial of health insurance coverage to an individual with a disability based solely on that person's disability.

First, Title V does not require special treatment in the area of health insurance or other insurance for individuals with disabilities. An employer is permitted to offer insurance policies that limit coverage for a certain procedure or treatment, even though this might have an adverse impact on the individual with a disability. The employer or insurance provider may not, however, establish benefit plans in order to evade the purposes of the ADA.

Second, Title V provides that the ADA will not limit or invalidate other federal or state laws that provide equal or greater protection to individuals with disabilities. Additionally, the ADA does not preempt medical or safety standards established by federal law or regulation, nor does it preempt state, county, or local public health laws. However, state and local governments and their agencies are subject to the provisions of the ADA, and courts may provide the same remedies (except punitive damages at this time) against the state or local governments as any other public or private covered entity.

In an effort to minimize litigation under the ADA, Title V promotes the use of alternative dispute resolution procedures to resolve conflicts under the ADA. As it states in section 514, "Where appropriate and to the extent authorized by law, the use of alternative means of dispute resolution, including settlement negotiations, conciliation, facilitation, mediation, factfinding, minitrials, and arbitration, is encouraged to resolve disputes under this Act." Safety and health professionals should note, however, that the use of alternative dispute resolution is voluntary and, if used, the same remedies must be available as provided under the ADA.

THE INJURED WORKER AND ADA

Safety and loss prevention professionals should be aware that many employees injured on the job may be afforded protection under the ADA if the injury or illness meets the definition of a "qualified individual with a disability." The fact that any employee is awarded workers' compensation benefits does not automatically establish that the person is protected under the ADA. However, if the injured employee possesses a permanent mental or physical disability that substantially affects one or more life functions, or is perceived as possessing a disability that affects a major life function, the employee may be afforded protection under the ADA in addition to receiving workers' compensation coverage.

The ADA allows an employer to take reasonable steps to avoid increased workers' compensation liability, while protecting persons with disabilities against exclusion from jobs that they can safely perform.

After making a conditional offer of employment, an employer may inquire about a person's workers' compensation history in a medical inquiry that is required of all applicants in the same job category. However, an employer may not require an applicant to have a medical examination because of a response to a medical inquiry (as opposed to results from a medical examination) that discloses a previous on-the-job injury, unless all applicants in the same job category are required to take the physical examination.

Safety and loss prevention professionals may use information from medical inquiries and examinations to perform several functions:

- Verify employment history.
- Screen out applicants with a history of fraudulent workers' compensation claims.
- Provide information to state officials as required by state law regulating workers' compensation and "second injury" funds.
- Screen out individuals who would pose a "direct threat" to the health or safety of themselves or others, which could not be reduced to an acceptable level or eliminated by reasonable accommodation.

An employer may not base an employment decision on the speculation that an applicant may cause increased workers' compensation costs in the future. However, an employer may refuse to hire or may discharge an individual who is not currently able to perform the job without posing a significant risk of substantial harm to the health or safety of himself, herself, or others, and the risk cannot be eliminated or reduced by reasonable accommodation.

Safety and loss prevention professionals should be aware that most injured employees who have received some percentage of disability rating may be able to qualify for protection under the ADA. Employees who return to work after a work-related injury or illness with some percentage of permanency may require an accommodation to be able to perform their jobs. The ADA possesses no requirements that an employer establish a "light duty" or "restricted duty" program. Employers should be prepared to address accommodation requests by these employees.

For applicants, the employer must make the conditional offer of employment before conducting any medical examinations or inquiring about the applicant's workers' compensation history. A prudent employer may wish to document the conditional offer of employment in order to safeguard against any questions of impropriety in this area.

Safety and loss prevention professionals should be aware that filing a workers' compensation claim does not bar an individual with a disability from filing a charge under the ADA. The "exclusivity" clause in most state workers' compensation law bars all other civil remedies related to the injury or illness that have been compensated by the workers' compensation system, but does not prohibit a qualified individual with a disability from filing a discrimination charge with the EEOC or filing an action under the ADA, if issued a "right-to-sue" letter by the EEOC.

In the area of health insurance, life insurance, and other benefit plans, the ADA does not limit insurers, hospitals, medical service companies, health maintenance organizations, or any agent or entity that administers such plans from underwriting risks, classifying risks, or administering such risks so long as these assessments are based on, and not inconsistent with, the individual state laws. This provision is a limited exemption that is only applicable to those organizations that sponsor, observe, or administer benefit plans. This exemption does not apply to organizations that establish, sponsor, observe, or administer plans not involving benefits, such as liability insurance programs.

Additionally, the safety and loss prevention professional should recognize that the ADA is not a regulation that affects only the personnel or human resource function, but definitely will require his/her expertise in safety and health in several strategic locations in order to achieve and maintain compliance. A prudent safety and loss prevention professional should become thoroughly knowledgeable in the requirements of the ADA and work to ensure compliance throughout his or her company or organization.

PROGRAM DEVELOPMENT

Given that many safety and loss prevention professionals have responsibilities in the development and management of the organizations' ADA compliance program, a general outline of the 25 key areas to be considered when developing a plan of action follows.

1. Acquire and read the entire Americans with Disabilities Act. Also review the rules and interpretations provided by the EEOC, the Department of Labor, and the Office of Federal Contract Compliance Programs (OFCCP).

Acquire the health and human services list of communicable diseases. Keep abreast with government publications and case law as published. It may be prudent to have the organization's counsel or designated agency representative review and identify pertinent issues.

2. Educate and prepare the organization's hierarchy. Explain in detail the requirements of the ADA and the limits of the "undue hardship" and "safety or health" exceptions. Ensure complete and total understanding regarding the ADA. Communicate the philosophy and express the organization's commitment to the achievement of the goals and objectives of the ADA.

3. Acquire necessary funding to make the necessary accommodations and acquire the auxiliary aids. If necessary, search for outside agency funding and assistance. Review possible tax incentives that are available with the appropriate department or agency.

4. Designate an individual(s) (either employee or consultant) who is well versed on the requirements of the ADA or establish an advisor group to serve as the ADA "expert" within the organization.

5. Establish a relationship with organizations serving individuals with disabilities for recruiting, advice, or other purposes.

6. Analyze operations and identify applicable areas, practices, policies, procedures, etc., requiring modification. Remember to include all public areas, parking lots, access-ways, and all equipment. Document this analysis in detail.

7. Develop a written plan of action for each required area under the ADA. Set completion dates for each phase of the compliance plan in accordance with the mandated target date.

8. Develop and publish a written organizational policy incorporating all of the provisions of the ADA.

9. Review and renegotiate employment agency contracts, referral contracts, and other applicable contractual arrangements. Document any and all modifications.

10. Acquire the posting information and make certain this document is appropriately posted within the facility. Develop and place notices of ADA compliance on all applications, medical reports, and other appropriate documents.

11. Develop an employee and applicant self-identification program and communication system to permit employees and applicants to identify themselves as qualified individuals with disabilities and to communicate the limitations of their disability. This program should be in writing and available for review by employees.

12. Implement the plan of action.

13. Review and modify selection policies and procedures including, but not limited to, the following items:
   - interview procedures
   - selection criteria
   - physical and psychological testing procedures
   - alcohol and controlled substance testing programs

- application forms
- medical forms
- file procedures
- disability and retirement plans
- medical examination policies and procedures
- physical and agility testing
- other applicable policies and procedures. (Develop procedures to ensure confidentiality of medical records. All new procedures or modifications should be documented.)

14. Review all current job descriptions and identify the essential functions for each position. Develop a written job description for each and every job in the organization. Remember, the written job descriptions are the evidence of the essential functions of the jobs and will be the first line of defense.

15. Review and modify the personnel and medical procedures and policies. Maintain all medical files separately and confidentially. Address the option of having a separate entity conduct medical reviews.

16. Plan and complete all physical accommodations to the workplace. Remember to analyze the complete work environment and the entire surrounding areas including parking lots, access-ways, doors, water fountains, rest rooms, etc. Document all physical accommodations made to the workplace. Provide documentation of all bids, reviews, etc., for any accommodation not made because of undue hardship.

17. Document all requests for accommodations made by job applicants or current employees. Document the accommodation provided or, if unable to accommodate, the reason for the failure to accommodate.

18. Analyze the workplace for the need for possible "auxiliary aids" and other accommodation devices. Prepare a list of vendors and services so that all possible auxiliary aids can be acquired within a reasonable time. Maintain documentation of all auxiliary aids that are requested and provided.

19. If the health or safety exception is relied upon for employment decisions or other situations, the safety director or other individual making this determination should document all reasonable accommodations that were explored and all other information used to make this determination.

20. If the undue hardship or burden exceptions are to be used, all financial records, workforce analysis, and other information used to make this decision should be documented and secured for later viewing.

21. Educate and train all levels within the organization's structure. Remember, if a member of the organization's team discriminates against a qualified individual with a disability, the organization will be responsible for his or her actions. Develop an oversight mechanism to ensure compliance.

22. Develop a mechanism to encourage employees to come forward (in confidence) and discuss their disabilities.

23. Evaluate and analyze the employee assistance programs, restricted duty or light-duty programs, and other related programs. Organizations should address and plan in advance for such situations to become permanent light-duty positions.

24. If necessary, enter into negotiations with any labor organizations to reopen, or otherwise modify, collective bargaining agreements to ensure compliance with the provisions of the ADA. Evaluate all insurance plans, retirement plans, contracts with employment agencies, or other contractual arrangements to ensure compliance with the ADA. Documentation of any agreement should be included in the written contract.
25. Develop a written evaluation or audit instrument in order to properly evaluate compliance efforts. Designate specific individual(s) to be responsible for the audit and corrective actions. Establish a schedule for conducting the ADA audit.107

ADA AMENDMENTS ACT OF 2008

Effective January 1, 2009, the new Americans with Disabilities Act Amendments Act of 2008 became effective. This new Act makes significant changes to the term "disability" as defined in the ADA by rejecting several Supreme Court decisions and portions of previous EEOC's ADA regulations. The new Act retains the basic definition of "disability" as defined in the ADA as being an impairment that substantially limits one or more major life activities, a record of such an impairment, or being regarded as possessing an impairment. More specifically, the ADA Act Amendments Act provided the following:

1. The Act requires the EEOC to revise the section of their regulations that define the term "substantially limits."
2. The Act expands the definition of "major life activities" by including two nonexhaustive lists. List 1 includes many activities that the EEOC has previously recognized (such as walking) as well as activities that the EEOC previously did not specifically recognize (such as reading, bending, and communicating). The second list includes major bodily functions, such as functions of the immune system, normal cell growth, digestive system, bowel, bladder, neurologic, brain, respiratory, circulatory, endocrine, and reproductive functions.
3. The Act states that mitigating measures, such as "ordinary eyeglasses or contact lenses," shall not be considered in assessing whether an individual possesses a disability.
4. The Act clarified that an impairment that is episodic or in remission is a disability if it would substantially limit a major life activity when active. The Act provides that an individual subjected to an action prohibited by the ADA (such as failure to hire) because of an actual or perceived impairment will meet the "regarded as" definition of disability, unless the impairment is transitory or minor.
5. The Act provides that individuals covered only under the "regarded as" prong of the ADA test are not entitled to reasonable accommodation.
6. The Act emphasizes that the definition of "disability" should be interpreted broadly.

Safety and loss prevention professionals should take careful note of the changes to the ADA provided under the ADA Amendment Act because this new law reverses several of the U.S. Supreme Court's decisions. Additionally, prudent safety and loss prevention professionals should become familiar with the requirements under the ADA Amendment Act and follow the numerous opinions and cases on the Equal Employment Opportunity Commission's website located at EEOC.gov to ensure that the safety and health programs are achieving and maintaining compliance and your actions or inactions do not discriminate in any way.

10 Legal Liabilities Under Workers' Compensation Laws

CHAPTER OBJECTIVES

1. Acquire an understanding of the concept of workers compensation.
2. Acquire an understanding of the benefit structure under most state workers compensation laws.
3. Acquire an understanding of the differences between the safety and workers compensation functions.
4. Acquire an understanding of the management of the workers compensation function.

OVERVIEW OF WORKERS' COMPENSATION SYSTEMS

Many safety and loss prevention professionals have found that the management and administration of their organization's workers' compensation program has fallen upon their shoulders because the end result of work-related accidents (i.e., an injured employee) encompasses many issues regarding safety and health. There are many similarities between the management of a workers' compensation program and a safety program such as program management, but there are also many significant differences. A workers' compensation program is generally a reactive mechanism to compensate employees with monetary benefits after an accident has occurred. However, safety and loss prevention programs are designed by nature to be proactive programs, designed to prevent employees from being injured in the first place. Safety and loss prevention professionals who wear the dual hats of safety and/or health as well as workers' compensation must be able to delineate which hat they are wearing at any given time. They must also effectively manage the individuals, the situations, and the potential liabilities surrounding the workers' compensation program. The rising cost of workers' compensation for most employers has resulted in a significantly increased focus by management in this area. Employers, always cognizant of the bottom line, have found that their workers' compensation costs have risen significantly because of many factors, including, but not limited to, increased injuries and illnesses, increased medical and rehabilitation costs, increased time loss and benefits, and other factors. With this increased focus, safety and health professionals are often thrust into the administrative world of workers' compensation with little or no training or education regarding the rules, regulations, and requirements. In the safety and health arena, many of the potential liabilities encountered in the area of workers' compensation are a direct result of acts of omission rather than commission. Safety and loss prevention professionals should understand the basic structure and mechanics of the workers' compensation system and the specific rules, regulations, and requirements under their individual state system.

Virtually all workers' compensations systems are fundamentally no-fault mechanisms through which employees who incur work-related injuries and illnesses are compensated with monetary and medical benefits. Either party's potential negligence is usually not an issue as long as this is the employer/employee relationship. In essence, workers' compensation is a compromise in that employees are guaranteed a percentage of their wages (generally two-thirds) and full payment for their medical costs when injured on the job. Employers are guaranteed a reduced monetary cost for these injuries or illnesses and are provided a protection from additional or future legal action by the employee for the injury.

The typical workers' compensation system possesses these features:

- Every state in the United States has a workers' compensation system. There may be variations in the amounts of benefits, the rules, administration, etc., from state to state. In most states, workers' compensation is the exclusive remedy for on-the-job injuries and illnesses.
- Coverage for workers' compensation is limited to employees who are injured on the job. The specific locations as to what constitutes the work premises and on the job may vary from state to state.
- Negligence or fault by either party is largely inconsequential. No matter whether the employer is at fault or the employee is negligent, the injured employee generally receives workers' compensation coverage for any injury or illness incurred on the job.
- Workers' compensation coverage is automatic, i.e., employees are not required to sign up for workers' compensation coverage. By law, employers are required to obtain and carry workers' compensation insurance or be self-insured.
- Employee injuries or illnesses that "arise out of and in the course of employment" are usually considered compensable. These definition phrases have expanded this beyond the four corners of the workplace to include work-related injuries and illnesses incurred on the highways, at various in- and out-of-town locations, and other such remote locales. These two concepts, "arising out of" the employment and "in the course of" the employment, are the basic burdens of proof for the injured employee. Most states require both. The safety and health professional is strongly advised to review the case law in his or her state to see the expansive scope of these two phrases. That is, the injury or illness must "arise out of," i.e., there must be a causal connection between the work and the injury or illness and it must be "in the course of" the employment; this relates to the time, place, and circumstances of the accident in relation to the employment. The key issue is a "work connection" between the employment and the injury/illness.

- Most workers' compensation systems include wage-loss benefits (sometimes known as time-loss benefits), which are usually between one-half and three-fourths of the employee's average weekly wage. These benefits are normally tax-free and are commonly called temporary total disability (TTD) benefits.
- Most workers' compensation systems require payment of all medical expenses, including such expenses as hospital expenses, rehabilitation expenses, and prosthesis expenses.
- In situations where an employee is killed, workers' compensation benefits for burial expenses and future wage-loss benefits are usually paid to the dependents.
- When an employee incurs an injury or illness that is considered permanent in nature, most workers' compensation systems provide a dollar value for the percentage of loss to the injured employee. This is normally known as permanent partial disability (PPD) or permanent total disability (PTD).
- In accepting workers' compensation benefits, the injured employee is normally required to waive any common law action to sue the employer for damages from the injury or illness.
- If the employee is injured by a third party, the employer usually is required to provide workers' compensation coverage, but can be reimbursed for these costs from any settlement that the injured employee receives through legal action or other methods.
- Administration of the workers' compensation system in each state is normally assigned to a commission or board. The commission/board generally oversees an administrative agency located within state government that manages the workers' compensation program within the state.
- The workers' compensation act in each state is a statutory enactment that can be amended by the state legislatures. Budgetary requirements are normally authorized and approved by the legislatures in each state.
- The workers' compensation commission/board in each state normally develops administrative rules and regulations (i.e., rules of procedure, evidence, etc.) for the administration of workers' compensation claims in the state.
- In most states, employers with one or more employees are normally required to possess workers' compensation coverage. Employers are generally allowed several avenues to acquire this coverage. Employers can select to acquire workers' compensation coverage from private insurance companies or state-funded insurance programs or become "self-insured" (i.e., after posting bond, the employer pays all costs directly from its coffers).
- Most state workers' compensation coverage provides a relatively long statute of limitations. For injury claims, most states grant between 1 and 10 years to file the claim for benefits. For work-related illnesses, the statute of limitations may be as high as 20 to 30 years from the time the employee first noticed the illness or the illness was diagnosed. An employee who incurred a work-related injury or illness is normally not required to be employed by the employer when the claim for benefits is filed.

- Workers' compensation benefits are generally separate from the employment status of the injured employee. Injured employees may continue to maintain workers' compensation benefits even if the employment relationship is terminated, the employee is laid off, or other significant changes are made in the employment status.
- Most state workers' compensation systems possess some type of administrative hearing procedures. Most workers' compensation acts have designed a system of administrative "judges" (normally known as administrative law judges or ALJs) to hear any disputes involving workers' compensation issues. Appeals from the decision of the administrative law judges are normally to the workers' compensation commission/board. Some states permit appeals to the state court system after all administrative appeals have been exhausted.

Safety and loss prevention professionals should be very aware that the workers' compensation system in every state is administrative in nature. Thus, there is a substantial amount of required paperwork that must be completed in order for benefits to be paid in a timely manner. In most states, specific forms have been developed.

The most important form to initiate workers' compensation coverage in most states is the first report of injury/illness form. This form may be called a "first report" form, an application for adjustment of claim, or may possess some other name or acronym like the SF-1 or Form 100. This form, often divided into three parts so that information can be provided by the employer, employee, and attending physician, is often the catalyst that starts the workers' compensation system reaction. If this form is absent or misplaced, there is no reaction in the system and no benefits are provided to the injured employee.

Under most workers' compensation systems, there are many forms that need to be completed in an accurate and timely manner. Normally, specific forms must be completed if an employee is to be off work or is returning to work. These include forms for the transfer from one physician to another, forms for independent medical examinations, forms for the payment of medical benefits, and forms for the payment of permanent partial or permanent total disability benefits. Safety and loss prevention professionals responsible for workers' compensation are advised to acquire a working knowledge of the appropriate legal forms used in their state's workers' compensation program.

In most states, information regarding the rules, regulations, and forms can be acquired directly from the state workers' compensation commission/board. Other sources for this information include the insurance carrier, self-insured administrator, or state-fund administrator.

Safety and loss prevention professionals should be aware that workers' compensation claims possess a "long tail," i.e., stretch over a long period of time. Under the OSHA recordkeeping system, which is familiar to most safety and loss prevention professionals, injuries and illnesses are totaled on the OSHA Form 300 log every year and a new year begins. This is not the case with workers' compensation. After an employee sustains a work-related injury or illness, the employer is responsible for the management and costs until such time as the injury or illness reaches maximum medical recovery or the time limitations are exhausted. When an injury reaches maximum medical recovery, the employer may be responsible for payment of permanent partial or permanent total disability benefits prior to closure of the claim. Additionally, in some states, the medical benefits can remain open indefinitely and cannot be settled or closed with the claim. In many circumstances, the workers' compensation claim for a work-related injury or illness may remain open for several years and, thus, require continued management and administration for the duration of the claim process.

Some states allow the employer to take the deposition of the employee claiming benefits, whereas others strictly prohibit it. Some states have a schedule of benefits and have permanent disability awards strictly on a percentage of disability from that schedule. Other states require that a medical provider outline the percentage of functional impairment due to the injury/illness, usually utilizing the American Medical Association (AMA) Guidelines. The functional impairment, as well as other factors such as the employee's age, education, and work history, are utilized by the ALJ to determine the amount of occupational impairment upon which permanent disability benefits are awarded. Still other states have variations on these systems.

In summation, safety and loss prevention professionals who are responsible for the management of a workers' compensation program should become knowledgeable in the rules, regulations, and procedures under their individual state's workers' compensation system. Safety and loss prevention professionals who possess facilities or operations in several states should be aware that although the general concepts may be the same, each state's workers' compensation program possesses specific rules, regulations, schedules, and procedures, which may vary greatly between states. There is no substitute for knowing the rules and regulations under your state's workers' compensation system.

POTENTIAL LEGAL LIABILITIES IN WORKERS' COMPENSATION

The potential liabilities for a safety and loss prevention professional in managing a workers' compensation program are many and varied. Above all, a safety and loss prevention professional should realize that most workers' compensation systems are no-fault systems that generally require the employer or the employer's insurance administrator to pay all required expenses whether the employer or employee was at fault, whether the accident was the result of employee negligence or neglect, or whether the injury or illness was the fault of another employee. Most workers' compensations systems are designed to be liberally construed in favor of the employee.

Many safety and loss prevention professionals who have been taught to use a proactive method of identifying the underlying causes of accidents and immediately correcting the deficiency may find that management of the workers' compensation function can often be very time consuming and frustrating and show little progress. In situations of questionable claims, that is, whether the injury or illness was actually work related, safety and health professionals should be aware that in many states the employee has the right to initiate a workers' compensation claim with the workers' compensation commission/board and initiate or continue time-loss benefits and medical benefits until such time as the professional can acquire the appropriate evidence to dispute the claim benefits. This administrative procedure is often foreign to many safety and loss prevention professionals and can be stressful and frustrating to a safety and loss prevention professional accustomed to a more direct management style. Above all, the safety and health professional must realize that he or she must follow the prescribed rules, regulations, and procedures set forth under each state's workers' compensation system. Any deviation thereof or failure to comply can place the company, the insurance carrier or administrator, and the safety and loss prevention professional at risk for potential liability.

In our modern litigious society, safety and loss prevention professionals should be aware that they will often be interacting with the legal profession when managing an injured employee's workers' compensation claim. Although most workers' compensation systems are designed to minimize the adversarial confrontations, in many states, attorneys are actively involved in representing injured employees with their workers' compensation claims. Safety and health professionals should be aware that the amount of money paid by the injured employee to the attorney, generally a contingent fee, is normally set by statute within the individual state's workers' compensation act.

Safety and loss prevention professionals should also be aware that when an injured employee is represented by legal counsel, often the direct lines of communication to the employee are severed and all communications must be through legal counsel. Circumvention of this communication method by safety and loss prevention professionals often leads to confusion, mismanagement, and adversarial confrontations. Safety and loss prevention professionals should be aware of the rules and regulations of the individual state regarding contact and communication with an employee who is represented by legal counsel. One of the major components in the management of a workers' compensation program is the communications with the medical professionals who are treating the injured or ill employee. Safety and loss prevention professionals should be aware that this can be an area of potential miscommunication and conflict. The goal of the safety and health professional and the medical professional is normally the same, i.e., making the injured employee well, but the methodology through which the goal is attained often conflicts. Although the potential liability in this area is not proscribed by statute, safety and health professionals should make every effort to ensure open and clear lines of communication to avoid any such conflicts. The potential liability in this area occurs when there is a loss of trust between the safety and loss prevention professional and the medical community, which can ultimately lead to additional benefit costs.

Given the many individuals who may be involved in a work-related injury situation (for example, the injured employee, the attorney, the physician, the administrator, and the safety and health professional, to name a few), the potential for conflict and litigation is relatively high. Safety and loss prevention professionals should know the rules and regulations of this administrative system and avoid areas of potential conflict.

The first and most common area of potential liability in the area of workers' compensation is simply not possessing or maintaining the appropriate workers' compensation coverage for employees. Often through error or omission, the employer either does not acquire the appropriate workers' compensation coverage or has allowed the coverage to lapse. In most states, the employer's failure to possess the appropriate workers' compensation coverage will not deny the employee the necessary benefits. The state workers' compensation program, through a special fund or uninsured fund, will incur the costs of providing coverage to the employee, but will bring a civil or criminal action against the employer for repayment and penalties. In several states, failure to provide the appropriate workers' compensation coverage can permit the individual employee to bring a legal action in addition to the legal action by the state workers' compensation agency. Often, the employer is stripped of all defenses.

Given the paperwork requirements of most workers' compensation systems, safety and loss prevention professionals can incur liability for failing to file the appropriate forms in a timely manner. In most states, failing to file the appropriate forms in a timely manner can carry an interest penalty. Additionally, safety and loss prevention professionals should be aware that it is the employee's right to file a workers' compensation claim and it is often the employer's responsibility to file the appropriate form(s) with the agency or party. Liability can be assumed by the safety and loss prevention professional for refusing to file the form or failing to file the form with the agency to initiate benefits. In most states, civil and criminal penalties can be imposed for such actions and additional penalties such as loss of self-insurance status can also be imposed on the employer.

Safety and loss prevention professionals may be confronted with situations where it is believed that the injury or illness is not work related. Safety and loss prevention professionals often assume liability by playing judge and jury when the claim is being filed and inappropriately denying or delaying payment of benefits to the employee. In most states, civil and criminal penalties can be imposed for such actions and other penalties, such as loss of self-insurance status, can additionally be imposed on the employer. Safety and loss prevention professionals should become knowledgeable in the proper method to appropriately petition for the denial of a non–work-related claim through the proscribed adjudication process.

In all states, an employee who files a workers' compensation claim possesses the right not to be harassed, coerced, discharged, or discriminated against for filing or pursuing the claim. Any such discrimination against an employee usually carries civil penalties from the workers' compensation agency, and often separate civil actions are permitted by the employee against the employer. In these civil actions, injunctive relief, monetary damages, and attorney fees are often awarded.

In most states, employees who file fraudulent workers' compensation claims are subject to both civil and criminal sanctions. The employer bears the burden of proving the fraudulent claim and can often request an investigation be conducted by the workers' compensation agency. Additionally, in some states, employees who intentionally fail to wear personal protective equipment or to follow safety rules can have their workers' compensation benefits reduced by a set percentage. Conversely, an employer who does not comply with the OSHA or other state safety and health regulations, thereby causing the injury or illness, can be assessed an additional percentage of workers' compensation benefits over and above the proscribed level. Safety and loss prevention professionals should also be aware that in a number of states, failure by the employer to comply with the OSHA or state plan safety and health regulations, thereby directly or indirectly resulting in the injury or death of an employee, can result in the employee, or his or her family, recovering workers' compensation benefits and being permitted to evade the exclusivity of workers' compensation to bring a civil action against the employer for additional damages.

With the burden of attempting to disprove that an injury or illness was incurred on the job, safety and loss prevention professionals are often placed in the position of an investigator, or as the individual responsible for securing outside investigation services, to attempt to gather the necessary information to deny a workers' compensation claim. The areas of potential liability with regard to surveillance, polygraph testing, drug testing, and other methods of securing evidence can be substantial. Prior to embarking on any type of evidence gathering that may directly or indirectly invade the injured individual's privacy, the safety and health professional should seek legal counsel to identify the applicable laws, such as common law trespass, invasion of privacy, federal and state polygraph laws, and alcohol and controlled substance testing perimeters. Potential sanctions for violations of these laws usually take the form of a civil action against the employer and individual involved, but criminal penalties can also be imposed for such actions as criminal trespass.

The previous list includes only a few of the areas of potential liability for a safety and loss prevention professional in the area of workers' compensation. Safety and loss prevention professionals should be aware that workers' compensation is an administrative system and any deviation from the proscribed procedures that may directly or indirectly affect the injured employees workers' compensation benefits is a potential minefield for liability. The assessment of criminal sanctions in the area of workers' compensation is infrequent and is usually reserved for egregious situations. However, assessment of civil penalties by the workers' compensation commission or agency for mismanagement of a workers' compensation claim are far too frequent and the potential of legal action by the injured employee inside and outside of the workers' compensation system is a growing area of potential liability.

QUESTIONABLE CLAIMS

One area in which safety and loss prevention professionals with workers' compensation responsibilities often encounter difficulty is in the area of questionable claims. These are the claims that involve unusual circumstances and require additional investigation prior to determining whether to accept the claim as compensable or pursue a denial of the claim. Some of the signs safety and loss prevention professionals may identify in order to pursue a more vigorous or extensive investigation include the following characteristics:

1. Friday and Monday claims: These are claims, usually back or soft tissue injuries, which allegedly occurred on Friday and were not reported until Monday or claims, again usually back or soft tissue injuries, which occurred previously. However, the employee does not go to the doctor until the last day of the work week.
2. No health insurance: For new employees or employees who do not possess health insurance coverage, one possible avenue to acquire payment for a healthcare issue is workers' compensation.

3. Holiday or hunting season claims: When an employee requests vacation or time off and is denied due to seniority or other reasons, one way to achieve the time off is to file a workers' compensation claim.

4. Late reporting of claim: Most companies possess policies and procedures requiring employees to report all injuries and illnesses immediately. A claim that is reported through the medical provider, through legal counsel, or substantially late may require additional investigation.

5. Before layoff, strike, or downsizing: Prior to any dramatic changes in the workplace, such as downsizing, layoffs, strikes, lockouts, or other events, many companies have experienced a sudden rise in the number of claims filed. Safety and loss prevention professionals should also note that claims can be filed on the rumor of a major event; thus, further investigation into the claim may be warranted.

6. Failing to recall details or change in the story: Unless a head injury is involved, employees can usually recall details of an accident. When the employee's story changes or the employee cannot recall the events surrounding the accident, this may be a sign that further investigation is needed.

7. There is no witness: In many industrial workplaces, it would be difficult to incur an injury without another employee witnessing the accident. Although a witness is not required, the lack of a witness, depending on the circumstances, may be a sign that further investigation is needed.

8. Upset employees: Employees who are upset with their supervisor or team leader may file a claim simply to avoid their supervisor or as retribution against their supervisor or team leader.

9. Employees receiving disciplinary action: An employee in the latter stages of a progressive disciplinary process may see a workers' compensation claim as an alternative avenue to avoid such actions.

10. Cannot make the payments: Employees sometimes view workers' compensation as a way to stop bill collection or a way to avoid a tough decision in their financial life.

11. Employee does not answer the phone: Many companies call employees receiving time-loss benefits to check on their condition. When the phone rings and no one answers or the employee calls back at a later time after a message is provided, this may provide a sign that further investigation is needed into the employee's activities.

12. Employee not attending medical appointments: When an employee is off work and receiving time-loss benefits, an employee regularly missing scheduled doctor's appointments, physical therapy, or other appointments may be a sign that further investigation is needed into the employee's "other" activities.

13. Outside activities: When the employee cannot work as a result of a back injury, but his or her photograph is in the newspaper for winning the local golf tournament last weekend, further investigation may be warranted.

14. Claim after termination: Any claim filed after an employee is terminated from employment should be carefully scrutinized.

Safety and loss prevention professionals should note that most state workers' compensation laws possess a method to pursue the investigation of fraudulent workers' compensation claims. Most state laws possess criminal sanctions for the filing of a fraudulent workers' compensation claim. Although the burden of proof is significantly higher when attempting to prove fraud, if the circumstances of your investigation identify such activities, appropriate internal and external parties should be informed in order to investigate such actions.

GENERAL GUIDELINES FOR EFFECTIVE MANAGEMENT OF WORKERS' COMPENSATION

Safety and loss prevention professionals responsible for the management of workers' compensation within the organization will find that an effective management system can control and minimize the costs related to this required administrative system while also maximizing the benefits to the injured or ill employee. Although the workers' compensation system is basically reactive in nature, safety and health professionals should develop a proactive management system to effectively manage the workers' compensation claims after they are incurred within the organization. The following basic 12-step guideline can be used to implement an effective workers' compensation management system:

1. Become completely familiar with the rules, regulations, and procedures of the workers' compensation system in your state. A mechanism should be initiated to keep the professional updated with all changes, modifications, or deletions within the workers' compensation law or regulations. A copy of these laws and rules can normally be acquired from your state's workers' compensation agency at no cost. Additionally, the state bar association, universities, and law schools in many states have published texts and other publications to assist in the interpretation of the laws and rules.

2. A management system should be designed around the basic management principles of planning, organizing, directing, and controlling. Given the fact that most state workers' compensation programs are administrative in nature, appropriate planning can include, but is not limited to, such activities as acquiring the appropriate forms, developing status tracking mechanisms, establishing communication lines with the local medical community, and informing employees of their rights and responsibilities under the workers' compensation act. Organizing an effective workers' compensation system can include, but is not limited to, selecting and training personnel who will be responsible for completing the appropriate forms, coordinating with insurance or self-insured administrators, acquiring appropriate rehabilitation and evaluation services, and developing medical response mechanisms. The directing phase can include, but is not limited to, implementing tracking mechanisms, coordinating on-site visitations by medical and legal communities, developing work-hardening programs, and installing return-to-work programs. Controlling can include such activities as establishing

an audit mechanism to evaluate case status and progress of the program, using injured worker home visitation, and acquiring outside investigation services, among other activities.

3. Compliance with the workers' compensation rules and regulations must be the highest priority at all times. Appropriate training and education of individuals working within the workers' compensation management system should be mandatory and appropriate supervision should be provided at all times.

4. When an employee incurs a work-related injury or illness, appropriate medical treatment should be top priority. In some states, the employee possesses the first choice of a physician; in other states, the employer has this choice. The injured or ill employee should be provided the best possible care in the appropriate medical specialty or medical facility as soon as feasible. Improper care in the beginning can lead to a longer healing period and additional costs.

5. Employers often fool themselves by thinking that if employees are not told their rights under the state workers' compensation laws there is less chance that an employee will file a claim. This is a falsehood. In most states, employees possess easy access to information regarding their rights under workers' compensation through the state workers' compensation agency, their labor organization, or even television commercials. A proactive approach that has proven to be successful is for the safety and loss prevention professional or other representative of the employer to explain to the employee his or her rights and responsibilities under the workers' compensation laws of the state as soon as feasible following the injury. This method alleviates much of the doubt in the mind of the injured employee, begins or continues the bonds of trust, eliminates the need for the involvement of outside parties, and tends to improve the healing process.

6. The safety and loss prevention professional should maintain an open line of communication with the injured employee and attending physician. The open line of communication with the injured employee should be of a caring and informative nature and should never be used for coercion or harassment purposes. The open line of communication with the attending physician can provide vital information regarding the status of the injured employee and any assistance the employer can provide to expedite the healing process.

7. Timely and accurate documentation of the injury or illness and appropriate filing of the forms to ensure payment of benefits is essential. Failure to provide the benefits in a timely manner, as required under the state workers' compensation laws, can lead the injured employee to seek outside legal assistance and cause a disruption in the healing process.

8. Appropriate, timely, and accurate information should be provided to the insurance carrier, organization team members, and others to ensure that the internal organization is fully informed regarding the claim. There is nothing worse than an injured employee receiving a notice of termination from personnel while lying in the hospital because personnel was not informed of the work-related injury and counted the employee absent from work.

9. As soon as medically feasible, the attending physician, the insurance administrator, the injured employee, and the safety and health professional can discuss a return to light or restricted work. A prudent safety and loss prevention professional may wish to use photographs or a video recording of the particular restricted duty job, written job descriptions, and other techniques in order to ensure complete understanding of all parties of the restricted job duties and requirements. After the injured employee has returned to restricted duty, the safety and loss prevention professional should ensure that the employee performs only the duties that were agreed upon and within the medical limitations prescribed by the attending physician. An effective return-to-work program can be one of the most effective tools in minimizing the largest cost factor with most injuries or illnesses, namely time-loss benefits.

10. In coordination with the injured employee and attending physician, a rehabilitation program or work-hardening program can be used to assist the injured employee's return to active work as soon as medically feasible. Rehabilitation or work-hardening programs can be used in conjunction with a return-to-work program.

11. Where applicable, appropriate investigative methods and services can be used to gather the necessary evidence to address fraudulent claims, deny non-work-related claims, or address malingering or other situations.

12. A prudent safety and loss prevention professional should audit and evaluate the effectiveness of the workers' compensation management program on a periodic basis to ensure effectiveness. All injured or ill employees should be appropriately accounted for, the status of each meticulously monitored, and cost factors continuously evaluated. Appropriate adjustments should be made to correct all deficiencies and to ensure continuous improvement in the workers' compensation management system.

WHAT TO EXPECT IN A WORKERS' COMPENSATION HEARING

Embedded within the framework of most workers' compensation systems, an arbitration system has been established to decide disputes in an informal and cost-effective manner. In most systems, the initial level of adjudication is a hearing before an administrative law judge, followed by an appeal stage before an appellate panel. Appeals from the appellate panel are normally made to the commission/board. In some states, the final appeal stage lies with the commission or board, but in other states, appeals to the state court system are allowed.

Safety and loss prevention professionals are normally involved during the initial hearing phase before the administrative law judge. In some organizations, the safety and loss prevention professional is responsible for the presentation of evidence at the hearing, but in other organizations, the safety and loss prevention professional assists legal counsel in the preparation of the case. In either circumstance, it is important that the safety and loss prevention professional be familiar with the rules and regulations of the individual state's workers' compensation system and the methods to prepare an effective case.

Workers' compensation hearings before an ALJ are often informal in comparison to a court of law. These hearings are often held in conference rooms in government buildings or may even be held in hotel conference rooms. Most ALJs are granted wide discretion about courtroom procedure, rules of evidence, and other procedural aspects of the hearing. Safety and loss prevention professionals should be prepared for the administrative law judge to be actively involved in the hearing and to ask questions of the parties and witnesses.

When preparing for the hearing, the safety and loss prevention professional should know the time limitations prescribed by the ALJ. Often, the parties are provided a limited time period to present each phase of their case. Additionally, appropriate preparation should be made regarding the recording, or lack thereof, of this hearing. There is great variation among jurisdictions as to whether this hearing is recorded and the type of recording method used by the ALJ.

In a hearing before an ALJ, a prudent safety and loss prevention professional should be prepared for the four major components of the hearing; namely the opening statement, presentation of testimony and evidence, cross-examination of opponent's witnesses, and a closing statement. Although many administrative law judges, in an effort to conserve time, expedite the opening and closing phases of the hearing, safety and loss prevention professionals should be prepared to present a concise, complete, and accurate account of their case.

In opening statements, the parties are normally afforded the opportunity to present their theory of the case. This is an opportunity to explain to the ALJ the theory of the case, outline the evidence to be presented to support the case, and request a decision in your favor. Normally, the employee presents first followed by the employer.

Following the employee's opening statement, the employee is provided an opportunity to call witnesses for direct examination and to present other documented evidence. In direct examination, open-ended questions are permitted, but leading questions are usually not permitted. The rules of evidence are often relaxed in this hearing. Cross-examination of witnesses is always allowed. Additionally, the ALJ often questions the witnesses. After the employee has called all his or her witnesses, the employer is normally provided an opportunity to call witnesses in support of its position.

In a cross-examination of the opponent's witnesses, leading questions (or "yes and no" questions) are normally permitted. A leading question is defined as, "one which instructs the witness how to answer or puts into his mouth words to be echoed back … ." Safety and health professionals should frame questions in a concise manner and "get to the point" of the examination as quickly as possible. Although this type of examination is intended to unearth discrepancies, bias, and credibility in the witness's testimony, safety and loss prevention professionals should bear in mind the issues in dispute and not permit this examination to evolve into a character assassination or personal attack.

Written documentation, diagrams, photographs, video recordings, and other evidence are normally presented to the ALJ for review and acceptance into evidence. This type of evidence can be provided at the end of the opening statement but prior to witness testimony or it can be provided in conjunction with witness testimony.

Both parties are provided time to examine the evidence or they are provided a copy of the documents.

In most states, the ALJ will not render an immediate decision in the case. The ALJ will conclude the hearing at the end of closing statements and provide a written decision to the parties via mail. Appeals from the written decision normally must be filed within a relatively short period of time from the receipt of the written decision (commonly 30 days).

Preparation is the key to success in a workers' compensation administrative hearing. Safety and loss prevention professionals should develop their theory of the case, assemble all necessary evidence and witnesses to support their theory, maintain an objective viewpoint, and prepare a file or manual containing all information and evidence prior to the hearing. Presentations in opening and closing statements should be concise and to the point, information and evidence should always be at the parties' fingertips for immediate location, and, above all, demeanor should always be professional during the hearing.

DENYING LIABILITY FOR A WORKERS' COMPENSATION CLAIM

In the event that an attorney to represent the employer is required or retained, the safety and loss prevention professional can be an enormous asset to the attorney by preparing the previously noted information and offering suggestions and questions for the attorney at the hearing.

Within most workers' compensation systems, a procedure is designated in the statute or administrative rules for an employer to deny liability for a particular claim. In virtually all states, this procedure will provide benefits to the employee filing the claim immediately, in accordance with the requirements, and place the burden of proving that the claim is not within the scope of workers' compensation coverage on the employer. With medical benefits being paid immediately and time-loss benefits normally being paid after a short waiting period, the safety and loss prevention professional is under a demanding time constraint to gather the necessary evidence to deny liability for the claim.

Procedurally, the initial step of denying liability for an alleged work-related injury or illness is completing the first report of injury form. On this form, there is a question that asks whether or not this injury or illness is work related and/or whether the employer wishes to petition for denial of the claim. If the safety and loss prevention professional possesses information or evidence that the injury or illness is not work related, indication through the marking of the appropriate box will place the administrator and workers' compensation agency on notice that this claim may be disputed. For employees, this designation has virtually no effect on their initial receipt of benefits.

With the "clock ticking" regarding the payment of benefits, most states require that the employer request a review by an administrative law judge or other representative of the agency in order to make an initial determination as to whether the claim is compensable and whether TTD benefits should be paid. This request for review, normally required in writing, is made to the administrative agency and normally either a review of written evidence is requested or a hearing date is designated.

In either case, the safety and loss prevention professional or other representative of the employer is required to present the evidence proving that the liability for the claim does not belong to the employer. The employee or representative for the employee may also present evidence to the contrary. After a review or hearing, the ALJ will make a determination regarding the initial compensability of the claim.

Denial of liability for a workers' compensation claim is significantly different from other litigation. Negligence by the employee is inconsequential in most circumstances. The primary theories to deny compensability in most states include the following possibilities:

- If the injury or illness was not work related; that is, the injury or illness did not arise out of or in the course of employment.
- The claim is fraudulent in nature.
- The employee incurring the injury or illness is excluded from coverage under the workers' compensation act. This exclusion may be voluntary (that is, opted out of coverage at an earlier date) or involuntary through the provisions of the specific act.
- The injured individual is not an employee within the definition of the act. The individual may be an independent contractor or subcontractor.
- The employee may have been injured while being lent to another employer.
- The employee may be a dual employee working for two or more employers.
- The employee did not file the claim within the specific time limitations set forth under the act.
- The death was a result of suicide, the injury was self-inflicted, or, in states where applicable, the employee was involved in horseplay or other misconduct directly resulting in the injury or illness.

The time and place of the accident is of utmost importance in evaluating a workers' compensation claim. Through established accident investigation procedures, safety and loss prevention professionals can normally begin to gather the necessary information and documentation to ascertain the status of the individual involved, the scope and type of injury or illness, and the specific information as to the factors leading up to the accident.

Safety and loss prevention professionals should acquire the necessary level of competency to conduct a complete and thorough accident investigation. Immediately after providing medical attention to the injured employee, the accident scene should be isolated and "frozen in time" while the investigation is conducted. Appropriate witness interviews should be conducted and documented. Photographs, video recordings, or other means of documenting the accident scene should be used. Appropriate sampling should be conducted and documented. In short, appropriate and accurate documentation is vital to any possibility of success in workers' compensation adjudication.

Upon completion and analysis of the accident investigation, if the evidence supports a petition for denial of workers' compensation benefits, the safety and loss prevention professional should assemble and prepare the evidentiary information for submittal to the agency or ALJ. Evidence should be prepared in a logical and organized manner, and all supporting information should be included for review and analysis prior to submittal. If specific and vital supporting information is absent, additional investigations should be initiated in an attempt to acquire this information.

In most states, the rules of evidence are relaxed in workers' compensation adjudications. Virtually any information related to the accident, the injury or illness, the employment status, or other related information may be heard by the ALJ. Hearsay evidence is admissible and can often be used to support a position. This is within the ALJ's discretion.

Other circumstances that the safety and loss prevention professional may be involved in are actions involving a liability issue under workers' compensation. These include the following situations:

- recovery of paid workers' compensation benefits from an employee who has recovered from a third party in situations such as auto accident, medical malpractice, and product liability
- denial of additional benefits after a claim has been settled
- a request for reopening of a claim by a previously injured employee
- payment of benefits through a special fund or second injury fund that can alleviate or minimize the employer's exposure for benefits

Safety and loss prevention professionals should be cognizant that denial of a workers' compensation claim must be based on an exception from legislative coverage rather than fault of the employer or employee. Responsibility for the injury or illness is presumed to be with the employer if the injury or illness occurred on the job. Immediate and appropriate investigation, information acquisition, and documentation are vital in order to attempt denial of liability for workers' compensation coverage.

11 Protecting Your Organization or Company

CHAPTER OBJECTIVES

1. Acquire an understanding of potential civil liabilities.
2. Acquire an understanding of potential criminal liabilities.
3. Acquire an understanding of protections for company documents.
4. Acquire an understanding of shifting liabilities and protections.

INTRODUCTION

Companies today face a myriad of potential civil and criminal liabilities from a wide variety of areas. In addition to the safety and loss prevention issues, corporate officials should also be aware of the potential liabilities with regard to other laws in the area of antitrust and trade regulations (Sherman Act, Clayton Act, Robinson–Patman Act, Federal Trade Commission Act, and state antitrust laws), employee benefit and wage laws, federal and state tax laws, and especially the federal and state environment laws (Resource Conservation and Recovery Act, Clean Air Act, etc.). The best protection that a company can acquire in order to avoid potential civil and criminal liability in all areas, including safety and loss prevention, is to ensure that its program is in compliance with the appropriate government regulations and is able to demonstrate or prove compliance if called upon to do so.

Specifically in the area of safety and loss prevention, the directors, officers, and management employees should be made aware of the potential monetary fines that can be imposed by OSHA or a state plan agency in addition to the potential of other civil damages if compliance with the appropriate regulations is not acquired and maintained. Although the monetary fines imposed by OSHA and state plan agencies are often relatively small, with the sevenfold increase in the maximum fines the possibility of six- or seven-figure fines for noncompliance is a distinct possibility. Monetary fines for noncompliance can now have a dramatic effect on the bottom line and possibly the financial future of the company.

With our current litigious society, companies should realize that the potential of civil damages, outside the realm of OSHA, is becoming commonplace. Companies face potential civil actions in a wide variety of areas including, but not limited to, product liability, discrimination in employment, and tort actions outside of the exclusivity of workers' compensation. In these types of actions against a corporation or company, the total monetary expenditure, whether the case is won or lost, can be astronomical. Companies should strive to educate their management team as to the "real" cost of litigation and how to avoid such litigation, through ensuring compliance with the government regulations and internal company policies and procedures.

All levels within a company's management hierarchy should be made aware of the increased potential for criminal sanctions being applied to a work-related injury or fatality under the OSH Act and by state prosecutors. Companies should prepare for such catastrophes and be ready to exercise their constitutional rights when necessary (see Chapter 6).

Acquiring and ensuring compliance is the key in avoiding much of the potential liability in the area of safety and loss prevention. Top-level management must be committed to creating and maintaining a safe and healthful workplace, no matter what the economic loss or other conditions that may create difficulties. Top-level management should be actively involved in the area of safety and health and provide all of the necessary resources including, but not limited to, acquisition of competent personnel, providing necessary financial support, and providing enforcement and moral support in order to achieve the safety and loss prevention compliance objectives. In essence, management commitment and support are necessary from the top down so that compliance can be achieved and maintained over the long run.

Companies should realize that the area of safety and loss prevention is constantly changing and evolving, and therefore companies must change in order to maintain compliance with OSHA and other regulations. On a daily basis, new standards are developed and promulgated, the OSHRC decides cases, and the courts decide issues that directly or indirectly affect companies. Companies are required to know the current status of a particular standard or law at any given time and to be in compliance. As has been said many times, ignorance of the law is no defense. Companies should be aware of the changes and make the appropriate modifications to ensure compliance within their operations.

Companies should be aware that, in the area of compliance, proper and appropriate documentation is essential in order to prove that the particular program is in compliance with the applicable standard. Although many OSHA and state plan programs do not require written programs, companies should be aware that the lack of supporting documentation can often affect credibility and lead to unnecessary citations. Appropriate documentation of compliance programs, required training, acquisition of equipment, and other required items eliminates doubt and increases the credibility of the company.

One method often used by companies to acquire and ensure compliance is the safety and loss prevention audit methodology. A safety and loss prevention audit instrument is designed to enable management to properly assess its current structure, assess adequacy of the safety and health program in numeric terms, and identify deficiencies within the program for immediate correction. This type of program is usually for in-house use only and should be conducted either by management team members educated in the OSHA standards or by an outside independent assessment team. Truthfulness and thoroughness are essential in developing and conducting the safety and loss prevention audit assessment.

The basic premise of the safety and loss prevention audit assessment is to compare the current position of the safety, health, and loss prevention programs to the ideal or optimum program status. This evaluation is often reduced to numeric terms by providing each question with a numeric value weighed according to importance. The percentage of effectiveness for the safety and loss prevention program is achieved by dividing the total amount of points earned during the audit by the total amount of points possible in a perfect program.

The safety and loss prevention audit methodology and mechanism utilized should be specifically designed for your organization and facilities. For example, some facilities will have confined spaces, but others may not have such hazards and would not be required to have a confined space program. The audit instrument provided at the end of this chapter should only be used as a guideline. The safety and health professional should design an instrument suited for the organization's operations and facilities. The development of the audit instrument should entail a complete and thorough identification and evaluation of all mandatory requirements and the required elements and sub-elements of each requirement. Each element and sub-element should be analyzed for applicability and effectiveness. Appropriate questions should be assembled on the audit instrument with a subjective evaluation of the numeric values. The audit should be conducted and the audit instrument modified to ensure effectiveness. Safety and loss prevention professionals should be aware that omission from the audit instrument of a required element or program is just as deficient as an improper element.

The safety and loss prevention audit assessment should be conducted on a periodic basis and a report generated for the upper management group. It should identify, at the very minimum, the current status, percentage of improvement, and deficiencies noted. Areas of deficiency should be addressed and corrective action taken immediately.

The safety and loss prevention audit assessment program is simply a tool that enables management to identify and evaluate its status in the area of safety, health, environment, and loss prevention. This tool additionally provides, on at least a yearly basis, a method of identifying deficiencies within the overall safety and loss prevention program, thus permitting the safety and loss prevention professional to properly focus his or her time and energy on the major areas of importance.

As with any compliance documentation, potential pitfalls exist that the safety and loss prevention professional should be aware of prior to establishing such a program. Safety and loss prevention audit documentation is a potential gold mine for the opposition in a civil action against a corporation because this document identifies all of the deficiencies within the safety and health program. Additionally, safety and loss prevention audit documentation has been requested by OSHA, as discussed in *Secretary of Labor v. Hammermill Paper*, and thus may be a source of litigation. Lastly, if deficiencies are identified and the safety and loss prevention professional or corporation willfully disregards this information after it is gathered, this documentation may be "smoking gun" evidence following future incidents.

SECRETARY OF LABOR V. HAMMERMILL PAPER

Of particular importance to companies in the area of documentation is the 1992 decision in *Secretary of Labor v. Hammermill Paper*. In this case, the secretary of labor authorized OSHA to require the employer, under the authority granted under section 8(b) of the OSH Act, to disclose its voluntary internal safety and health compliance audits for a three-year period pursuant to an administrative subpoena duces tecum (i.e., bring your records with you to the trial or deposition). The employer refused to provide these documents and the secretary of labor brought action against the employer to compel disclosure of these documents.

The employer argued that the subpoenaed materials were beyond the secretary of labor's statutory authority as provided under section 8(b) of the OSH Act. In addition, this subpoena of voluntary internal documents was contradictory to OSHA's established policy of encouraging employers to conduct voluntary self-audits in order to improve their compliance efforts. Although the court sympathized with the employer's position, noting in addition that "the secretary of labor wrote to the defendants (employer), along with the CEO's of other Fortune 500 companies, calling upon them to act in a leadership role in protecting the American workers from injury by implementing a safety and health audit, and produce a safer and more healthful work environment." The court additionally noted that there was no legal requirement for the employer to have safety and health audits or any requirement to keep records of voluntary audits. The court went even further in advising that the "secretary of labor should not undertake this action." However, the court found that the secretary of labor possessed the authority to require complete disclosure, which made the subpoena enforceable.

This decision, in essence, now permits OSHA, and possibly other government agencies, to acquire access to internal safety and health audits through the subpoena authority under section 8(b) of the OSH Act. Internal safety and health audits are normally prepared for the purpose of identifying deficiencies within the safety and health management system in order to initiate corrective action. Because OSHA is permitted to acquire access to these internal audit documents that identify the deficiencies within a safety and health program, many employers have decided to either forgo the development of such potentially damaging documents or engage in other methods to protect these documents from OSHA.

WORK PRODUCT RULE

Although companies may prevent acquisition of sensitive documents, such as internal safety and health audits, by simply not developing these documents in the first place, this method may be detrimental to the overall safety and health effort. One method under which employers may be able to protect these internal documents is through the use of the "work product" doctrine as prescribed in section 503 of the Federal Rules of Evidence, dealing with attorney–client privilege. Simply put, if the internal safety and health audit is prepared in anticipation of litigation and provided to legal counsel, this document may be considered privileged information and thus not accessible to the opposition (OSHA). The leading case under this theory is *Hickman v. Taylor*. In this U.S. Supreme Court decision, the "work product privilege" was created to protect pretrial preparation materials. Ultimately, the Supreme Court utilized the Federal Rule of Civil Procedure 26(b) in the following statement.

> Subject to the provisions of subsection (b)(4) of this rule, a party may obtain discovery of documents and tangible things otherwise discoverable under subsection (b)(1) of this rule and prepared in anticipation of litigation or for trial by or for another party or by or for that other party's representative (including his attorney, consultant, surety, indemnitor, insurer or agent) only upon a showing that the party seeking discovery has substantial need of the materials in the preparation of his case and that he is unable without undue hardship to obtain the substantial equivalent of the materials by other means.

In *Secretary of Labor v. Bally's Park Place Hotel & Casino*, it was decided that the document in question, a consultant's report that was prepared in response to the OSHRC contacting the employer about toxic emissions, was in fact protected by the work product rule because it was prepared for the prospect of litigation in the future.

An employer seeking to invoke the work product privilege must establish that: (1) the party is invoking the privilege in the right type of proceeding (i.e., most statutes apply the privilege to civil actions only; however, the privilege has been extended in some criminal proceedings), (2) the party is asserting the right type of privilege (the work product privilege is a personal right of refusal), (3) the party claiming the privilege is the proper holder (i.e., the client or the attorney), and (4) the information the party seeks *to suppress is work product material (i.e., the information is the work product of the attorney,* is derivative rather than primary materials, and is in preparation of litigation). Although this theory is often utilized in civil litigation, use of this privilege with regard to an administrative subpoena under the OSH Act is virtually untried.

Companies should also be aware that under the Freedom of Information Act (FOIA), most documents that are acquired by OSHA or other government agencies are normally accessible to the public and press. A prudent company may want to discuss available protection of internal safety and health audits and related documents with legal counsel in order to develop the appropriate protections.

In addition to safety audits, safety and loss prevention professionals should be aware that many of the documents that are produced as part of performing the job function, such as safety inspections, personal protective equipment purchases, etc., are normally discoverable by outside parties through the subpoena power of the agency or court. These documents can be used as evidence by your company as part of a defense or can be used by the opponents to support or prove their case. In many civil actions against companies for work-related accidents, the safety and loss prevention professional is often a named party in order that the safety and loss prevention professional can be available for deposition and his or her records can be acquired under a subpoena duces tecum. After the safety and loss prevention professional has been deposed, the safety and loss prevention professional is often released as a named party, and the "deep pocket" company remains as the defendant.

Safety and loss prevention professionals should prepare all documents with surgical preciseness in a defensive manner. In any legal action, all documents will be viewed with hindsight because an accident or injury has already occurred and the issue of liability is now being placed on the appropriate parties. Safety and health professionals should prepare all documents, especially written compliance programs, company policies, and related documents, in preparation of a challenge and to ensure that the appropriate evaluations have been made of these documents by the legal department, personnel or human resource department, and any other applicable department prior to publication.

SHIFTING LIABILITY, THE SCAPEGOAT, AND THE LIAR'S CONTEST

Safety and loss prevention professionals should realize that the basic instinct of self-preservation normally overtakes all levels of the management hierarchy following a serious incident in which criminal charges or substantial civil liability may be involved. Although companies do not intend to become involved in such incidents, when a serious incident such as a multiple fatality accident or a million-dollar OSHA citation occurs, often the management team concept is discarded and the situation becomes every man or woman for himself or herself.

In these types of circumstances, companies should be aware that many levels of the management hierarchy tend to shift the responsibility, and thus the consequential potential liability, for the incident to another level and often disavow any knowledge of previously accepted responsibilities. For example, following a serious incident, the president may pass the buck and responsibility to the vice-president, stating he or she delegated the responsibility to the vice-president. The vice-president may claim that he or she delegated the responsibility to the safety and loss prevention manager. The safety and loss prevention manager may claim he or she is only a corporate function acting only in an advisory capacity to the plant operations. The plant operations management may shift the potential liability to the lower levels through delegation until no one individual can be "pinned" with the liability for the incident. No individual wants to be the designated scapegoat to bear the burden for the deficiencies that ultimately led to the incident.

Companies should be aware that when management-level employees place this cloak of self-preservation upon their shoulders in order to protect their job, freedom, or position, appropriate documentation is vital in order to ascertain the precise extent of responsibility and any deficiency in the management system that ultimately led to the incident. However, companies should be aware that documentation can be shredded, individuals can shade or twist the truth, and finger-pointing can be initiated in order for individuals to attempt to shift the responsibility and thus the liability elsewhere for self-preservation.

Appropriate documentation can sort through this type of situation to ascertain the underlying causes for the incident. Protection should be afforded to these documents in order to avoid the possibility of unexplained disappearances or alteration. As discussed in Chapter 12, many authors have promoted the use of "Pearl Harbor" files for safety and loss prevention professionals in order to protect the safety and loss prevention professionals from being designated as the scapegoat following an incident. Companies, as well as individuals within the management hierarchy, should protect appropriate documentation in order to be able to reassemble the circumstances following an incident and protect the company.

As shown in the *Film Recovery* case and other related cases in which criminal liability has attached, all levels of the management hierarchy, from the president to the first-line supervisor, are at risk for potential criminal liability for failure to follow the prescribed safety and health regulations. In the past, lower and middle management bore the major burden of deficiencies, but now every level of the management hierarchy can and is being held liable for the deficiencies of the organization. Each company should ensure that all levels of its management team are committed to the safety and health effort. As the saying goes, "The chain is only as strong as the weakest link." In safety and health, the management team is only as strong as its weakest member. If a team member falters and an incident results, all levels within the management hierarchy may incur the risk of liability.

The corporation may also be liable for criminal acts. Although a corporation itself is essentially a legal fiction created by statute, courts have found a corporation itself to be criminally liable for the acts of its agent when the agent performed the illegal act within the scope of his or her office or authority. Several states, such as Texas and New York, have broad-based provisions in their criminal codes that permit criminal actions to be brought against a corporation for the actions of its agents, directors, officers, and even employees simply because of the position that they hold within the corporation.

Safety and loss prevention professionals should be aware that the incidents of imposition of criminal sanctions by OSHA and state prosecutors are usually reserved for only the most egregious and willful situations. Also, the safety and loss prevention professionals must have been knowingly involved in the criminal acts resulting in the death or injury. The number of incidents in which OSHA has referred a matter to the Justice Department for criminal action or state prosecutors have filed criminal charges are relatively few in relation to the number of incidents that happen in the United States each year. Although these cases make the headlines, if a safety and loss prevention professional is performing his or her job appropriately and to the best of his or her ability, the potential of criminal liability is remote.

Safety and loss prevention professionals are often a named party in civil actions against a company because of the visibility of their position, the management structure of the corporation, and the name recognition by employees. Bearing in mind that the ultimate goal of a civil action is to acquire monetary damages, the naming of the safety and loss prevention professional is often only a method to ensure that the safety and loss prevention professional will be available during the discovery phase to "pick his or her brain" for information rather than actually seeking monetary damages from the individual. The ultimate "deep pocket" is the company rather than the individual, but the safety and loss prevention professional is often a vital link to the information and documents.

Civil liability for safety and loss prevention professionals is available and is often used in situations involving negligence, willful disregard, or other similar circumstances. In most situations, the company indemnifies the individual from personal liability as long as the safety and loss prevention professional was performing within the scope of his or her employment. If a safety and loss prevention professional is performing his or her job in an appropriately professional manner, neither civil nor criminal liability is likely. That is not to say the loss prevention professional may not be named in a lawsuit. In many circumstances, the safety and loss prevention professional is named to acquire information and documents in discovery and often released later in the action. However, if the safety and loss prevention professional should breach his or her duty and an injury or damage should result, liability may be present. As noted earlier in this chapter, safety and loss prevention professionals should assess their personal risks and take appropriate action to protect against the potential risk.

Safety and loss prevention professionals should be aware that most workers' compensation laws bar civil recovery by individuals who have sustained injuries arising out of and in the course of their employment. With most injuries that occur as a result of a safety systems failure, workers' compensation is the sole remedy, and civil actions against the safety and loss prevention professional or company are usually barred. However, in situations where willful negligence is the cause of the injury, some jurisdictions provide a separate cause of action in addition to workers' compensation. In addition, if a piece of equipment was involved in the accident, a product liability suit may be filed against the manufacturer, distributor, or supplier. Then there is a good chance that the employer will be named in a third-party liability suit. Any negligence on the part of the employer or its agents may expose the company to significant liability.

Safety and loss prevention professionals should also be aware that the appearance or perception of fault can often culminate in a legal action against the individual or corporation. Even if no civil or criminal liability is ultimately found against the safety and loss prevention professional or corporation, the individual or corporation can sustain immense damage in terms of legal costs, damage to reputation, and other efficacy harms. Safety and loss prevention professionals should be aware of these perceptions and do everything feasible to maintain the appropriate appearance to avoid these legal actions in the first place.

In essence, if a safety and loss prevention professional is performing his or her job in a professional manner and is doing everything possible to safeguard the employees working for the company and comply with the various laws, the potential risks in the areas of criminal or civil liability are usually minimal. However, where the safety and loss prevention professional and/or the company is not willing or able, for whatever reasons, to provide this safe and healthful work environment, the potential risks of liability exist and accumulate until an incident results in the liability attaching to the individual or corporation.

SUBCONTRACTORS

A major area of potential liability for companies is in the area of subcontractors. There are two basic tactics being utilized by companies today, namely a complete hands-off approach with regard to the safety and health efforts of the subcontractors or exercising complete control over the subcontractors.

The reason for these two distinct approaches is due to the decisions made by the OSHRC in 1976 in two companion cases. In *Anning-Johnson Company*, a subcontractor who engaged in the installation of acoustical ceilings, drywall systems, and insulation was cited along with the general contractor for the total lack of perimeter guards. Although the subcontractor had complained to the general contractor about the lack of guards, no abatement had occurred prior to the OSHA inspection. The OSHRC found that both the subcontractor and the general contractor possessed responsibility for the violation. With regard to the subcontractor, the OSHRC stated, "What we are holding in effect is that even if a construction subcontractor neither created nor controlled a hazardous situation, the exposure of its employees to a condition that the employer knows or should have known to be hazardous, in light of the authority or 'control' it retains over its own employees, gives rise to a duty under section 5(a)(2) of the Act [29 U.S.C. § 654(a)(2)]. This duty requires that the construction subcontractor do what is 'realistic' under the circumstances to protect its employees from the hazard...."

With regard to the responsibility of the general contractor, the OSHRC held that liability for the OSHA violations would attach despite the fact that the general contractor did not have any employees at the worksite. It held that the general contractor possessed sufficient control to give rise to a duty to correct the situation.

In a companion case, *Grossman Steel & Aluminum Corp.*, the OSHRC reached the same result with respect to liability for OSHA violations for general contractors and subcontractors. In this case, the OSHRC found liability for the general contractor based upon the general contractor's supervisory authority and control. OSHRC stated that, "[T]he general contractor normally has responsibility to assure that the other contractors fulfill their obligations with respect to employee safety that affects the entire site. The general contractor is well situated to obtain abatement of hazards, either through its own resources or through its supervisory role with respect to other contractors."

The *Anning-Johnson/Grossman Steel* analysis derived from these companion decisions still represents the position of the OSHRC with regard to general contractors and subcontractors. Since 1976, four basic categories of cases have evolved from this decision:

1. the employer's affirmative defense that the hazard was not created by nor did the employer control the worksite
2. the employer did not know of the hazard, did not possess control with due diligence, and could not have noticed the hazard
3. the employer either created or failed to control the hazard
4. the employer was found to possess control over the hazard

The *Anning-Johnson/Grossman Steel* analysis has been endorsed, in whole or in part, by five circuit courts of appeal.

In light of the *Anning-Johnson/Grossman Steel* analysis, OSHA has adopted rules for apportioning liability between the general contractor and subcontractor on a multi-employer worksite. OSHA will hold each employer primarily responsible for the safety of his or her own employees, and employers are generally held responsible for violations to which their own employees are exposed, even if another contractor is contractually responsible for providing the necessary protection. However, OSHA has created a two-pronged affirmative defense through which an employer or general contractor might avoid liability (see Selected Case Study). If the general contractor or employer did not create or control the hazard, liability can be avoided either by proving that the general contractor or employer took the steps that were reasonable under the circumstances to protect the employees or that the general contractor or employer lacked the expertise to recognize the hazard.

Given these decisions, the two basic approaches by most companies to minimize the risk of liability are (1) asserting direct control over the worksite and ensuring compliance through the development and management of a safety and health program in compliance or (2) attempting to eliminate this control or management of the worksite by shifting the liability solely to the subcontractor. With either approach, companies normally require the subcontractors to possess a functioning safety and health program meeting the requirements and standards and to ensure compliance on the worksite. Companies should be aware, however, that relinquishing control over the worksite may create other difficulties and may not serve to protect the company with regard to other areas of potential liability.

INSURANCE PROTECTION

In addition to standard insurance coverage (i.e., property, casualty, automobile, etc.) and required insurance coverage by statute (i.e., workers' compensation), companies may want to consider other types of protection for the directors, officers, and other key personnel. Although there is a wide variety of insurance protection that can be obtained, a common type of protection is directors' and officers' liability insurance, also known as "D & O insurance."

D & O insurance is a type of property and casualty coverage similar in many respects to professional liability insurance for physicians or lawyers (also known as malpractice insurance). D & O insurance can serve to protect the directors and officers directly against losses in which the corporation cannot or will not indemnify them, but can also serve to protect the corporation or company from potentially sizable losses because the company or corporation did elect to indemnify an officer or director. D & O insurance, like indemnification discussed earlier in this book, generally protects directors and officers against the consequences of honest mistakes or omissions. Reckless, willful, or criminal misconduct are not insurable as a matter of public policy. The advantages of D & O insurance include the following:

- it provides an independent contractual source of indemnity for officers and directors that eliminates the potential catastrophic consequences of large damage awards during a period when the company or corporation is in an economic slump
- it provides a source of reimbursement for the company or corporation for any amounts paid while indemnifying an officer or director

- it may and can cover some areas, such as payment in derivative actions against a director or officer, that the company is not permitted to indemnify under the particular state's statutes

The negative aspects of D & O insurance include the following:

- it is a contractual agreement with generally strict requirements and limitations
- it tends to be expensive
- it generally has limits on coverage

Variations such as "fronting arrangements" (the director enters into an arrangement that is in all respects a standard D & O contract but, unlike most contracts, the company agrees to reimburse the insurer in full for all losses it pays out in excess of premiums received, and the insurance company receives a fee for the underwriting and claim services) and other forms of D & O insurance have also developed.

With insurance costs escalating, companies often search for alternate methods for providing protection for themselves as well as their officers and directors. One method that has become more common in recent years is an indemnification trust. With this method, a trust is established to fund the indemnification provisions in the corporate articles or bylaws or under separate contract. This may benefit one or all of the directors or officers. This arrangement often offers greater flexibility and some insulation for the directors and officers in case the corporation should go bankrupt or otherwise become insolvent. Additionally, because the trust is administered by a trustee, the trust can be designed to provide greater flexibility than other methods. An indemnification trust can be used as part of an integrated protection program that enhances the standard indemnification or can be separate indemnification contracts with each officer or director. An indemnification trust can be established, like any other trust, with a bank or other appropriate entity and is usually structured as irrevocable and nonamendable in order to avoid the reach of corporate creditors. Legal counsel should be consulted prior to attempting to structure such an indemnification trust arrangement. The negative aspects of an indemnification trust include specific procedures and amounts for funding the trust, required claims procedures, challenges under state fraudulent conveyance statutes, and tax and accounting issues.

With any insurance or other method of protection, companies are attempting to protect themselves and their officers and directors against the unknown and unforeseen cataclysmic event or circumstance that could cause financial ruin. In the area of safety and health, directors and officers can protect themselves and their companies from most civil liability, but normally cannot acquire protection against criminal liability or administrative fines. The most efficient and cost-effective method of minimizing the risks in the area of safety and loss prevention is to develop and maintain a safety and loss prevention program that complies with all government regulations and that creates a work environment free of hazards.

CORPORATE COMPLIANCE PROGRAM CHECKLIST

To assess whether your company is in need of a comprehensive compliance program in the area of safety and health and other areas of potential risk, the following list of questions can assist you to assess your current position. Every "no" answer should signal a potential risk for which a program is needed.

1. Does the board of directors put a high priority on safety, health, loss prevention, environment, and other regulatory compliance requirements?
2. Has the company adopted policies with regard to compliance with OSHA regulations and other laws having a direct bearing on the operations?
3. Has the company established and published a code of conduct and distributed copies to employees?
4. Has the company employed an individual(s) who will be directly responsible for safety and loss prevention and OSHA compliance? Are these individuals properly educated and prepared to manage the safety and loss prevention function?
5. Has the company formally developed a safety and health or loss prevention committee?
6. Does the company possess all of the necessary resources to develop and maintain an effective safety and loss prevention compliance program?
7. Are required safety and loss prevention compliance programs documented in writing?
8. Are corporate officers, managers, and supervisors sensitive to the importance of safety, health, and loss prevention compliance?
9. Are employees involved in the safety and health efforts?
10. Are corporate officers and managers committed to the safety and loss prevention compliance efforts? Do they provide the necessary resources, staffing, etc., to effectively perform the safety and loss prevention function successfully?
11. Does the company conduct periodic safety and loss prevention and other legal audits to detect compliance failures? Are deficiencies or failures corrected in a timely manner?
12. Has the company established "hotlines" or other mechanisms to facilitate reporting of safety and health, environment, etc., compliance failures?
13. Are all employees properly trained in the required aspects of the safety and health compliance programs?
14. Is training properly documented (i.e., will the documentation prove beyond a shadow of a doubt that a particular employee was trained in a particular regulatory requirement)?
15. Does the company provide a new-employee orientation program?
16. Does the orientation program for new employees include review of safety and loss prevention policies, codes of conduct, and other policies and procedures?
17. Are employees provided hands-on safety, health, and loss prevention training? Are employees provided other on-the-job training?

18. Does the company conduct compliance-training sessions to sensitize managers and rank-and-file employees to their legal responsibilities?
19. Does the company provide information and assistance to employees regarding their rights and responsibilities under the individual state's workers' compensation laws?
20. Does the company communicate safety, health, and loss prevention compliance issues to employees with postings, newsletters, brochures, and manuals?
21. Does the company go beyond the "bare bones" compliance requirements to create a safe and healthful work environment?
22. Does the company keep up with the new and revised OSHA standards and other regulatory compliance requirements?
23. Does the company discipline employees for failure to follow prescribed safety rules and regulations? Is this discipline fair and consistent?
24. Is safety, health, and loss prevention a high priority for officers and directors of the company?
25. Is the company proactive in the area of safety, health, and loss prevention?

In developing a corporate code of conduct for use within a company, special care should be taken to properly structure the code to encompass a broad spectrum of potential risks while also providing guidance to employees as to the expected behaviors. The following items should be considered:

1. Publish the content of a corporate code of conduct. The most common laws covered in a code of conduct include labor law, antitrust law, business ethics, conflicts of interest, corporate political activity, environmental law, safety and health laws, employee relations law, securities laws, etc. Special care should be provided in the area of individual and personal rights.
2. Distribute the code to officers, directors, and employees. Most companies distribute a code of conduct to directors, officers, and employees when they join the company and on a yearly basis. The company may want to document that the directors, officers, and employees have read, have understood, and will adhere to the code.
3. Provide training and education to promote compliance. In-house seminars can be targeted to small groups of key managers to sensitize them to their legal responsibilities. Some companies use guidebooks as memos to communicate the importance of compliance to their employees. Newsletters and memos can be distributed to managers on a periodic basis to remind them about their legal responsibilities, advise them of the developments of the law, and give them preventive law tips.
4. Enforce the code of conduct. Compliance programs should include disciplinary procedures to punish violations of the code of conduct. Compliance programs are not effective if they are not enforced. Sanctions for violation can include verbal warnings, written warnings, suspension, demotion, discharge, and referral to law enforcement agencies. In addition to sanctions, the disciplinary procedures should include provisions for protecting whistle-blowers and for investigating allegations of illegal conduct.

5. Monitor compliance through legal audits. Legal audits can include, but are not limited to, the following activities:
   - assemble legal audit team
   - educate and train legal audit team
   - assign legal team responsibilities
   - develop a site-specific audit instrument
   - conduct site visitation and facility inspection
   - conduct employee interviews
   - conduct records search
   - develop and use employee questionnaires
   - review record retention procedures and policies
   - develop an audit report
   - present audit report to board of directors and officers
6. The legal audit mechanism, like the internal safety and loss prevention audit discussed previously, produces documentation that may be the subject of discovery requests in civil or criminal litigation. This type of evaluation may also produce sensitive data that the company seeks to keep confidential. Extreme caution should be exercised to preserve privilege and confidentiality.

In summation, corporations and other legal entities should assess their potential risks and balance these risks against the possible benefits and possible detriments. With the appropriate preparation, the vast majority of potential risks can be properly managed to safeguard not only your company but also your personnel and yourself. Identifying the risk, assessing the risk, and adopting a proactive approach before something happens can minimize or eliminate the impact of the risk or even the risk itself.

SELECTED CASE STUDY

Note: Case summarized for the purpose of this text.
United States Court of Appeals, Seventh Circuit.

UNITED STATES OF AMERICA, PLAINTIFF-APPELLEE, v. L.E. MYERS COMPANY, DEFENDANT-APPELLANT

No. 07-2464.

Argued Feb. 26, 2008.
Decided April 10, 2009.

Background

After electrical contractor was convicted of willfully violating Occupational Safety and Health Administration (OSHA) regulations, causing the death of one of its employees, the United States District Court for the Northern District of Illinois, Brown, United States Magistrate Judge, 2005 WL 3875213, denied contractor's motion for new trial. Contractor appealed. The United States District Court for the Northern District of Illinois, James B. Zagel, J., affirmed. Contractor appealed.

L.E. Myers Company, a large electrical contractor, was convicted of willfully violating Occupational Safety and Health Administration ("OSHA") regulations, causing the death of one of its employees. *See* 29 U.S.C. § 666(e). An apprentice linesman in the early stages of his training with L.E. Myers was killed while working on a repair assignment atop a transmission tower owned by Commonwealth Edison ("ComEd"). The "static" wire on the tower was not in fact "static" (i.e., a grounded "dead" wire) but instead was energized; the apprentice came into contact with it and was electrocuted.

On appeal L.E. Myers argues that the magistrate judge who presided at trial improperly instructed the jury on the issues of corporate knowledge and conscious avoidance. The company also argues that the judge erroneously excluded evidence of a 1979 fatality involving a ComEd linesman who was electrocuted by contact with an energized static wire. Finally, the company claims it is entitled to a new trial on the basis of a proposed OSHA regulation creating new duties for "host employers" like ComEd regarding hazards at their transmission facilities. The proposed rule, published for notice and comment after the trial was concluded, was accompanied by an explanatory comment from OSHA describing energized static wires as one such hazard.

We reverse. The magistrate judge improperly instructed the jury on corporate knowledge and conscious avoidance. Corporate knowledge in this context includes knowledge of hazards acquired by the corporation's employees provided the employees in question are responsible for reporting such hazards to the corporate hierarchy. This important proviso was omitted from the jury instruction on corporate knowledge. Furthermore, there was insufficient evidentiary support for the conscious avoidance or "ostrich" instruction; it should not have been given. Because the statute's willfulness requirement was the central point of contention in this criminal OSHA case, we are not convinced that these instructional errors were harmless. Remand for retrial is required.

We reject L.E. Myers's evidentiary argument, however; the evidence of the 1979 ComEd fatality was properly excluded. Finally, because we are reversing for a new trial based on instructional error, we need not address whether L.E. Myers is entitled to a new trial on the basis of the proposed OSHA regulation.

12 Personal Liability for Safety and Loss Prevention Professionals

CHAPTER OBJECTIVES

1. Acquire an understanding of potential civil liabilities.
2. Acquire an understanding of the potential criminal liabilities under the OSH Act.
3. Acquire an understanding as to how to assess risk in a safety job function.
4. Acquire an understanding of the protective measures to minimize risk.

INTRODUCTION

Safety and loss prevention professionals are becoming increasingly aware of the pressures placed upon them and the increased potential of personal liability for errors and omissions in the decision-making process. Unlike most managerial positions, safety and loss prevention professionals are making decisions on a daily basis that can directly or indirectly affect the health, safety, and personal welfare of the employees within their charge. Within this decision-making process, safety and loss prevention professionals are usually required to juggle many issues when making these decisions as well as balancing the internal and external pressures to make a correct decision on an issue. Although safety and loss prevention professionals are only human and mistakes can be made, when a mistake is made in the area of safety and loss prevention, the employees, customers, contractors, or other individuals are placed at risk for incurring an injury, illness, or even death.

Because of the unique managerial position in which most safety and loss prevention professionals are now being placed (i.e., on the management team but closely aligned with the needs of the employees), conflicts often occur that require the professional to make tough decisions. Safety and loss prevention professionals are often pressured by upper management to minimize cost expenditures and make do with the status quo. However, pressure is also placed upon the safety and loss prevention professional from employees and labor organizations to make improvements in the work environment. In addition, OSHA and other government agencies continue to produce new and modified standards and regulations that require further development, change, and implementation. In short, safety and loss prevention professionals are normally pulled in many directions by varying forces. They also often possess minimal resources and personnel to handle their current workload, let alone expanding regulatory burdens. Added to this is the pressure to make tough decisions and to be right every time. Naturally, the probability of a safety and loss prevention professional making a mistake is high.

The current status of safety and loss prevention programs can be summarized as: "the game is the same, but the risks have increased!" Try to imagine being transported to Las Vegas and standing next to the $2 blackjack table. For most individuals, the risks are minimal. We then move to the $100 blackjack table. The game is the same, but the risks have increased. If we move again to the $1,000 blackjack table, the game is still blackjack, but the risks have been increased greatly. This is similar to the safety and loss prevention profession. From 1970 to 1985, the game was compliance with the OSHA standards. Personal liability risks were minimal. If an accident happened or a citation was issued, the greatest risk to a safety and loss prevention professional would be the loss of a job. From 1985 to 1990, the game was still safety and loss prevention, but the risks increased with the sevenfold increase in the OSHA penalties and the use of state criminal sanctions. Beginning in the late 1990s, an even higher-risk situation exists because of the increased utilization of criminal sanctions by OSHA and the use of state criminal code sanctions by state prosecutors. The game is still safety and loss prevention, but the risks for the company and for the safety and loss prevention professional personally have increased dramatically.

Above all, safety and loss prevention professionals must realize that there is no substitute for a safety and health program that is in compliance with the OSHA standards. Safety and loss prevention professionals should utilize all of their skills and knowledge to ensure a safe and healthful work environment for their employees. Taking shortcuts and placing programs on the back-burner as a result of cost difficulties, internal pressures, and other influences places the safety and health professional at greater risk of making a mistake that could dramatically change his or her personal and professional life.

PERSONAL RISK ASSESSMENT INSTRUMENT

Safety and loss prevention professionals may want to evaluate and assess their personal risk in their current capacity. The following general personal risk assessment instrument will assist in this evaluation:

1. Are all of your programs in compliance with OSHA, EPA, and other applicable regulations?
2. Do you possess upper management support?
3. Are you provided with the appropriate resources to complete the work in a timely and appropriate manner?
4. Do you possess the necessary education, training, and education resources to adequately perform the function?
5. Are you provided the resources to appropriately train your employees?
6. Do you possess outside resources to assist you when necessary?
7. Is your company a target company?

8. Do you possess adequate personnel to appropriately manage the safety and health function?
9. Are you responsible for other functions besides safety and health, such as security, environment, workers' compensation, and so on? How much of your time is spent on safety and health?
10. What is your company's safety, health, and loss prevention history? Is your company at high risk for serious injuries or fatalities?
11. Is everything (i.e., OSHA 200, training logs, etc.) documented appropriately? Are all of your compliance programs in writing?
12. If push came to shove, would your company support you or would you be the designated scapegoat?
13. Do you document your recommendations to your management?
14. Do you carry separate professional liability insurance?

Safety and loss prevention professionals who are working under adverse conditions often burn out or give up the fight after a period of time. The safety and health function is not for the perfectionist or the faint of heart. The safety and loss prevention profession is often a battle to acquire funding, to achieve compliance, to maintain compliance, and to effectively manage a number of subfunctions over a long period of time. Although the perfection of achieving a zero accident goal and achieving compliance with all regulatory requirements is the supreme goal, the struggle to attain this goal and, if attained, to maintain the status is extremely difficult and often fleeting. In a vacuum, the ultimate goals can be achieved and maintained, but in our world of change, a variety of outside factors, such as labor strife, a change in personnel, and a modification in workload, can affect the status quo of the safety and health effort. The constant building and rebuilding of the safety and health program can lead to enormous frustration and possible mistakes.

Although most safety and loss prevention professionals view their work as a quest to protect their employees, long work hours and efforts over and above the call of duty take a toll. Safety and loss prevention professionals may find that they are at risk, but cannot make a change, internally because of the corporate structure or externally because of personal financial considerations such as a house mortgage and car payment. What can be done to protect the individual from potential personal liability?

As far back as the 1950s, many safety and health scholars, such as Dan Petersen in his text *Safety Management* and C. Everett Marcum in *Modern Safety Management,* stated that a "Pearl Harbor" file may be needed. Such a file documents every detail of every action to protect a safety and loss prevention professional in a high-risk situation where management commitment has evaporated or the safety and health program is not or cannot achieve compliance. Like the attack on Pearl Harbor in World War II, the safety and loss prevention professional will not know when a decision he or she has made will evolve into a disaster and when the bombs might be dropped on him or her. This type of documentation can support the safety and loss prevention professional and lend credibility when under fire. A Pearl Harbor file includes such items as appropriate correspondence, documentation of telephone conversations, denial of funding at meetings, and other appropriate information.

Hindsight is 20/20! Safety and loss prevention professionals should realize that in any type of litigation the jury or judge is evaluating the accident situation from a hindsight perspective. When the pressures of litigation are applied to the various individuals involved in the case, the number of individuals who contract instant amnesia is amazing. Without the supporting information that can be provided through a Pearl Harbor file, the judge or jury will have to make a determination about credibility, identifying which individual is lying (sometimes called "a liars' contest"). Appropriate documentation can lend support and credibility to the safety and loss prevention professional and often serves as a refresher to the memory of other witnesses. Like proving compliance with OSHA and other regulatory agencies, documentation is vital in protecting yourself and avoiding being the designated scapegoat for the situation.

PERSONAL PROTECTIVE THEORIES

Safety and loss prevention professionals often ask how they can protect themselves and their families from risks involving personal liability. Only one theory is available in the area of personal criminal liability, but three basic personal protective theories can be utilized in the area of personal civil liability. For personal criminal liability, the only method of protecting yourself and your company from potential liability is to be in compliance with all government regulations and to develop the best safety and health program possible. If the injury or fatality that serves as the catalyst does not occur, the potential of criminal liability is minimal.

The three protective theories involving personal civil liability in the area of safety and loss prevention include indemnification by the company, acquisition of personal professional liability insurance, and "going naked" or possessing no coverage but possessing minimal assets. These broad theories are general in nature and each safety and loss prevention professional should assess his or her individual situation to determine the level of personal risk involved in the job and the amount of protection that can be afforded.

PERSONAL CIVIL LIABILITY AND INDEMNIFICATION BY THE COMPANY

In many larger companies, the company, through charter provisions, bylaws, or separate contracts, indemnifies their managerial employees for errors and omissions made while performing within the scope of their employment. Indemnity, by definition, is "a collateral contract or assurance, by which one person engages to secure another against anticipated loss or to prevent him from being indemnified by the legal consequences of an act or forbearance on the part of one of the parties or of some third party. Terms pertaining to liability for loss are shifted from one person held legally responsible to another person." In a safety and loss prevention type situation, indemnification means that the company will bear the costs of legal fees, any civil penalties, and damages for the safety and loss prevention professional if he or she is named personally in the action that is basically against the company.

In our current litigious society, individuals performing the safety and loss prevention function can expect to be named in most related suits against the company. From a plaintiff's viewpoint, the naming of the individual responsible for the safety and loss prevention function is a method of ensuring that the individual will be available for deposition. As often happens, after the safety and loss prevention professional provides testimony under oath at the deposition and the plaintiff has picked his or her brain, the safety and loss prevention professional is dropped from the action because he or she is not the "deep pocket." With the high costs of personal liability insurance, safety and loss prevention professionals often assume that their employer will indemnify them for their actions on the job. This may not always be the case because of the types of charges or action brought, the individual state laws, the public policy limitations, or even the individual corporation bylaws. A prudent safety and loss prevention professional may want to check the scope of the company's indemnification with corporate counsel prior to an incident. Assumption of indemnification by the company may lead to problems after an incident has happened.

The scope and parameter of indemnification can vary from state to state and from company to company. In the area of state law, historically, certain states, such as New York and Tennessee, provide that defined indemnification rights were granted to officers, directors, and "others." These were exclusive and occupied the entire field of permissible indemnification. Similarly, states that enacted the Model Business Corporation Act require consistency in the indemnification provisions with respect to directors, but not for officers, employees, or agents. While the Model Business Corporation Act approach has largely been abandoned in favor of the nonexclusivity provisions, close review of the individual state's incorporation statutes with regard to indemnification should be closely evaluated. Additionally, it is not always sufficient to check the statutes of the incorporating state. Several states have sought to apply their public policy exceptions with regard to indemnification to corporations doing business in the state but organized in another state, whereas other states have permitted bargaining for indemnification.

Some corporations, in an attempt to resolve uncertainties in state laws or for other reasons, have incorporated into their corporate bylaws certain provisions regarding indemnification. In most circumstances, the bylaw clause provides additional rights not granted under the statute or clarifies the indemnity rights of the directors, officers, and others. Although this extension of the nonexclusivity language is often controversial, several courts have permitted corporations to extend or vary the state statutory scheme of indemnification by contract and by article or bylaw amendments. Safety and loss prevention professionals should also be aware that both federal and state public policy limitations may prevent indemnification, even if the corporation possesses a nonexclusivity clause and the corporate directors wish to indemnify the safety and loss prevention professional.

In assessing whether your company is permitted to indemnify you as the safety and loss prevention professional, the following questions should be raised with corporate counsel prior to any action:

1. Am I, in my current capacity, considered a covered person under the provisions of the incorporating state statute?
2. What is the extent of my indemnification, if any?
3. Are attorney fees and other peripheral expenses covered through my indemnification?

4. Are there any federal or state public policy exceptions that could prohibit my indemnification?
5. Are there any bylaws or amendments in the corporate charter prohibiting indemnification?
6. Is the indemnification provided to me different from the indemnification provided to directors and officers of the company?
7. What are the scope and parameters of such indemnification?
8. As noted previously, indemnification under virtually all state statutes and corporate bylaws is for civil actions only. If criminal charges are brought under the OSH Act or by state prosecutors, is the company permitted under state law and through the charter and bylaws to provide financial support? What is the company's position regarding criminal actions?

A prudent safety and loss prevention professional should attempt to address potential areas of personal liability in advance with the legal staff and acquire the company's position regarding indemnification. If the company is permitted to indemnify the safety and loss prevention professional, the safety and loss prevention professional should define the extent of the provided indemnification coverage and identify the areas in which the safety and loss prevention professional is on his or her own.

This is normally not a subject that corporate officers and legal counsel discuss with their managers and agents until after an incident has happened. Given the potential risks that exist in the safety and loss prevention profession, a prudent safety and loss prevention professional may want to know the parameters of any such indemnification prior to an incident so that a valid assessment can be made of the personal civil risks and appropriate decisions can be made regarding other possible coverage.

PERSONAL CIVIL LIABILITY AND PROFESSIONAL LIABILITY INSURANCE

The second basic theory for personal protection of a safety and loss prevention professional is the acquisition of professional liability insurance. This theory has many obvious benefits but contains several pitfalls.

Professional liability insurance in the safety and loss prevention profession is widely utilized by consultants, individuals in high-risk positions, and other similar positions. Like other insurance coverage, professional liability insurance provides varying levels of coverage, normally ranging from $100,000 to several million dollars. The coverage is usually for errors and omissions that result in civil actions against the individual. The rates for professional liability insurance usually vary with the education and experience of the individual, the type of work performed, prior liability history, and other factors. At this point in time, there are only a few specialized insurance carriers that will write liability insurance for safety and loss prevention professionals.

The positive benefits of professional liability insurance are that if an incident occurs where civil liability could attach, the insurance policy will pay for the cost of the defense as well as any damages, up to the policy limits. A safety and loss prevention professional deciding to purchase professional liability insurance should read the policy closely to determine the parameters of the coverage and closely scrutinize whether such peripheral costs as legal fees and expert witness fees are covered and whether the safety and health professional relinquishes authority and control over the case to the carrier. Professional liability insurance does provide peace of mind for safety and loss prevention professionals at risk and often protects personal assets if an error or omission is made by mistake and is not willful.

The downside of professional liability insurance is the cost. This type of specialized coverage is relatively expensive and the scope of coverage is normally limited. In many circumstances, the company will not provide such coverage and the safety and loss prevention professional wishing to acquire this coverage would be required to pay the premiums personally. It should also be noted that most professional liability insurance does not cover the individual for criminal actions.

A peripheral consideration for professional liability insurance in a civil action is whether possession of the insurance coverage will make the safety and loss prevention professional another deep pocket. In civil actions, the individual injured or damaged (known as the plaintiff) is seeking monetary damages. In some circumstances, the plaintiff will name the safety and loss prevention professional to the action in the beginning in order to ensure that all parties are included in the action. Upon acquisition of records and deposition testimony, the safety and loss prevention professional is dropped as a named party to the action and the company is pursued because the company is the possessor of the money, that is, the deep pocket. However, if the safety and health professional possesses a substantial amount of insurance coverage, would the plaintiff drop the safety and loss prevention professional from the action or would the safety and loss prevention professional be maintained in the action as another deep pocket? This may be a consideration to be evaluated by the safety and loss prevention professional depending on the risks and circumstance.

PERSONAL CIVIL LIABILITY WITHOUT PROTECTION

The third and most common theory occurs when the safety and loss prevention professional has no protection. Although most safety and loss prevention professionals are afforded some protection from civil actions under the indemnification theory through their company, most safety and loss prevention professionals possess no insurance coverage and, in the event of a failure to acquire protection from the company, would be "left in the cold."

The argument may be made that if the safety and loss prevention professional has few appreciable personal assets, going naked may be the best avenue. The concept is that they cannot take something that you do not have. Remember, in a civil action, the plaintiff is seeking money damages. If the safety and loss prevention professional does not possess personal or real property of any appreciable value, there is nothing to take. The negative aspect of this concept is that if the safety and loss prevention professional owns a home, a car, a bass boat, or anything of appreciable value, these items could be attached and sold to pay any damage awards. In a criminal action, monetary fines can be assessed in addition to incarceration.

The negative aspect of going naked is that everything owned is at risk in the event of a judgment. If the company does not indemnify the safety and loss prevention professional in a civil action, the safety and loss prevention professional would be responsible for all costs, including but not limited to, attorney fees, court fees, and expert witness fees, in addition to any money damage awards. Legal fees can be a substantial sum of money, even if the safety and loss prevention professional is successful in defending against the action. If the action is criminal, most safety and loss prevention professionals will not be able to qualify for public legal services because of their incomes.

Several alternatives have been tried in order to minimize the exposure to civil liability. Individuals have attempted to transfer assets to a wife or family member following an incident that resulted in civil liability. In most circumstances, this "shell game" tactic is easily exposed by the courts. In essence, the courts disallow the transfer and return the transferred assets to the defendant. If assets are to be transferred legitimately, in most states they are required to have been transferred several years prior to the incident.

The establishment of a corporation for the purposes of performing safety and loss prevention "work," such as consulting activities, is another method gaining wide acceptance. Under this method, a corporation (usually either a "C" or "S" corporation) is formed and incorporated within a particular state. The corporation would possess, but not be limited to, a board of directors (normally at least two individuals), possess articles of incorporation and bylaws, hold at least annual meetings, file separate tax forms, and possess separate assets. In essence, a "corporate veil" would be developed around the safety and loss prevention professional and only corporate assets would be at risk. If a civil action was filed, the corporation would be named (possibly along with the director and/or officers) and liability would be to the corporation, which could indemnify the officers and directors. Personal liability for the officers, directors, and agents would apply only if the "corporate veil" could be pierced. Because of the tax and liability issues involved in this methodology, evaluation by legal counsel about the type of corporation and by an accountant for tax ramifications is critical.

In many states, you have the ability to form and utilize a limited liability company. This legal entity possesses many of the protections of a corporation as well as the simplicity and tax considerations of a partnership. Additionally, limited liability companies often provide additional legal protections for individuals.

This method of potential protection works well for consultants and other independent parties. However, for a safety and loss prevention professional working for a corporation, most corporations require the safety and loss prevention professional to possess the status of employee within the management team rather than a contractor.

Individual Employment Contract

A possible method of protection for safety and loss prevention professionals working for a corporation is the use of an individualized employment contract. Although this method usually has been reserved for the highest echelons within the corporate world, an individualized employment contract may be a method through which a safety and loss prevention professional can set the parameters of any indemnification protection prior to the start of employment. A standard employment contract usually provides the terms and conditions of employment, the compensation and fringe benefit package, and length of the agreement. One important aspect of the employment contract is that it can spell out the duties of the professional. Consequently, it can also, by implication, indicate what is not the responsibility of the professional.

An employment contract additionally removes the safety and loss prevention professional from the at-will employment status (can be terminated for good cause, bad cause, or no cause at all) and places the individual in a contractual relationship where termination, demotion, layoff, or other changes in the employment relationship must be in accordance with the written agreement. Most employers do not wish to enter into such a binding relationship with their low- or middle-level management employees. In essence, most employers wish to maintain the at-will relationship with these employees where the control over the relationship is maintained at the highest levels and the employment relationship can be terminated at any time without penalty to the company.

In conclusion, there is no one right method or theory that is applicable to all situations. A prudent safety and loss prevention professional may want to evaluate and assess the personal risks that he or she is shouldering in the safety and loss prevention capacity and explore some of the options that may afford additional protection. The simple solution is to develop and maintain a program in compliance with all government regulations and one that protects your employees from injuries or illnesses. If the accidents are prevented before they happen and your facilities are in compliance, the potential of personal liability is minimal.

WHAT TO EXPECT WHEN THE UNEXPECTED HAPPENS

A civil action can be based on many different laws, both statutory and common, but basically follows the same path through the steps prescribed by the Federal Rules of Civil Procedure and/or individual state rules and regulations regarding civil procedure. Unlike a criminal action, there is usually no initial investigation conducted by the plaintiff or legal counsel for the plaintiff. The first notice that a safety and loss prevention professional will have that a civil action has been filed is when he or she is served with a complaint. The complaint is, by definition, "The original or initial pleading by which an action is commenced under the codes or Rules of Civil Procedure. The pleading (complaint) sets forth the claim for relief. ... and shall include: (1) a short and plain statement of the grounds upon which the court's jurisdiction depends ... (2) a short and plain statement of the claim ... (3) a demand for judgment for the relief to which he deems himself entitled. ..." A complaint is often served to the safety and health professional by a marshal or process server. Upon receipt of a complaint, the safety and loss prevention professional should seek legal counsel. A relatively short time frame, 20 to 30 days, is available for answering the complaint or filing some type of responsive pleading or objection.

Unlike a criminal charge where a speedy trial is required, a civil action may take many months or even years to get to trial. The next step in which a safety and loss prevention professional is involved is called the discovery phase. In discovery, the plaintiff sends interrogatories to the parties (written questions) and will likely take the deposition of the safety and loss prevention professional. Outside of the trial itself, the discovery phase can be the most grueling and stressful period during the case.

In a normal deposition, a time, date, and location are agreed upon by your counsel and counsel for the opposition. The location can be anywhere from a conference room to your office. A court reporter is usually present to record the testimony, and the individual being deposed must swear or affirm to tell the truth. The deposition may be stereotypically recorded and/or videotaped. In most situations, the plaintiff's attorney will ask most of the questions and the counsel for the defendant may make objections on the record. It is highly advisable to review testimony with legal counsel prior to the deposition. The rule of thumb for being deposed is to stick to the facts and always answer truthfully. Do not volunteer information and do not answer following an objection by your attorney unless instructed to do so thereafter.

Safety and loss prevention professionals should be aware that, in general, anything and everything is fair game in a deposition. As part of the discovery process, the plaintiff's attorney is attempting to add substance and evidence to support his or her case. Depositions can be as short as a couple of minutes to as long as several weeks. However, most depositions are between 30 minutes and two hours long. Safety and loss prevention professionals should maintain their composure throughout the deposition and not become upset by the various tactics and challenges offered by the plaintiff's attorney. Additionally, safety and loss prevention professionals should be prepared to answer the same question, although phrased or couched differently, numerous times during the deposition. In essence, the plaintiff is attempting to evaluate whether the answer is changing throughout the testimony. Your attorney will likely object after a few times of repetitive questions by saying "objection, asked and answered."

Safety and loss prevention professionals should be aware that the testimony provided in a deposition is being recorded and can be used at trial if the answer provided to the question asked has substantially changed. In essence, a deposition is locking in your testimony for future use at trial. It is also giving the plaintiff answers to important questions and leads to other discoverable information, documents, persons, or witnesses.

Another important part of discovery is the obtaining of documents through a motion to produce documents. Generally, notes, memos, logs, and other documents kept by the safety and loss prevention professional are discoverable. That is, the professional would have to make copies or allow the plaintiff to review these documents. The professional should always seek legal counsel prior to handing over documents. Throughout and after discovery, the parties will likely engage in what is called motion practice. This is where the case is generally narrowed or discussed, or additional parties or causes of action are added or deleted. Following the discovery phase, the case is usually set for trial. Normally, the safety and loss prevention professional will assist legal counsel in preparation for trial. It should be noted that the vast majority of civil cases (more than 90 percent) are settled prior to trial. Assistance may come in the form of specialized expertise, assisting legal counsel in deposing the opposition, evaluating expert witnesses, and other activities.

The format for a civil trial is basically uniform in most courts. The trial can be before a jury (if permitted) or a judge only. The plaintiff begins by providing an opening statement that summarizes the case from the plaintiff's point of view. The defendant then is provided an opportunity to give an opening statement. The plaintiff then calls witnesses to support his or her case and the defendant is provided an opportunity to cross-examine the witnesses. Following all of the plaintiff's witnesses and after the plaintiff rests (finishes), the defendant is then provided the opportunity to present his or her witnesses and the plaintiff is permitted to cross-examine these witnesses. Upon completion of all testimony, both sides are permitted closing statements. The judge will provide the jury with instructions on the law and standards of proof. The jury will then deliberate the case and render a verdict.

If the safety and loss prevention professional is a named defendant, he or she will be expected to sit at the defendant's table in the courtroom. If the safety and loss prevention professional is to be called as a witness, the witnesses usually are seated in the gallery or just outside the courtroom and called in by name by the court.

In preparing to testify, safety and loss prevention professionals should properly and fully prepare themselves and review their direct testimony with their legal counsel prior to the hearing. The direct questions by your legal counsel are usually open-ended questions that permit the individual testifying to verbalize the situation. Safety and loss prevention professionals should be prepared to give short "yes and no" answers on cross-examination and possibly receive verbal attacks from the plaintiff. The plaintiff, in cross-examination, is usually attempting to either discredit the testimony or attack the credibility of the individual testifying, if different from the plaintiff's perspective of the case.

Specific documents to be used as reference are often permitted while testifying, but should be reviewed by legal counsel prior to trial. Charts, graphs, photographs, and other demonstrative evidence should also be reviewed by legal counsel prior to trial.

The mental anguish, sleepless nights, and emotional stress that can occur during civil or criminal litigation can take a heavy toll on an individual irrespective of whether the case is won or lost. The best method to avoid potential lawsuits is by avoiding accidents, but if litigation should arise, the best advice is to remain calm, think before you speak or act, and always tell the truth. Remember: document, document, document.

SELECTED CASE STUDY

From OSHRC website and the case modified for the purposes of this text.

Secretary of Labor, Complainant, v. The Barbosa Group, Inc., d/b/a Executive Security, Respondent

OSHRC Docket No. 02-0865

Decision

Before: Chairman; Commissioners.

The Barbosa Group, Inc., d/b/a Executive Security (Barbosa), a Texas-based sole proprietorship, supplied security personnel under contract to a detention facility operated by the Immigration and Naturalization Service (INS) in Batavia, New York. In May 2002, following an inspection of the Batavia facility by the Occupational Safety and Health Administration (OSHA), OSHA issued Barbosa two citations—one serious and one willful—each alleging two violations of 29 C.F.R. § 1910.1030, OSHA's bloodborne pathogens (BBP) standard. The late Administrative Law Judge Michael H. Schoenfeld affirmed all four violations as alleged and assessed OSHA's total proposed penalty of $132,750.

On review, Barbosa does not dispute the existence of the violative conditions. Indeed, as the judge noted, Barbosa admits that "its contract employees at the Batavia facility were unlawfully denied adequate protection against blood borne pathogens." The issues we decide today include (1) whether Barbosa is the employer responsible for the cited conditions; and (2) whether two of the citation items—those alleging violations of the BBP standard's provisions on the hepatitis B virus (HBV) vaccine and post-exposure follow-up treatment—are not willful because Barbosa reasonably believed it had no duty and, in fact, was powerless to provide these required protections to its security personnel under its contract with the INS.

As indicated in this opinion, and in the separate opinions of Chairman Railton and Commissioner Rogers, we determine that Barbosa was responsible for the cited conditions as an employer of the contract security personnel it provided to the INS's Batavia facility and that it failed to effectively delegate its compliance responsibilities to the INS or any other entity. We, therefore, affirm the four citation items at issue. However, for the reasons stated herein, Chairman Railton and I agree to recharacterize one of the willful violations as serious, group the three affirmed serious violations for penalty purposes, and assess a total penalty of $69,300.

Background

The INS contracted with Barbosa to provide approximately sixty-five security personnel who worked alongside an equal number of INS security personnel at its Batavia detention facility. It is undisputed that the INS had control over the Batavia facility as well as all security personnel physically on the premises, including those provided by Barbosa. The INS's control even extended to duty assignments, as well as to discipline and removal.

Barbosa hired all of the security personnel it provided to the INS's Batavia facility and paid their salaries and benefits. Barbosa's security personnel also received day-to-day instructions, assignments, work schedules, promotions and pay from two Barbosa managers located at the facility. However, regular on-site supervision was also provided by Barbosa "shift" supervisors who were hourly employees. Barbosa and the Service Employees International Union were parties to a collective bargaining agreement covering nonmanagement personnel at the Batavia facility. Under the Barbosa/INS contract, the INS was required to provide site-specific job training, including BBP training, to Barbosa security personnel while Barbosa was required to provide separate training to its on-site supervisors. The U.S. Public Health Service conducted the BBP training for all hourly security personnel at the Batavia facility.

Discussion

I. Employer under the OSH Act

The first question presented by this case is that of Barbosa's status as an employer of the security personnel it provided to the INS at its Batavia facility. In determining whether an employer–employee relationship exists under the Occupational Safety and Health Act of 1970, 29 U.S.C. §§ 651–678 (OSH Act), the Commission applies the common-law agency doctrine enunciated in Nationwide Mut. Ins. Co. v. Darden, 503 U.S. 318, 322–23 (1992) ("Darden"). See Froedtert Mem. Lutheran Hosp., Inc., 20 BNA OSHC 1500, 1506, 2002 CCH OSHD 32,703, p. 51, 733 (No. 97-1839, 2004) ("Froedtert").

In Froedtert, the Commission applied the Darden analysis to a case involving co-employment issues that is factually similar to this one. There, OSHA cited a hospital for violations of the BBP standard based on the exposure to workplace hazards of housekeepers supplied to the hospital by two temporary help agencies. Applying Darden, the Commission concluded that the hospital was properly cited under the OSH Act as an employer of the housekeepers because the hospital directed and controlled the means, methods, location, and timing of their work, and also provided sole on-site supervision and on-the-job instruction. Froedtert, 20 BNA OSHC at 1505-07, 2002 CCH OSHD at pp. 51,732–35.

Barbosa maintains that the Commission's holding in Froedtert dictates that the INS is solely responsible for the cited conditions. This contention is rejected. Application of the Darden factors clearly establishes that Barbosa had an employment relationship with its security personnel and, therefore, OSHA could properly cite it under the OSH Act. Barbosa's managers and supervisors provided regular on-site supervision to Barbosa security personnel at the Batavia facility. Consistent with the terms of its contract with the INS, Barbosa supervisors provided first-line direction and meted out discipline to its contract security personnel at the Batavia facility, unless contravened by INS personnel. Indeed, these contract security personnel considered Barbosa to be their employer due, in no small part, to the fact that Barbosa informed them of their daily work assignments and schedules, provided their pay and promotions, and entered into a collective bargaining agreement with their union covering the terms and conditions of their employment.

Unlike the unskilled manual work the housekeepers performed at the hospital in Froedtert, the duties performed by the contract security personnel at the Batavia facility required some degree of skill and prior experience, as evidenced by the Barbosa/INS contract provision that "[a]ll contract employees shall have a minimum one year's experience as a law enforcement officer or military policeman or six months experience as a security officer engaged in functions related to maintenance of civil order." While the INS provided site-specific job training to Barbosa's security guards, Barbosa provided separate supervisory training to its on-site "shift" supervisors, as required by its contract with the INS. Thus, regardless of whether the INS had any sort of employment relationship with the security personnel supplied by Barbosa, the degree of control Barbosa retained over its contract security personnel compels the conclusion that Barbosa remained their employer in these circumstances and was properly cited as such under the OSH Act.

II. Delegation of Duty

On the basis of the record in this case, there is no evidence that Barbosa effectively delegated its compliance responsibilities under the OSH Act to the INS or any other entity. As the Commission recognized in Froedtert, "[a]n employer may carry out its statutory duties through its own private arrangements with third parties, but if it does so and if those duties are neglected, it is up to the employer to show why he cannot enforce the arrangements he has made." Froedtert, 20 BNA OSHC at 1508 (quoting Central of Georgia R.R. Co. v. OSHRC, 576 F.2d 620, 624 (5th Cir. 1978)). See also Baker Tank Co./Altech, 17 BNA OSHC 1177, 1180, 1993-95 CCH OSHD ¶ 30,734, p. 42,684 (No. 90-1786-S, 1995) (an employer cannot "contract away its legal duties to its employees or its ultimate responsibility under the Act by requiring another party to perform them"). Here, the question of delegation arises only with regard to the BBP training provided by the INS to Barbosa's security personnel pursuant to their contract.

Barbosa claims on review that the BBP training "included an 'overview of communicable diseases and use of universal precautions,'" and that "the INS trained the contract employees along side [sic] federal employees on blood borne [sic] pathogens." Barbosa neglects to mention, however, that the training provided by the INS clearly lacked procedures for employees to follow in the event that an exposure incident occurred, including how to obtain post-exposure follow-up medical treatment. These omissions, which would render any delegation ineffective, could have been discovered by Barbosa had it exercised reasonable diligence. See Automatic Sprinkler Corp., 8 BNA OSHC 1384, 1387, 1980 CCH OSHD 24,495, p. 29,926 (No. 76-5089, 1980) (employer "must make a reasonable effort to anticipate the particular hazards to which its employees may be exposed in the course of their scheduled work"). Indeed, Barbosa was more than familiar with the contents of a comprehensive BBP training program, having provided such training to its personnel located at other facilities. According to Barbosa's operations manager, Jeanne McMichael, Barbosa brought in its own certified trainer for a BBP training program provided to Barbosa employees working for the federal government in New Jersey. McMichael sat in on four of these BBP training classes within a six-month period and described the trainer Barbosa hired as "one of the best" with regard to BBP training. As for the training provided to Barbosa's contract security personnel at the Batavia facility, McMichael simply testified that the INS "said that they did" the training. Yet, neither McMichael nor any other Barbosa manager attended this training nor made any other inquiries prior to the OSHA investigation to determine whether the training complied with the BBP standard.

Under these circumstances, it is clear that Barbosa not only failed to delegate its compliance duties with regard to these specific requirements under the BBP standard but also failed to show why those duties were not carried out with regard to its contract security personnel located at the Batavia facility. See Central of Georgia, 576 F.2d at 624 (effective delegation of responsibilities to third parties requires that employer show why it cannot enforce its own arrangements). Accordingly, all four violations at issue are affirmed.

III. Willfulness

Willful violations are "characterized by an intentional or knowing disregard for the requirements of the Act or a 'plain indifference' to employee safety, in which the employer manifests a 'heightened awareness' that its conduct violates the Act or that the conditions at its workplace present a hazard." Weirton Steel Corp., 20 BNA

OSHC 1255, 1261, 2003 CCH OSHD 32,672, p. 51,451 (No. 98-0701, 2003) (citations omitted). Willfulness may be obviated by a good faith, albeit mistaken, belief that particular conduct is permissible. E.g., Gen. Motors Corp., Electro-Motive Div., 14 BNA OSHC 2064, 2068-69, 1991-93 CCH OSHD 29,240, pp. 39,168-69 (No. 82-630, 1991) (consolidated).

Here, as indicated in his separate opinion, Chairman Railton and I find that the contrasting approaches Barbosa took in addressing the conditions covered by the two BBP provisions under which it was cited for willful violations clearly differentiate these citation items for the purposes of characterization. With regard to the violation of 29 C.F.R. § 1910.1030(f)(3), the post-exposure evaluation and follow-up treatment item, Barbosa paid—either directly or through workers' compensation—for the initial post-exposure evaluation obtained by its injured security personnel at a local hospital. All of these personnel also obtained the post-exposure evaluation and follow-up treatment required by 29 C.F.R. § 1910.1030(f)(3) pursuant to Barbosa's employer-provided health care coverage. However, Barbosa not only failed to cover the co-pay associated with this treatment but it also charged leave to the injured personnel for the work-time spent obtaining this treatment. While Barbosa's conduct does not fully comply with the requirements of the cited provision, its personnel did receive the treatment required by the standard. Under these circumstances, Chairman Railton and I find no evidence in the record that Barbosa demonstrated an intentional disregard rising to the level of willfulness and, therefore, affirm the violation as serious. See Beta Constr. Co., 16 BNA OSHC 1435, 1444-45, 1993–95 CCH OSHD 30,239, pp. 41,652-53 (No. 91-102, 1993) (employer's efforts to prevent violation sufficient to negate willfulness, even if efforts are insufficient to fully eliminate hazardous conditions), aff'd without published opinion, 52 F.3d 1122 (D.C. Cir. 1995).

The willful characterization of the violation based on Barbosa's failure to provide the HBV vaccine to its security personnel is another matter altogether. Vaccination is one of the critical ways of preventing the harmful effects of exposure to bloodborne pathogens. See Occupational Exposure to Bloodborne Pathogens, 56 Fed. Reg. 64,004, 64,152, 64,154 (Dec. 6, 1991) ("OSHA believes that the risk of infection is sufficient to require that the employer make Hepatitis B vaccination available to all employees who have occupational exposure"). The cited standard requires that "[h]epatitis B [or HBV] vaccination be made available within 10 working days of initial assignment to all employees who have occupational exposure." 29 C.F.R. § 1910.1030(f)(2)(i).

13 Other Legal Considerations

CHAPTER OBJECTIVES

1. Acquire an understanding of privacy issues in the workplace.
2. Acquire an understanding of privacy issues involving social media and e-mail.
3. Acquire an understanding tort law and negligence actions.
4. Acquire an understanding of workplace violence.

INTRODUCTION

In the previous chapters, the major issues surrounding the responsibility and potential liability for safety and loss prevention professionals have been thoroughly addressed. In this chapter, other collateral legal considerations that may affect safety and loss prevention professionals are addressed in a summary fashion. This chapter is not meant to set forth the complete ramifications of these issues, but it will give a quick overview for the safety and loss prevention professional.

WORKPLACE PRIVACY

A general outline for the various forms of workplace privacy issues can be found in Restatement (Second) of Torts Section 652A(2) (1977). In employment-based invasion of privacy actions, the allegations usually center upon one of the following elements:

- access to personal information in the possession of the employer
- unreasonable collection of information by an employer
- retaliation by an employer for an employee's refusal to provide personal information
- unreasonable means used by an employer to collect information
- personnel decisions based upon a person's off-duty activity
- unwarranted disclosure of personal information about an employee by an employer
- employer insults and affronts to the dignity of an individual

The issue of workplace privacy is a growing concern. There has been an apparent increase in incidents of people staging accidents to bilk insurance companies, sales scams with special emphasis on targeting elderly citizens and poorly educated individuals, acts of embezzlement, theft of valuable and confidential information, resume fraud to secure a job or promotion, and theft by employees. An NBC *Nightly News* report (May 21, 1996) indicated that a survey showed one-third of all employees steal from their employers, and in the United States alone, an estimated $4 billion is lost from fraud and abuse.

Employers are faced daily with decisions about the honesty and reliability of employees who seek to occupy or already enjoy positions of trust. Thus, many employers have resorted to various investigative techniques in both the pre- and post-hiring stages of employment. These include personal surveillance, security cameras, and monitoring calls, as well as background investigations. A relatively new concern is the rapid increase and use of technology. E-mail and social networking are also causing numerous problems and concerns for employers.

What legal issues may a dismissed employee maintain in tort against an employer for invasion of privacy? Basically, there are four accepted variations of the tort of invasion of privacy. These include:

- intrusion upon seclusion;
- appropriation of name or likeness;
- publicity given to private life; and
- publicity placing a person in false light.

A traditional invasion of privacy action probably would not be brought for the dismissal of an employee itself, but for other acts connected with the dismissal. The appropriation of name or likeness is unlikely to be involved in the dismissal situation, and the false light variant is difficult to distinguish from defamation. Accordingly, a dismissed employee is most likely to benefit from the intrusion and publicity given to private life variance.

INTRUSION UPON SECLUSION

The intrusion upon seclusion issue consists of an intentional interference with the plaintiff's private affairs in a manner "that would be highly offensive to a reasonable man;" it does not depend on any publicity given to the information collected about the plaintiff. Most of the cases accepting the intrusion variant of the privacy tort have involved intrusion into a physical area with respect to which the plaintiff had a reasonable expectation of privacy. It is as simple as going through the desk or locker of an employee who believes he or she has a personal right of privacy in these places. In several circumstances, the courts have agreed.

Most privacy cases involved the acquisition or misuse of information. Thus, if an employee was dismissed for reasons related to his or her private life and the employee could prove that the employer had, in some manner, unreasonably investigated his or her private life, a cause of action could possibly be made by the employee. For example, the illegal wiretapping of an employee's telephone has been held to constitute an invasion of privacy. Similarly, polygraph examinations may constitute invasion of privacy, depending upon the permissibility of testing and extent of the inquiry.

In the employment context, intrusion into privacy areas may involve an employer's testing of employees, gathering of medical information, obtaining credit records, electronic surveillance, and obtaining background information on an employee's suitability for employment. Many other issues involving workplace investigations also address this subject. Complaints of sexual harassment require employers to investigate the allegations. Growing concerns involving workers' compensation and whether injuries are bona fide may require an investigation and/or surveillance. In addition, investigations may be required when employees complain of other discriminatory acts at the hands of their supervisors.

Surveillance of employees in plain view at the workplace as part of a work-related investigation is a permissible practice. However, this does not permit the employer to spy on employees while they are in the bathroom or other private settings. There is absolutely no employer protection to place surveillance cameras, one-way mirrors, or other forms of surveillance in bathrooms or other private settings.

Courts have been inclined to grant employers latitude with respect to home surveillance if done as part of a claims investigation. However, there is an increased likelihood that surveillance of an employee's nonwork-related activities may be deemed by a court or jury to cross the line of acceptable activities. The key is the intrusiveness of the activity. As long as the surveillance is conducted on public property and does not interfere with the daily activities of the individuals being monitored, some latitude is given. In 1995, a Kentucky Court of Appeals decision provided some insight into the extent of proof necessary for a plaintiff to get a verdict involving a surveillance claim.

The underlying facts in the May case involved the investigation of an employee who claimed a work-related injury. The employer was informed that the employee, Mr. May, was engaged in outside work while assigned to light duty. The employer hired an investigation service to determine the truth of the allegations. Videotape of Mr. May's activities was taken from a van on a public road. Neighbors were interviewed and Mrs. May was followed on several occasions. The Mays were unaware of the surveillance until the videotape was played during a workers' compensation proceeding.

The court stated that:

> [I]t is not uncommon for defendants and employers to investigate personal injury and workers' compensation claims. Because of the public interest in exposing fraudulent claims, a plaintiff or claimant must expect that a reasonable investigation will be made after a claim is filed. It is only when an investigation is conducted in a vicious or malicious manner not really limited and designated to obtain information needed for the defense of a legal claim or deliberately calculated to frighten or torment that the courts will not countenance it.

APPROPRIATION OF NAME OR LIKENESS

One who appropriates for his or her own use or benefit the name or likeness of another is subject to liability to the other for invasion of his or her privacy. This tort reserves to the employee the exclusive use of his or her name or photograph, usually for commercial gain.

Publicity Given to Private Life

Under this third category, publicity given to private life, one can be subject to liability for invasion of privacy when the matter publicized is one that would be highly offensive to a reasonable person and is not of legitimate concern to the public. This tort of public exposure of private facts typically involves disclosure of private facts without the consent of the employee. In the context of employment, it may involve attempts to gain background information about an applicant or disclosure of medical information. Unlike defamation (i.e., libel or slander), truth is not a defense. The disclosure of private facts is generally made to a wide audience. Republication of a private fact already known by the employee to fellow employees does not generally provide a cause of action. However, if an employer communicated to a larger number of people private information about the plaintiff–employee in connection with the employee's dismissal, a claim might be established under this theory.

In *Bratt v. International Business Machines Corp.*, the court, applying Massachusetts law, held that the disclosure of information obtained when an employee used IBM's open-door internal grievance policy was not an invasion of privacy because the information disclosed was not "intimate" or "highly personal." The court affirmed summary judgment for the employer on this allegation. It held that disclosure of mental problems to supervisors was not an invasion of privacy because they had a legitimate need to know. It reversed summary judgment respecting disclosure of psychiatric problems by the company doctor to supervisors. It held that the expectation of privacy was much greater with respect to information disclosed in the doctor–patient setting, particularly when company policy reinforces the employee's expectation that such communication would not be divulged. The court noted that the privacy interest of the employee might be outweighed by the legitimate interest of the employer. A balancing test should be employed by the fact finder.

Publicity Placing a Person in False Light

This claim involves both inaccurate portrayals of private facts and accurate portrayals where disclosure would be highly objectionable to the ordinary person. Such a claim generally has been difficult to maintain. The key defense is whether the plaintiff has truly been placed in a false light.

Other Privacy Issues

Sexual privacy is another topic encompassing a variety of employment-related issues such as dating and marriage between employees, dating and sexual relationships with outsiders such as employees in competing companies or customers, extramarital relationships, sexual orientation, and even dress codes. In general, the courts have granted employees wide latitude in adopting policies in addressing these issues.

Generally, employees can be discharged as related to marital status. There are few states that provide statutory protection regarding marital status, and it is non-existent under federal law. Obviously, the employer cannot use this as a protected basis for discrimination or in retaliation for an employee exercising his or her rights as recognized under public policy. However, anti-nepotism policies have generally been upheld.

The theory behind barring spouses from working together is that it prevents conflict in the workplace, that is, complaints of favoritism by co-workers, interference with workplace productivity, and so on. Employers generally prevail in these types of claims. In addition, many employers have policies forbidding dating among employees, especially between employees and supervisors. Discipline, including termination, has been upheld by the courts for violations of these policies, even in the face of invasion of privacy suits.

Privacy claims have also failed when employees were fired after continuing friendships with former company officers or employees. Most states, excluding California, have sided with employers when employees are discharged for dating or marrying a competitor's employee.

E-mail

An issue that presents itself to most employers is e-mail and the potential for an invasion of privacy claim. Many employees assume their e-mail is private and cannot be accessed by anyone else. When a company then reads their e-mail—either as part of an investigation or for some other reason—the employees might sue for invasion of privacy.

As of this date, the author knows of no such successful suit, but a growing number of employees are bringing such lawsuits for invasion of privacy. Even so, they can become expensive for companies to fight even if they win, and an e-mail policy may go a long way toward preventing the suits from being brought in the first place.

The following examples identify some of the cases where employees brought suit:

- An employee sued after being fired for sending an e-mail in which he said he wanted to "kill the back-stabbing bastards" who managed the sales department.
- Two employees at Nissan Motor Corporation were fired for sending e-mail that was critical of the manager.
- A California employee sued after she discovered that employees' e-mail was being monitored.

Even though these cases were brought under state law and dismissed, companies could soon face a rash of suits under a federal statute—the Electronic Communications Privacy Act of 1986. This act prohibits the intentional interception or disclosure of wire, oral, or electronic communications. It does not apply if the interception is made by "the person or entity providing a wire or electronic communication service," so it would probably allow a company to read messages on its own internal e-mail system.

However, as a growing number of company e-mail systems are linked to the internet, it is not clear whether the exception would apply in such a case. The Act does allow e-mail to be monitored if one of the parties to the e-mail has consented to the monitoring. Therefore, it would be important for companies who want to monitor their employees' e-mail to protect themselves by getting employees to sign off on the monitoring in advance.

Sexual Orientation

With regard to sexual orientation, although not a protected class under federal law, it is conceivable that the employer inquiries of this nature, which lead to an employee's termination, may become a basis for both an invasion of privacy claim and an ADA claim. Several states and cities have adopted laws and ordinances that address this area.

Personal Grooming

Dress and grooming policies have also been generally upheld in favor of employers. The key issue is that employers have a right to ensure a "proper" public image and customers and co-workers should not be put off by how a co-worker is dressed. There have been a number of cases regarding sex discrimination on this issue, but invasion of privacy has generally not been upheld in these types of cases.

Drug Testing

Another major issue regarding invasion of privacy concerns is substance abuse/drug and alcohol testing. Obviously, it is undisputed that employers have a number of legitimate work-related reasons for wanting and needing to know whether employees are using illegal drugs, alcohol, or other potentially harmful substances. The reasons include having a good public corporate image, reducing medical costs, lost productivity, and possible theft incidental to supporting such a habit.

Generally, U.S. constitutional restrictions against drug testing apply only to public sector employees, because the requisite "state action" is not present for private employers. However, in the future, constitutional claims may increasingly be asserted against private sector employers in industries subject to government-imposed drug testing requirements.

Under the Fourth Amendment of the U.S. Constitution, courts have found that urinalysis infringes upon one's reasonable expectation of privacy and thereby constitutes a search and seizure within the meaning of the Fourth Amendment. The courts then balance the competing interest of the individual's right to privacy against the government's right to investigate misconduct. In *National Treasury Employee's Union*, the U.S. Supreme Court applied a reasonableness requirement of the Fourth Amendment and approved tests performed on employees seeking promotion into highly sensitive areas of the U.S. Customs Service. The courts found the reasonableness standard met because of three criteria:

1. Advanced notice was provided to the employees.
2. Elaborate chain of custody and quality-control procedures were employed.
3. Individuals were given the opportunity to resubmit a positive test to a lab of their own choosing.

In another case, where railroad labor organizations filed suit to enjoin regulations promulgated by the Federal Railroad Administration, which governs drug and alcohol testing of railroad employees, the Supreme Court found several concepts:

- The Fourth Amendment was applicable to drug and alcohol testing.
- Because of the compelling government interests served by the regulations, which outweighed the employees' privacy concerns, the drug and alcohol tests mandated or authorized by the regulations were reasonable under the Fourth Amendment even though there was no requirement of a warrant or a reasonable suspicion that any particular employee might be impaired.
- Suspicionless post-accident testing of train crews pursuant to a 1985 Federal Railroad Administration regulation is valid.

Some states and at least one municipality have enacted laws that place limits on drug testing in employment. Generally, the issues include reasonable suspicion that an employee is under the influence, chain of custody issues, and guarantees of privacy. In *Wilkinson*, the California appellate court held that state constitutional right to privacy applied to private sector employees, but that the drug testing program did not violate that right because the program was reasonable and the employer had an interest in a drug- and alcohol-free workplace.

Federal statutes have been enacted such as the Omnibus Transportation Employee Testing Act of 1991 and the Drug Free Workplace Act of 1988.

The public has the right to be secure in the knowledge that individuals employed in industries such as aviation, railroads, and trucking are not human time bombs waiting to go off while they fly an airplane, operate a train, or drive down the interstate in a heavy tractor trailer.

Drug testing must be done as quickly as possible and as accurately as possible. There are testing requirements set forth in the mandatory guidelines for federal drug testing programs, 53 Fed. Reg. 11, 1979 (April 11, 1988), and these should be followed to the letter. Employers should find a company with a well-established reputation for such testing and set up procedures with guidelines from experts in the field to avoid or minimize liability.

For tort claims premised upon invasion of privacy for drug testing, courts have centered their inquiry as to whether or not there has been an unreasonable intrusion into an employee's seclusion. Factors include:

- the type of job the employee performs;
- whether objective evidence of probable cause exists that the employee is under the influence;
- the methods used to conduct the testing (i.e., does a person watch an employee provide a urine specimen or does the person wait outside the bathroom door).

However, at least the Sixth Circuit has ruled the right of privacy not to be implicated if the employer has a bona fide right to investigate.

Social Media

Employees generally perceive social media as being protected while employers generally have legitimate reasons to monitor their employee's use of social media, the cloud, and other uses of company-provided technology. With the expansion of the "e-workplace", companies are faced with new federal and state laws in monitoring and limiting employee internet activities, security breaches, use of personal devices on the job, and a growing number of related issues.

Safety and loss prevention professionals should be aware of the federal and state laws as well as company policies regarding use of personal technology (such as a phone with a camera) on the job as well as employees' use of company-provided technology. Additionally, safety and loss prevention professionals should be cognizant of other related laws, such as the FLSA, regarding possible overtime for phone calls, text, or e-mail after working hours as well as possible work-related expenses for use of personal phones. Additionally, issues involving the consequences of a serious breach of security of the employer's data when employees are accessing and exchanging data over unsecured networks are increasing.

Of particular importance for safety and loss prevention professionals is the use of technology while on the job. In addition to the security concerns, the distraction of cell phones in the workplace can distract employees creating safety hazards. With personal phones or tablets in the workplace, employees are constantly tempted to text, make telephone calls, or even play games on their phones. As with texting while driving, the distraction of the phone or tablet in the workplace can create unforeseen hazards that OSHA has recognized and cited under the general duty clause.

Given the changing landscape of social media and the law, safety and loss prevention professionals should exercise caution when investigating internet, phone, or other devices misuse or other legitimate reasons to monitor or examine internet activities. Safety and loss prevention professionals should be aware that issues such as requiring an applicant provide his/her social media password are now addressed under a number of state laws. Additionally, safety and loss prevention professionals should be aware that some states, such as California, as well as several courts have addressed e-discovery issues. Although there is no generally accepted privilege or protection of privacy for all social media, safety and loss prevention professionals should be cognizant of this emerging area and be aware that information placed not only in e-mails but also on social media may be discoverable in litigation.

DEFAMATION

Because safety and loss prevention professionals are increasingly involved in issues such as drug testing, the potential for defamation actions are increasing. Defamation occurs when an untrue statement is communicated to a third party that tends to harm the reputation of another to lower him or her in the estimation of the community or to deter third persons from dealing with him or her. As stated by the Kentucky Supreme

Court in *McCall*, defamation is a statement or communication to the third person that tends to:

- bring a person into public hatred, contempt, or ridicule;
- cause him or her to be shunned or avoided;
- cause injury to him or her in his or her business or occupation.

The following prima facie elements of defamation are needed in most jurisdictions:

- the statement is false and defamatory
- the statement is about the plaintiff
- the statement is published
- the publication is due to negligent or reckless fault of the defendant
- the publication was not privileged
- the publication causes injury to reputation

Publication is an important element of defamation. The publication must be shown to have been done either negligently or intentionally. Unless the employee's communication to the third party was privileged, no actual malice must be proven. In another case, *Hay*, a hotel manager informed his entire staff that they were suspects following a robbery because evidence indicated the crime was an "inside job." Because the accusation was made before the entire group, the statement was considered published. The hotel manager then subjected the entire staff to polygraph examinations.

In some circumstances, publication of the allegedly defamatory statement may encompass more than oral or written statements communicated to a third person. Some courts recognize that "acts" can constitute publication of a defamatory statement. In a Pennsylvania case, the court refused to grant summary judgment because an issue remained as to whether or not defamatory meanings could be inferred from an employer's actions in terminating the employee, such as packing up the employee's belongings and changing the locks on the office door.

Another important aspect is the nature of the words used, which have a bearing on the damages in a defamation case. Words that are harmful by themselves are considered defamatory per se. Injury may be presumed if defamation per se is involved. Most causes of action based on defamation in the employment relationship concern statements impugning the character of an individual or his or her abilities as an employee.

In *O'Brien v. Papa Gino's of America*, a jury found that the employer's statement that the plaintiff was terminated for drug use was not completely true. The jury also found that the employer had a retaliatory motive as well. It awarded the plaintiff damages for both defamation and wrongful termination.

Truth is an absolute defense in a defamation action, even when the plaintiff asserts that the alleged defamatory statements were inspired by malice and the alleged defamation is per se defamatory.

Probably the most common affirmative defense asserted in defamation claims arising from the employment relationship is qualified privilege. The publication is

qualified when circumstances exist that cast on the defendant the duty to communicate to certain other parties information concerning the plaintiff. For example,
managers within the corporation may disclose to other managers rumors or comments made about employees that are defamatory. However, due to the potential for
harm within the workplace setting, courts have found qualified privilege in these
situations. If the publication is qualified, the presumption of malice is lost and must
be proven by the plaintiff.

WORKPLACE NEGLIGENCE

NEGLIGENT HIRING

One of the newest tort theories being developed is that of negligent hiring. This theory has its foundations as an exception to the fellow servant rule and operates to find
liability against the employer where an employee is improperly hired and ultimately
causes injury to another employee. The general rule under the fellow servant doctrine is that the employer would be exempt from liability because of the negligence,
carelessness, or intentional misconduct of a fellow employee. However, the courts
in at least 28 states and the District of Columbia have recognized exceptions to this
general rule under the theory of negligent hiring.

The foundation for the theory of negligent hiring can be traced back to the case
of *Whalen*. In this case, the Illinois Supreme Court found that an employer had a
duty to exercise reasonable and ordinary care in the employment and selection of
careful and skillful co-employees. In recognition of this exception, the tort of negligent hiring has been expanded significantly by the courts to find that an employer
may be liable for the injurious acts of an employee if these acts were within the
scope of the employment. This theory was expanded even further when courts
began finding the employer liable, even when the employee's acts were outside the
scope of the workplace or the employment setting. In the early cases, the theory
of negligent hiring developed into what we would today call negligent security.
Many of the cases dealt with maintenance personnel or rental property managers
with access to individuals' dwellings through master keys and other means. In
these cases, the court generally found that if the owner or employer knew that the
duties of the job required these employees to go into the personal residences of
the individuals, the employer possessed a duty to use reasonable care in selecting
an employee reasonably fit to perform these duties. In more recent cases, the doctrine of negligent hiring has been significantly expanded to cover a wide variety
of areas. For example, employers have been found liable in cases where they have
employed truck drivers with known felony backgrounds who ultimately assaulted
individuals, cases involving sexual harassment charges, and situations where off-
duty management personnel assaulted others. The basis for the vast majority of
these cases involved the employer's failure to properly screen and evaluate the
individual before offering employment.

In the area of workplace violence, the case of *Yunker v. Honeywell, Inc.*, 496 N.W.
2d 419 (Minn. Ct. App. 1993) appears to be one of the first to address this issue. In
this case, the Minnesota Court of Appeals reversed the lower court's finding that

Honeywell, Inc., as a matter of law, did not breach its duty in hiring and supervising an employee who shot and killed a co-worker off the employer's premises. In reversing the summary judgment ruling for the employer, the court not only applied a negligence theory, but also made a distinction between the negligent hiring theory and negligent retention theory.

In this case, an individual worked at Honeywell from 1977 until his conviction and imprisonment for the strangulation death of a co-worker in 1979. On his release from prison, the employee reapplied and was re-hired as a custodian by Honeywell in 1984. In addition, the individual befriended a female co-worker assigned to his maintenance crew. The female employee later severed the relationship, stopped spending time with the individual, and requested a transfer from the particular Honeywell facility.

The individual began to harass and threaten the female employee, both at work and at her home. On July 1, 1988, the female employee found a death threat scratched on her locker door at work. The individual did not report to work after that date and Honeywell accepted his formal resignation on July 11, 1988. On July 19, 1988, the individual killed the female co-worker in her driveway at close range with a shotgun. The individual was convicted of first-degree murder and sentenced to life imprisonment.

The estate of the female employee brought a wrongful death action against Honeywell on the basis of the theories of negligent hiring, negligent retention, and negligent supervision of a dangerous employee. The district court dismissed the negligent supervision theory because it derives from the *respondeat superior* doctrine, which the court recognized relied on the connection to the employer's premises or chattels. The court additionally found negligent hiring as predicted upon the negligence of the employer in placing a person with known propensities, or propensities that should have been discovered by reasonable investigation, in an employment position in which it should have known the hired individual posed a threat of injury to others. The court went further in distinguishing the doctrine based on the scope of the employer's responsibility associated with the particular job. In this case, the individual was a custodian, which did not expose him to the general public and required only limited interaction with fellow employees.

The appeals court, in upholding the summary judgment for Honeywell, stated:

> To reverse the district court's determination on duty as it relates to hiring would extend *Ponticas* and essentially hold that ex-felons are inherently dangerous and that any harmful act they commit against persons encountered through employment would automatically be considered foreseeable. Such a rule would deter employers from hiring workers with a criminal record and offend our civilized concept that society must make reasonable effort to rehabilitate those who have erred so that they can be assimilated into the community.

Additionally, the court made the distinction between negligent hiring and negligent retention as theories of recovery. The court noted that negligent hiring focuses on the adequacy of the employer's preemployment investigation of the employee's background. The court found that there was a record of evidence of a number of episodes in which the individual's post-imprisonment employment at Honeywell demonstrated propensity for abuse and violence toward fellow employees, including sexual harassment of females and threatening to kill a co-worker during an angry confrontation after a minor car accident. The *Yunker* case exemplifies the general trend in the U.S. courts to permit theories of recovery for victims of workplace violence incidents. Employers should be cautious and take the appropriate steps in the hiring and screening phases to possibly avoid this potential area of legal liabilities. The trend to permit recovery under the theory of negligent hiring appears to be expanding in the courts, and employers can no longer rely upon the doctrine of the fellow servant rule to protect them in this area.

NEGLIGENT RETENTION

Closely allied with the tort theory of negligent hiring is that of negligent retention. In general terms, the theory of negligent retention involves an employer who possesses knowledge that an employee has a propensity toward violence in the workplace, but permits the employee to retain his or her employment status despite this knowledge. In *Yunker v. Honeywell, Inc.*, set forth previously, the Minnesota Court of Appeals defined negligent retention as focused "on when the employer was on notice that an employee posed a threat and failed to take steps to ensure the safety of third parties."

   Looking at the general theory of negligence, four basic elements are required to establish a prima facie case, that is, duty, breach, causation, and damage. Under the negligent retention theory, the duty would be created when the employer possessed knowledge of an individual with propensity toward workplace violence, the breach would apply when the employer failed to act or react to this knowledge, the causation would attach when the individual with a propensity actually assaulted or otherwise harmed fellow employees, and the damages would stem from this causation. The pivotal issue in most negligent retention cases involves whether the employer possessed knowledge of the propensity of the individual. In actuality, this is a catch-22 situation for many employers. If the employer did not properly screen the individual prior to hiring and the individual performed workplace violence, the negligent hiring theory would apply. In the event that the employer did not acquire the knowledge during the hiring phase and permitted the employee to continue to work and later acquired information regarding the propensity and failed to react, the theory of negligent retention would apply.

   Prudent employers should take the appropriate steps to properly screen and evaluate employees during the preemployment phase of the operation in order to avoid the possibility of liability in the area of negligent hiring. After an individual is employed, the employer appears to possess an affirmative duty to take appropriate steps to safeguard employees in the workplace after the employer possesses knowledge about the employee's propensity toward workplace violence. In essence, the employer must react after the knowledge is acquired in order to safeguard other employees in the workplace from the particular individual's propensity toward violence.

Negligent Supervision

The theory of negligent supervision has been gaining strength in various courts in the United States. Under this theory, the employer may assume liability when a management person fails to properly supervise an employee who ultimately inflicts harm on fellow employees or co-workers. In the negligent supervision cases, the proximate cause issues are the primary focus in most cases.

In the case of *St. Paul Fire and Marine Insurance v. Knight*, the issue of negligent supervision, as well as negligent hiring and negligent retention, was brought before the court. In this public sector case, a claim was made that an adolescent stress center had improperly hired, supervised, and retained an employee who sexually assaulted a young patient. The particular incident occurred off premises and the party knew that the meeting with the ex-supervisor was not part of the center's "after care" program. The court reversed the lower court, holding that there was no evidence that the employer should have known of its employees' sexual activities and that the incident did not arise out of the employment because the employer possessed specific policies prohibiting contact with former patients.

This novel theory of negligent supervision is gaining ground in the courts. The general rule is that the employer has an affirmative duty to supervise its employees in the workplace. When an employer fails to properly supervise or take appropriate actions, which could ultimately lead to some negative behavior such as sexual harassment or workplace violence, the potential of liability exists. This theory, in most circumstances, is applied in combination with the negligent hiring and negligent retention theories.

Negligent Training

The theory of negligent training involves the employer's failure to provide necessary training, to provide appropriate training, or to provide proper information within the training function. This theory has limited application and is primarily focused on the specific facts of the situation. This theory may be applicable in situations where an employer failed to provide the necessary training or where the training was improper. For example, an employer possesses an affirmative duty to train individuals going into confined space areas under the OSHA standards. The employer fails to provide this training in violation of the OSHA standard. The employee enters the confined space area and becomes injured. The only remaining issue is that of damages. Although this particular scenario would probably be covered under workers' compensation, several states have permitted tort recovery outside of workers' compensation in areas where the accident was caused by the willful negligence of the employer.

The theory of negligent training has also surfaced in the public sector in dealing with firearm safety for police officers, workplace safety, and other areas. The principal elements in a negligent training action would involve the employer's duty to provide appropriate training to employees. Prudent employers should pay special attention when an affirmative duty is created, such as with the OSH Act, EPA, or state laws where mandatory training is required by law. Employers should also evaluate any special relationships that have been created, such as contractors.

NEGLIGENT SECURITY

The theory of negligent security is often invoked in areas where the employer possessed an affirmative duty to safeguard employees or the general public. The theory of negligent security is most applicable in situations where an affirmative duty has been created to safeguard the public and the employees, such as the lighting in parking lot areas, the ability of outside individuals to enter the employment setting, and related areas. The duty is created for the employer through applicable laws or through knowledge because of past incidents. The pivotal issues in a negligent security case normally involve the issue as to whether or not a duty was created and whether or not that duty was breached, rather than issues of causation and damages.

For example, an employer has a large employee parking lot in which employees are required to park their vehicles; the employee parking area has substandard lighting and the employer possessed knowledge that there had been several attempted assaults in the parking area. A female employee, working the late shift, leaves the facility and walks to her vehicle, where she is sexually assaulted. Utilizing this example, the employer possessed knowledge of past assaults and the issue is whether this knowledge created a duty to safeguard this employee in the parking area. Was this duty breached when the employer failed to provide adequate lighting or security for the female employee leaving the plant? Was the failure to provide adequate security or lighting the cause of the sexual assault? The only remaining issue would be the extent of the damages and whether the particular workers' compensation statute applied to the parking lot areas.

Another area where the theory of negligent security may be applicable is the issue of domestic violence filtering into the workplace. Does the employer possess an affirmative duty to safeguard employees from the potential of outside violence from family members or significant others? Knowledge of the problem and a reasonable duty to protect appear to be the key issues. As a general rule, employers possessing knowledge of past incidents or being cognizant of the potential risks should safeguard employees from hazards created by outside forces. Although employers do not want to invade the privacy of domestic relations situations, the potential risk to the employee and co-workers may precipitate the need for additional security in the workplace. With the issue of ex-employees, again the employer should be cognizant of the potential risk of ex-employees returning to the workplace.

WORKPLACE VIOLENCE

Workplace violence has fast become the leading cause of work-related deaths in the United States and has opened an expanding area of potential liability against employers who fail to safeguard their workers. According to the statistics from the National Institute of Occupational Safety and Health (NIOSH), more than 750 workplace killings each year were reported in the 1980s. Additionally, according to the National Safe Workplace Institute, there were approximately 110,000 incidents of workplace violence in the United States in 1992. A common misconception is that violence incidents are a fairly new phenomenon; however, incidents of workplace violence have been happening for a substantial period of time. The primary reason for the emphasis in this area at this time is the increased frequency and severity of the incidents of workplace violence.

According to the U.S. Bureau of Labor Statistics, there were 1,063 homicides on the job in 1993 and of these deaths 59 were killed by co-workers or by disgruntled ex-employees. This report also noted that 22,396 violent physical acts occurred on the job in 1993 and approximately 6% of these incidents were committed by present or former co-workers. In addition to the incidents of workplace violence among and between employees and ex-employees, incidents of other individuals entering into the workplace, such as disgruntled spouses, have drastically increased also. Other areas that should be considered within the realm of workplace violence are the incidents of sabotage and violence directed at a company by outside organizations. Examples of such incidents are the World Trade Center bombing and the bombing of the federal building in Oklahoma City.

Incidents of workplace violence have been highly publicized. The most visible organization with a substantial number of workplace violence incidents is the U.S. Postal Service, which recorded some 500 cases of workplace violence toward supervisors in an 18-month period in 1992 and 1993. Additionally, the U.S. Postal Service also recorded 200 incidents of violence from supervisors toward employees. The following examples are just a few of the highly publicized incidents that resulted in injury or death to individuals:

- the shooting spree at the Chuck E. Cheese restaurant in Denver, Colorado, in which a kitchen worker killed four employees and wounded a fifth
- the ex-employee of the Fireman's Fund Insurance Company killed three individuals, wounded two others, and killed himself in Tampa, Florida
- the 1986 Edmond, Oklahoma, shooting during which a letter carrier killed 14 and wounded 6 others
- the disgruntled postal worker in Dearborn, Michigan, who shot another employee in May 1993
- the former postal worker who killed four employees and injured another in the Montclair, New Jersey, post office

So what exactly is workplace violence? Generally, workplace violence is defined as "physical assaults, threatening behavior, or verbal abuse occurring in the work setting." Although incidents of threatening behavior, such as bomb threats or threats of revenge, are not statistically available, there is a substantial likelihood that these types of incidents are also on the upswing.

Many companies and organizations in the United States have taken steps to safeguard their employees in the workplace through a myriad of security measures, policy changes, and other methods. The potential legal liabilities in this particular area have drastically increased for employers. In most circumstances, the employer would be responsible for any costs incurred by the employee through the individual state workers' compensation system. Now, however, new and novel theories such as negligent retention, negligent hiring, and negligent training, as well as the potential of government monetary fines, such as fines by OSHA, have also emerged to increase the potential risk.

Most experts concede that there are no magic answers when it comes to addressing problems in the area of work-related violence. Given the fact that the potential of violence exists on a daily basis and the method in which the violence can be precipitated can come from a wide variety of areas, the intangibles lend themselves to the fact that workplace violence is a very complicated issue. Is this a new issue? Absolutely not. Incidents of workplace violence have been occurring since the Industrial Revolution in the United States. The frequency of incidents has substantially increased, as well as the severity of these types of workplace incidents. This may correlate to a variety of reasons, including, but not limited to, increased violence in our society, availability of weapons, downsizing of the workplace, management style, and numerous other reasons. Additionally, when the different types of workplaces in America are considered there is no simple answer to this multifaceted question.

OSHA has provided guidelines for specific industries, such as the retail industry and health care operations. Many employers have taken proactive steps to develop a general strategy to protect their employees and thus reduce the potential legal risks as well as providing ancillary efficacy benefits to employees and management. In addition to the proactive strategy, many employers have developed a reactive plan and have implemented stringent employee screening and monitoring processes to identify and address potential incidents of workplace violence in order to minimize potential risks.

In most circumstances, employers are better able to combat the potential risk of workplace violence when the threat is initiated by an employee rather than an ex-employee or outside individual. Researchers have provided a general profile of individuals with a propensity toward workplace violence. The profile lists states such as depression (sometimes with suicidal threats), poor health, and other traits. Incidents precipitated by ex-employees, spouses of employees, and individuals outside the organization are substantially harder for the employer to address, given the comparative lack of control outside the workplace.

As employers attempt to address the potential risks of workplace violence and the correlating legal risks and costs, employers must be very cautious to avoid trampling upon the individual's rights and freedoms. As employers develop and implement more stringent activities and programs to curtail or minimize the potential risks of workplace violence, they must be extremely cautious not to create additional legal risks through their actions. Privacy laws, acquisition of information laws, and discrimination laws provide avenues of potential redress in this area.

Companies now walk a legal tightrope because of the expanding emphasis on workplace violence. To a great extent, this area of law is still expanding, and employers should attempt to maintain an approach that provides the maximum protection to employees but does not affect employees' privacy rights or other individual rights. This is a difficult endeavor but one that is becoming a necessity in the American workplace.

SELECTED CASE STUDY

Case modified for the purpose of this text.
United States District Court, N.D. New York.

New York State Electric and Gas Corporation, Plaintiff, v. SYSTEM COUNCIL U-7 OF THE INTERNATIONAL BROTHERHOOD OF ELECTRICAL WORKERS, Defendant

No. 3:04CV194FJSDEP.
Aug. 4, 2004.

Background

Employer sought to vacate and enjoin enforcement of arbitration award relating to termination of employee's employment, and union counterclaimed for interest, fees, and back pay, alleging employer violated collective bargaining agreement (CBA) by refusing to comply with arbitration award, which ordered reinstatement of employee without back pay or benefits on last-chance basis. Parties brought cross-motions for summary judgment.

Holding

The District Court, Chief Judge, held that arbitrator's award of reinstatement was not clearly contrary to public policy, though employee had expressed a desire to harm certain managerial and supervisory employees at employer's facility.

Employer's motion denied; union's motion granted in part.

Memorandum—Decision and Order

Chief Judge.

I. Introduction

Plaintiff filed its complaint on February 23, 2004, requesting that this Court vacate an arbitration award and enjoin enforcement of the award under 29 U.S.C. § 185. FN1 Defendant counterclaimed and sought interest, fees, and backpay, asserting that Plaintiff has violated the collective bargaining agreement ("CBA") between the parties by refusing to comply with the arbitration award.

Presently before the Court is Defendant's motion for summary judgment and Plaintiff's cross-motion for summary judgment pursuant to Rule 56 of the Federal Rules of Civil Procedure. Defendant also seeks attorneys' fees on the ground that Plaintiff's complaint is frivolous. The Court heard oral argument in support of, and in opposition to, these motions on April 23, 2004, and reserved decision at that time.

14 Safety and Labor

CHAPTER OBJECTIVES

1. Acquire an understanding of the interaction between safety and labor laws.
2. Acquire an understanding of employee rights under the labor laws.
3. Acquire an understanding of unfair labor practices.
4. Acquire an understanding how a labor union is formed.

INTRODUCTION

One of the major activities that safety and loss prevention professionals do on a daily basis is interact with employees and, in unionized operations, the employee representatives, regarding safety-related issues. This interaction, and the methods therein, can be significantly different depending on whether the workplace is unionized or the employees are not represented by a labor organization. Safety and loss prevention professionals working in a unionized environment are often required to negotiate or bargain over safety-related issues because safety is often a "term and condition" of the collective bargaining agreements wherein safety and loss prevention professionals directly manage the function in non-union operations. Safety and loss prevention professionals in unionized operations are often required to manage the safety functions within the terms agreed upon in the "union contract" or collective bargaining agreement wherein safety and loss prevention professionals in non-union operations possess no collective bargaining agreements. In essence, safety and loss prevention professionals should be aware that although the basic elements are the same, the way in which safety, health, and loss prevention is managed within a unionized operation can be significantly different from in a non-union operation.

Over the recent decades, unionization has been on the decline in the American workplace. However, the "elephant in the room" that could dramatically change this status is the proposed legislation titled the Employee Free Choice Act. This proposed legislation, if passed, would amend the National Labor Relations Act and substantially modify the methods and procedures through which a union could organize a company. For safety and loss prevention professionals, whether you are pro-union or anti-union, this proposed change could have a major impact on your workplace and could create situations that could directly or indirectly affect the development and management of your safety and loss prevention efforts.

CURRENT LABOR LAWS

In 1935, Congress passed the National Labor Relations Act (NLRA) in order to "encourage a healthy relationship between private-sector workers and their employers, which policymakers viewed as vital to the national interest." The NLRA created rights for workers who wished to form, join, or support labor organizations including the following examples:

- the right of self-organization
- the right to form, join, or assist labor organizations
- the right to bargain collectively for wages and working conditions through representatives of their own choosing
- the right to engage in other protected concerted activities with or without a union (usually two or more employees)
- the right to refrain from any of these activities

Each of the previous rights possesses a specific definition, substantial regulation, and case law. The governing agency is the National Labor Relations Board. For safety and loss prevention professionals who are not familiar with the extensive process of forming a union, the following basic summary identifies the various steps utilized when forming a union:

1. Employees must want to form a union. An employee usually contacts a labor organization to start the organizing process.
2. Upon selection of a labor organization, the determination as to the "bargaining unit" is made by the union and employees to pursue organization. A "bargaining unit" is the department, area, or facility to be included in the organizing efforts. "Bargaining units" are usually hourly personnel (supervisors and managers are usually excluded) with common work activities, rate of pay, or other activities in common.
3. The union usually provides a card or petition for employees to sign. Upon the signing of 30% of the cards from the employees in the bargaining unit, the union usually submits the cards or petition to the NLRB. The union often sends a certified letter to management identifying the union's wish to be recognized. Very seldom will management recognize a union at this stage.
4. After the NLRB is notified there are very specific rules and requirements. Safety and loss prevention professionals should become familiar with the requirements through the human resource department or legal counsel.
5. The NLRB will determine the exact "bargaining unit" and establish "laboratory conditions" within the operations prior to the secret vote.
6. Safety and loss prevention professionals should be aware that certain activities, communications, and conversations may be prohibited during this time period. Certain actions or inactions can be considered "unfair labor practices" that can result in the NLRB ordering the union to be recognized. Union representatives may also be prohibited from certain actions or inactions that also can constitute "unfair labor practices."

7. Safety and loss prevention professionals should be aware that safety is often a major issue and it is not uncommon for OSHA or other government agencies to be contacted to conduct inspections or investigations.

8. Safety and loss prevention professionals should be aware that there are often numerous hearings before the NLRB addresses various issues involved in the election.

9. The NLRB schedules a date for the secret election and a representative of the NLRB usually oversees the election and counts the ballots.

10. If the union wins the election, the company is usually required to begin contract negotiations toward the achievement of a collective bargaining agreement. If the company wins, the employees and union must wait at least one year before trying again.

Safety and loss prevention professionals should be aware that the previous list is a basic framework; however, a myriad of different rules, regulations, requirements, and strategies impact any union-organizing campaign. Given the often adversarial nature of these situations, prudent safety and loss prevention professionals should acquire guidance from human resources or legal counsel before attempting to make any modifications or changes in their safety efforts, as well as providing any communications or information to employees or the union.

UNFAIR LABOR PRACTICES

An unfair labor practice (ULP) is a charge filed with the NLRB, which can be brought by a union or management, alleging a violation of rights or failure to meet a required obligation. Specifically, Section 158 of the National Labor Relation Act lists the employer's actions, and thus the safety and loss prevention professional as an agent of the employer, that constitute unfair labor practices. Generally, this list of ULPs includes interfering with the rights of employees to establish, belong to, or aid labor organizations as well as to conduct collective bargaining through the union and to participate in "concerted activities" that include strikes. Specifically, the NLRB possesses the authority to investigate and remedy ULPs that are defined in Section 8 of the NLRA. Generally, safety and loss prevention professionals should be aware that it is unlawful to interfere with, restrain, or coerce employees in the exercise of their rights; to interfere with the formation or administration of a labor organization; to discriminate against employees for engaging in concerted activities; to discriminate against an employee for filing a charge with the NLRB; and to refuse to bargain with a union.

Although most, if not all, of these activities generally appear to be beyond the scope of the authority and activities involving most safety and loss prevention professionals, prudent professionals should exercise caution in all functions, activities, and conversations during the period of time in which an election is planned and acquire guidance from the human resource department or legal counsel. Safety and loss prevention professionals should be aware that the NLRB possesses a proscribed procedure for the filing of ULP charges, as well as a hearing process. The NLRB possesses the authority to remedy any ULPs, up to and including ordering the recognition of a union.

Although the previous description is simply a broad overview of this very technical area of the law, safety and loss prevention professionals should be aware of the potential implications of their actions or inactions and acquire guidance during this important period of time from their human resource department or legal counsel. Safety and loss prevention professionals, who are often one of the most visible and vocal members of the management team, can easily and inadvertently create an ULP situation with a simple "slip of the tongue" in a casual conversation in the operations.

EMPLOYEE FREE CHOICE ACT

Safety and loss prevention professionals should be cognizant of prior legislation, known as the Employee Free Choice Act or "Card Check," which could substantially change the federal laws regarding employee rights to form or join a union. In essence, this proposed legislation would permit employees to form a union by signing the authorization cards for union representation, would establish substantially harsher penalties for ULPs, and would institute a new mediation/arbitration process for first-time contract disputes. This bill was initially introduced in the House and Senate in the early 2000s and passed the House in March 2007. The Senate Republicans filibustered the bill in June 2007. With the change in the executive branch and majority in Congress during recent elections, this bill in its current form possesses little chance of passage. However, safety and loss prevention professionals should be aware of the same or similar legislation being brought forward again in the near future.

Safety and loss prevention professionals should be aware that if the Employee Free Choice Act or similar legislation becomes law, the NLRA will be amended and permit unions with a majority of employees signing the authorization cards to be certified as the bargaining representative. The current election process, which was described previously, will be eliminated. Additionally, if a union is certified, the employer and union would immediately begin bargaining over "wages, hours and conditions of employment," which often includes various aspects of an employer's safety program. If the employer and union cannot reach an agreement within 90 days, either party can request mediation through the Federal Mediation and Conciliation Service. If the Federal Mediation and Conciliation Service is not able to achieve an agreement within 30 days, the dispute would be referred to arbitration and the decision would be binding on the parties for a period of two years. Lastly, safety and loss prevention professionals should be aware that the civil penalties for violation of the National Labor Relations Act would be increased to $20,000 per violation for a willful or repeat violation and the amount paid to employees if discharged or discriminated against would be increased to "three times back pay."

LABOR ACTIONS AND IMPACT ON SAFETY

Whether you are pro-union or anti-union, safety and loss prevention professionals should be aware that any time a union campaign is underway, this time period may be very contentious and may dramatically impact your overall safety program and efforts. Safety and loss prevention professionals are often limited in their activities and communications with employees through the direction of human resources or legal counsel in order to avoid the potential of unfair labor practice charges. Additionally, safety and loss prevention professionals should be aware that these situations can often transform in adversarial situations wherein additional safeguard may be required to protect personnel, equipment, and operations. If/when the Employee Free Choice Act is passed by Congress and signed into law, safety and loss prevention professionals should be prepared for a substantial change in the labor dynamics and take the appropriate steps to being their safety programs and efforts in line with this new law.

SELECTED CASE STUDY

Case modified for the purposes of this text
United States Court of Appeals, Tenth Circuit.

RAMSEY WINCH, INC.; AUTO CRANE COMPANY; CONOCOPHILLIPS;
NORRIS, A DOVER RESOURCES COMPANY; DP MANUFACTURING,
INC., A DELAWARE CORPORATION; TULSA WINCH, INC.,
A DELAWARE CORPORATION, PLAINTIFFS-APPELLEES,

V.

C. BRAD HENRY, GOVERNOR OF THE STATE OF OKLAHOMA; W.A. DREW B.
EDMONSON, ATTORNEY GENERAL OF THE STATE OF OKLAHOMA, AND THEIR AGENTS
AND SUCCESSORS, DEFENDANTS-APPELLANTS, NATIONAL RIFLE ASSOCIATION; THE
BRADY CENTER TO PREVENT GUN VIOLENCE; THE AMERICAN SOCIETY OF SAFETY
ENGINEERS; ASIS INTERNATIONAL; SOCIETY OF HUMAN RESOURCES MANAGEMENT;
HR POLICY ASSOCIATION; EQUAL EMPLOYMENT ADVISORY COUNCIL; NATIONAL
FEDERATION OF INDEPENDENT BUSINESS LEGAL FOUNDATION, AMICI CURIAE

No. 07-5166.
Feb. 18, 2009.
Circuit Judge.
A number of Oklahoma businesses forbid their employees from bringing firearms onto company property. In March 2004, the Oklahoma legislature amended its laws to narrow the reach of such company policies. These new laws hold employers criminally liable for prohibiting employees from storing firearms in locked vehicles on company property. Various Oklahoma businesses subsequently filed suit seeking to enjoin the enforcement of the new Oklahoma laws, alleging they were (1) unconstitutionally vague; (2) an unconstitutional taking of private property, as well as a violation of Plaintiffs' due process right to exclude others from their property; and (3) preempted by various federal statutes. The district court for the Northern District of Oklahoma held that the challenged laws were preempted by the Occupational

Health and Safety Act (OSH Act) of 1970 and permanently enjoined enforcement of the new laws. We have jurisdiction under 28 U.S.C. § 1291, and reverse.

I.

Numerous Oklahoma businesses maintain a policy of absolute prohibition on employees' possession of firearms on company property, a violation of which may serve as grounds for termination. After several Oklahoma employees were, in fact, discharged for storing firearms in their vehicles on company parking lots, the Oklahoma legislature amended its firearms laws. Specifically, the legislature amended the Oklahoma Firearms Act (OFA) of 1971 and the Oklahoma Self-Defense Act (OSDA) of 1995 to prohibit property owners from banning the storage of firearms locked in vehicles located on the owner's property. See 21 Okla. Stat. §§ 1289.7a & 1290.22.

Whirlpool Corporation filed the initial action in this case seeking an injunction against enforcement of the Amendments.

In November 2004, the district court entered a temporary restraining order (TRO) against enforcement of the Amendments, finding they were probably preempted by various federal laws. Before deciding whether to issue a permanent injunction, the district court certified to the Oklahoma Court of Criminal Appeals the question of whether the Amendments were criminal statutes. At the time, the status of the Amendments was uncertain. The district court was concerned that if the Amendments were civil in nature, the Oklahoma Governor and Attorney General might not have enforcement authority over the Amendments, thereby making them improper parties to this action. The Court of Criminal Appeals alleviated the district court's concerns, ruling that the Amendments were, in fact, criminal statutes. See *Whirlpool Corp. v. Henry*, 110 P.3d 83, 86 (Okla.Crim. App.2005).

Following this ruling, the district court moved forward with Plaintiffs' request for a permanent injunction and ordered extensive briefing by the parties on the issue of preemption, in particular whether the Amendments conflict with the OSH Act.

In October 2007, the district court ruled the Amendments were not an unconstitutional taking and did not violate Plaintiffs' due process rights. The district court further ruled Plaintiffs lacked standing to assert a facial vagueness challenge. Lastly, the district court held the Amendments were preempted by the OSH Act's general duty clause. Accordingly, the district *1204 court permanently enjoined enforcement of the Amendments.

II.

[1] [2] [3] Congress derives its power to preempt state law under the Supremacy Clause in Article VI of the United States Constitution. See *Choate v. Champion Home Builders Co.*, 222 F.3d 788, 791 (10th Cir.2000). Determining whether Congress intended to preempt state law is the ultimate touchstone of preemption analysis. See *Gade v. Nat'l Solid Wastes Mgmt. Ass'n*, 505 U.S. 88, 96, 112 S.Ct. 2374, 120 L.Ed.2d 73 (1992). Three types of preemption exist. See *Choate*, 222 F.3d at 792. First, Congress can explicitly preempt state law, also known as "express preemption." *Id.* Second, courts infer preemption where Congress

extensively regulates conduct in an entire field, or where the federal interest clearly dominates. *See id.* This is known as "field preemption." *Id.* Express and field preemption do not apply to the present case. The third category, known as "conflict preemption," occurs "where it is impossible for a private party to comply with both state and federal requirements, or where state law stands as an obstacle to the accomplishment and execution of the full purposes and objectives of Congress." *Id.* Conflict preemption requires that the state law materially impede or thwart the federal law or policy. *See id.* at 796.

The district court enjoined enforcement of the Amendments based upon conflict preemption, ruling that (1) gun-related workplace violence is a recognized hazard under the general duty clause; and (2) the Amendments impermissibly conflict with Plaintiffs' ability to comply with the general duty clause, thereby thwarting Congress' overall intent in passing the OSH Act. See *ConocoPhillips*, 520 F.Supp.2d at 1330. In support of its ruling, the district court relied on various studies and scholarly works outlining the growing problem of workplace violence. The district court also cited published statements from the Occupational Safety and Health Administration (OSHA) and prior cases concerning the OSH Act's general duty clause. We review the district court's preemption determination de novo. See *Mount Olivet Cemetery Ass'n. v. Salt Lake City*, 164 F.3d 480, 486 (10th Cir.1998).

A.

[4] [5] Courts do not "lightly attribute to Congress or to a federal agency the intent to preempt state or local laws." *Nat'l Solid Wastes Mgmt. Ass'n v. Killian*, 918 F.2d 671, 676 (7th Cir.1990). In fact, we begin "with the assumption that the historic police powers of the States [are] not to be superseded by the Federal Act unless that was the clear and manifest purpose of Congress." *Altria Group, Inc. v. Good*, ___ U.S. ___, ___, 129 S.Ct. 538, 543, ___ L.Ed.2d ___, ___ (2008) (quoting Rice v. Santa Fe Elevator Corp., 331 U.S. 218, 230, 67 S.Ct. 1146, 91 L.Ed. 1447 (1947)). This assumption applies with greater force when the alleged conflict is in an area traditionally occupied by the states. *See id.* Here, we are faced with "public crimes" meant "to protect the health, safety, and public welfare of citizens and to deter crime." *Whirlpool*, 110 P.3d at 86. The Amendments, therefore, implicate Oklahoma's police powers, an area traditionally controlled by the states. See, e.g., *United States v. Lopez*, 514 U.S. 549, 561 n. 3, 115 S.Ct. 1624, 131 L.Ed.2d 626 (1995) (noting in its preemption review of the federal Gun-Free School Zones Act of 1990 that defining and enforcing criminal law primarily rests with the states); *Richmond Boro Gun Club, Inc. v. City of New York*, 97 F.3d 681, 687 (2d Cir.1996) (recognizing in its preemption review of a city gun ordinance that areas of safety and health are traditionally occupied by the states). Accordingly, our analysis is guided by the assumption that Congress did not intend the OSH Act to preempt the Amendments. See *Altria Group*, 129 S.Ct. at 543.

B.

Congress's declared "purpose and policy" in enacting the OSH Act was "to assure so far as possible every working man and woman in the Nation safe and healthful working conditions and to preserve our human resources." 29 U.S.C. § 651(b). To effect its stated purpose, Congress authorized the Secretary of Labor and OSHA to set and enforce occupational safety and health standards for businesses. See 29 U.S.C. § 651(b)(3); see also OSHA's Role, http://www.osha.gov/oshinfo/mission.html. In addition to requiring employers' compliance with OSHA's promulgated standards, see 29 U.S.C. § 654(a)(2), Congress imposed upon employers a general duty to "furnish to each of his employees employment and a place of employment which are free from recognized hazards that are causing or are likely to cause death or serious physical harm." 29 U.S.C. § 654(a)(1). This provision of the OSH Act, known as the general duty clause, was not meant to "be a general substitute for reliance on standards, but would simply enable the Secretary to insure the protection of employees who are working under special circumstances for which no standard has yet been adopted." S.Rep. No. 91-1282, at 5186 (1970).

The original impetus behind the OSH Act was danger surrounding traditional work-related hazards. See 29 U.S.C. § 651(a) (noting the OSH Act arose from concern surrounding "personal injuries and illnesses arising out of work situations"); S. Rep. 91-1282, at 5178 (describing at length the problems of industrial accidents and occupational diseases, without referencing workplace violence). In recent years, however, OSHA has recognized workplace violence as a serious safety and health issue. See, e.g., Workplace Violence, http://www.osha.gov/SLTC/workplaceviolence/index.html (a section of OSHA's website devoted to workplace violence). To that end, OSHA has issued voluntary guidelines and recommendations for employers seeking to reduce the risk of workplace violence in at-risk industries. See Guidelines for Preventing Workplace Violence for Health Care and Social Service Workers and Recommendations for Workplace Violence Prevention Programs in Late-Night Retail Establishments, both available at http://www.osha.gov/SLTC/workplaceviolence/solutions.html. OSHA has not, however, promulgated any mandatory standards regarding workplace violence.

C.

> [6] Because the absence of any specific OSHA standard on workplace violence is undisputed, the district court correctly recognized that the only possible area of OSH Act preemption was under the general duty clause and the OSH Act's overarching purpose. Thus, in finding preemption, the district court held that gun-related workplace violence was a "recognized hazard" under the general duty clause, and, therefore, an employer that allows firearms in the company parking lot may violate the OSH Act. We disagree. OSHA has not indicated in any way that employers should prohibit firearms from company parking lots. OSHA's website, guidelines, and citation history do not speak at all to any such prohibition. In fact, OSHA declined a request to promulgate a standard banning firearms from the workplace. See Standards Interpretations Letter, September 13, 2006, available at 2006 WL 4093048. In declining this request, OSHA stressed reliance on its voluntary

guidelines and deference "to other federal, state, and local law-enforcement agencies to regulate workplace homicides." *Id.* OSHA is aware of the controversy surrounding firearms in the workplace and has consciously decided not to adopt a standard. Thus, we are not presented with a situation where the general duty clause applies because OSHA has been unable to promulgate a standard for an "unanticipated hazard." *Teal v. E.I. DuPont de Nemours and Co.*, 728 F.2d 799, 804 (6th Cir.1984) (recognizing the purpose of the general duty clause was to cover unanticipated hazards that were not covered by a specific regulation); see also *Reich v. Arcadian Corp.*, 110 F.3d 1192, 1196 (5th Cir.1997) ("Courts have held that enforcement through the application of standards is preferred because standards provide employers notice of what is required under the OSH Act.").

The district court's conclusion is also belied by the only opinion issued by an Administrative Law Judge (ALJ) concerning a general duty clause violation due to workplace violence. See *Megawest Fin., Inc.*, 1995 OSAHRC Lexis 80 (May 8, 1995). In *Megawest*, the Secretary of Labor cited the operator of an apartment community located in a rough neighborhood for failing to take steps to prevent residents' violent acts. *See id.* at *1-2, *6-7. The ALJ reversed the Secretary's citation, ruling that potential violent behavior by residents did not constitute a "recognized hazard" within the meaning of the general duty clause. *Id.* at *32. In reversing the citation, the ALJ expressed the difficulties associated with requiring employers to abate hazards of random physical violence. *See id.* at *28 (recognizing that the "hazard of physical assault ... arises not from the processes or materials of the workplace, but from the anger and frustration of people"). The ALJ stressed that an employee's general fear that he or she may be subject to violent attacks is not enough to require abatement of a hazard under the general duty clause. *See id.* at *27; see also *Pa. Power & Light Co. v. Occupational Health and Safety Review Comm'n*, 737 F.2d 350, 354 (3d Cir.1984) (recognizing that an employer's "duty does not extend to the abatement of dangers created by unforeseeable or unpreventable employee misconduct"); *Pratt & Whitney Aircraft v. Sec'y of Labor*, 649 F.2d 96, 104 (2d Cir.1981) (indicating the OSH Act only requires employers to "guard against significant risks, not ephemeral possibilities"); *Nat'l Realty and Construction Co., Inc. v. Occupational Safety and Health Review Comm'n*, 489 F.2d 1257, 1266 (D.C.Cir.1973) (noting that "[a] demented, suicidal, or willfully reckless employee may on occasion circumvent the best conceived and most vigorously enforced safety regime").

Undeterred by OSHA's and *Megawest's* express restraint in policing social behavior via the general duty clause, the district court held firearms stored in locked vehicles on company property may constitute a "recognized hazard." In so finding, the district court relied heavily on OSHA's general statement that employers may be cited for a general duty clause violation "[i]n a workplace where the risk of violence and serious personal injury are significant enough to be 'recognized hazards.'" Standard Interpretations Letter, December 10, 1992, available at http://www.osha.gov/SLTC/workplaceviolence/standards.html.

The district court also relied on the ALJ's indication in *Megawest* that it might be possible to violate the general duty clause for failure to prevent workplace violence. *See id.* at *29 (noting a high standard of proof is necessary to show that an employer recognized the hazard of workplace violence). Despite these general statements, OSHA's action (or inaction) regarding this matter undermines the district court's conclusion. The broad meaning of "recognized hazard" espoused by the district court is simply too speculative and unsupported to construe as the "clear and manifest purpose of Congress." *Altria Group*, 129 S.Ct. at 543; see also *Oil, Chemical & Atomic Workers v. Am. Cyanamid Co.*, 741 F.2d 444, 449 (D.C.Cir.1984) (refusing to apply a broad meaning of "hazard" under the general duty clause and instead "confin[ing] the term 'hazards' under the general duty clause to the types of hazards [the Court] kn[e]w Congress had in mind").

D.

[7] [8] The district court further reasoned the Amendments thwart the overall purpose and objective of the OSH Act. We disagree. The OSH Act is not meant to interfere "with states' exercise of police powers to protect their citizens." *Lindsey v. Caterpillar, Inc.*, 480 F.3d 202, 208 (3d Cir.2007) (citation omitted); see also *Gade*, 505 U.S. at 96, 112 S.Ct. 2374 (noting "[f]ederal regulation of the workplace was not intended to be all encompassing"); *Florida Retail Federation, Inc. v. Attorney General*, 576 F.Supp.2d 1281, 1298 (N.D.Fla.2008) (stating in its rejection of a nearly identical challenge to the Florida "guns-at-work" statute that "[t]he OSH Act is not a general charter for courts to protect worker safety"); *Megawest*, 1995 OSAHRC Lexis 80, at *4 (recognizing that "enforcement in [the] arena [of workplace violence] could place extraordinary burdens on an employer requiring it to anticipate the possibility of civic disorder"). As such, "state laws of general applicability ... that do not conflict with OSHA standards and that regulate conduct of workers and non-workers alike [are] generally not ... preempted." *Gade*, 505 U.S. at 107, 112 S.Ct. 2374 (emphasis added).

Here, the Amendments conflict with no OSHA standard. Moreover, the Oklahoma Court of Criminal Appeals defined the Amendments as "public crimes" of general applicability "concern[ing] protection of the community as a whole rather than individual citizens." *Whirlpool*, 110 P.3d at 86. Thus, while the Amendments may "have a 'direct and substantial effect' on worker safety, they cannot fairly be characterized as 'occupational' standards, because they regulate workers simply as members of the general public." *Gade*, 505 U.S. at 107, 112 S.Ct. 2374. The district court's decision interferes with Oklahoma's police powers, see *Lindsey*, 480 F.3d at 208, and essentially promulgates a court-made safety standard—a standard that OSHA has explicitly refrained from implementing on its own. Such action is beyond the province of federal courts. See *Chevron, U.S.A., Inc. v. Natural Resources Def. Council*, 467 U.S. 837, 843-44, 104 S.Ct. 2778,

81 L.Ed.2d 694 (1984) (holding that deference must be given to an administrative agency in filling any gaps in regulations).

In sum, the facts before us do not approach the level necessary to overcome "the assumption that the historic police powers of the States [are] not to be superseded by the Federal Act." *Altria Group*, 129 S.Ct. at 543. We understand Plaintiffs may disagree with the wisdom of the Amendments. Our task, however, is not to second-guess the Oklahoma legislature, but rather to interpret the Congressional intent behind the OSH Act and its general duty clause. Accordingly, we hold that Congress did not clearly intend the OSH Act to preempt the Amendments.

III.

The district court rejected Plaintiffs' argument that the Amendments are an unconstitutional taking of private property and a violation of Plaintiffs' due process right to exclude others from their property. Plaintiffs raise these arguments as an alternative grounds for affirmance, however, and we address them accordingly. See *Medina v. City and County of Denver*, 960 F.2d 1493, 1495 n. 1 (10th Cir.1992) ("[W]e are free to affirm a district court decision on any grounds for which there is a record sufficient to permit conclusions of law, even grounds not relied upon by the district court."). As a matter of law, we review Plaintiffs' challenge to the constitutionality of the Amendments de novo. See *Powers v. Harris*, 379 F.3d 1208, 1214 (10th Cir.2004).

A.

[9] Regulation of private property may be so onerous that it violates the Takings Clause of the Fifth Amendment and requires the government to provide compensation. See *Lingle v. Chevron U.S.A., Inc.*, 544 U.S. 528, 536-37, 125 S.Ct. 2074, 161 L.Ed.2d 876 (2005). Regulatory acts requiring payment are either (1) a per se taking, *id.* at 538, 125 S.Ct. 2074, or (2) a taking as characterized by the standards set forth in *Penn Central Transp. Co. v. City of New York*, 438 U.S. 104, 98 S.Ct. 2646, 57 L.Ed.2d 631 (1978). One category of per se takings is "where the government requires an owner to suffer a permanent physical invasion of her property." *Lingle*, at 538, 125 S.Ct. 2074. Such regulatory action is often referred to as a "physical" taking. *Id.* at 548, 125 S.Ct. 2074. A sub-category of physical per se takings is a "land-use exaction" in which the "government demands that a landowner dedicate an easement allowing public access to her property as a condition of obtaining a development permit." *Id.* at 546, 125 S.Ct. 2074 (citing *Nollan v. Cal. Coastal Comm'n*, 483 U.S. 825, 107 S.Ct. 3141, 97 L.Ed.2d 677 (1987) and *Dolan v. City of Tigard*, 512 U.S. 374, 114 S.Ct. 2309, 129 L.Ed.2d 304 (1994)). Such demands by the government are "so onerous that, outside the exactions context, they would be deemed per se physical takings." *Id.* at 547, 125 S.Ct. 2074.

[10] [11] Recognizing that a permanent physical invasion by the government has not occurred here in the traditional sense, Plaintiffs argue the Amendments are a physical per se taking because they require Plaintiffs

to provide an easement for individuals transporting firearms. Thus, the argument goes, the Amendments constitute a permanent physical invasion akin to the "land-use exaction" takings in *Nollan* and *Dolan*. We do not find Plaintiffs' per se taking argument persuasive. A per se taking in the constitutional sense requires a permanent physical occupation or invasion, not simply a restriction on the use of private property. See *Loretto v. Teleprompter Manhattan CATV Corp. et al.*, 458 U.S. 419, 426-34, 102 S.Ct. 3164, 73 L.Ed.2d 868 (1982). Here, the Amendments are most accurately characterized as a restriction on Plaintiffs' use of their property. In *Nollan* and *Dolan*, specific, individual landowners were forced to dedicate portions of their privately owned land for public use in exchange for a development permit from the local governing authority. See *Lingle*, 544 U.S. at 546, 125 S.Ct. 2074. In contrast to the regulatory acts in *Nollan* and *Dolan*, the Amendments (1) apply to all property owners, not just Plaintiffs, (2) merely limit Plaintiffs use of their property, and (3) do not require Plaintiffs to deed portions of their property over to the state for public use. See *Dolan*, 512 U.S. at 385, 114 S.Ct. 2309. Thus, the specific set of circumstances present in *Nollan* and *Dolan* are simply not applicable here. See, e.g., *Lingle*, 544 U.S. at 546-47, 125 S.Ct. 2074 (describing the specific land-use exaction facts present in *Nollan* and *Dolan*); *City of Monterey v. Del Monte Dunes at Monterey, Ltd.*, 526 U.S. 687, 702-03, 119 S.Ct. 1624, 143 L.Ed.2d 882 (1999) (noting that the rough proportionality test used to find a taking in *Nollan* and *Dolan* is restricted to the "special context" of land-use exactions).

Rather, the facts here are more analogous to *PruneYard Shopping Center v. Robins*, 447 U.S. 74, 100 S.Ct. 2035, 64 L.Ed.2d 741 (1980). In *PruneYard*, California's constitutional protection of free speech rights prevented owners of a private shopping center from prohibiting the circulation of petitions on the owner's property. *See id.* at 77-78, 100 S.Ct. 2035. Despite the fact that individuals circulating petitions may have "physically invaded" the owner's property, *id.* at 84, 100 S.Ct. 2035, the Supreme Court held that California's requirement that property owners recognize state-protected rights of free expression and petition "clearly [did] not amount to an unconstitutional infringement of appellants' property rights under the Takings Clause." *Id.* at 83, 100 S.Ct. 2035. As in *PruneYard*, Plaintiffs have not suffered an unconstitutional infringement of their property rights, but rather are required by the Amendments to recognize a state-protected right of their employees. *See id.* at 81, 100 S.Ct. 2035 (noting that the state may exercise its police power to adopt individual liberties more expansive than those conferred by the Federal Constitution). As such, we conclude that Plaintiffs have not suffered a per se taking.

[12] Plaintiffs argue that, even if the Amendments are not a per se taking, a taking has nonetheless occurred under the standards set forth in *Penn Central*. *Penn Central* establishes that while a regulatory act may not constitute a per se taking, it can be "functionally equivalent to the classic taking in which government directly appropriates private property or ousts the

owner from his domain." *Lingle*, 544 U.S. at 539, 125 S.Ct. 2074. The major factors under the Penn Central inquiry are (1) "[t]he economic impact of the regulation on the claimant," (2) "the extent to which the regulation has interfered with distinct investment-backed expectations," and (3) "the character of the governmental action." *Penn Central*, 438 U.S. at 124, 98 S.Ct. 2646. In essence, *Penn Central* focuses on "the magnitude of a regulation's economic impact and the degree to which it interferes with legitimate property interests." *Lingle*, 544 U.S. at 540, 125 S.Ct. 2074.

Plaintiffs' takings argument also fails under the Penn Central inquiry. First, the only economic impact cited by Plaintiffs is the general claim (located in a footnote of their brief) that allowing firearms onto an employer's property inevitably increases costs linked to workplace violence. A constitutional taking requires more than an incidental increase in potential costs for employers as a result of a new regulation. *See id.* at 538, 125 S.Ct. 2074 ("Government hardly could go on if to some extent values incident to property could not be diminished without paying for every such change in the general law.") (quoting *Pa. Coal Co. v. Mahon*, 260 U.S. 393, 413, 43 S.Ct. 158, 67 L.Ed. 322 (1922)). Second, Plaintiffs do not assert any interference with their investment-backed expectations, and, therefore, "have failed to demonstrate that the 'right to exclude others' is so essential to the use or economic value of their property that the state-authorized limitation of it amount[s] to a 'taking.'" *PruneYard*, 447 U.S. at 84, 100 S.Ct. 2035. Third, the governmental action at issue here involves "public crimes" of general applicability "concern[ing] protection of the community as a whole rather than individual citizens." *Whirlpool*, 110 P.3d at 86. Plaintiffs must expect "the uses of [their] property to be restricted, from time to time, by various measures newly enacted by the state in legitimate exercise of its police powers." *Clajon Prod. Corp. v. Petera*, 70 F.3d 1566, 1579 (10th Cir.1995); see also *Penn Central*, 438 U.S. at 125, 98 S.Ct. 2646 (noting that laws meant to support the health, safety, morals, and general welfare of the entire community are generally upheld even if they destroy or adversely affect private property interests).

B.

[13] [14] In reality, Plaintiffs are less concerned about "compensation for a taking of [their] property ... but rather [seek] an injunction against the enforcement of a regulation that [they] allege[ ] to be fundamentally arbitrary or irrational." *Lingle*, 544 U.S. at 544, 125 S.Ct. 2074. As such, Plaintiffs' due process claim, i.e., the Amendments deprive Plaintiffs of the right to exclude others from their property, is more apt than their takings argument. A government regulation "that fails to serve any legitimate governmental objective may be so arbitrary or irrational that it runs afoul of the Due Process Clause." *Id.* at 542, 125 S.Ct. 2074. The Supreme Court, however, has "long eschewed ... heightened scrutiny when addressing substantive due process challenges to government regulation." *Id.* at 545, 125 S.Ct. 2074. Accordingly, we review the Amendments under a "rational

basis" standard. See *Powers*, 379 F.3d at 1215 (regulations not subject to heightened scrutiny require rational basis review); *Crider v. Bd. of County Comm'rs*, 246 F.3d 1285, 1289-90 (10th Cir.2001) (regulations restricting the use of property are subject to rational basis review). Under rational basis review, "we look only to whether a 'reasonably conceivable' rational basis exists." *Id.* at 1290 (citation omitted). We are not allowed to second-guess the wisdom of legislative policy-determinations. *Powers*, 379 F.3d at 1217.

One professed purpose of the Amendments is the protection of the broader Oklahoma community. We need not decide the long-running debate as to whether allowing individuals to carry firearms enhances or diminishes the overall safety of the community. The very fact that this question is so hotly debated, however, is evidence enough that a rational basis exists for the Amendments. See *Village of Euclid v. Ambler Realty Co.*, 272 U.S. 365, 388, 47 S.Ct. 114, 71 L.Ed. 303 (1926) (noting that if a regulation is fairly debatable, the legislative judgment must control). In addition to the Amendment's purpose of increasing safety, one could argue that the Amendments are simply meant to expand (or secure) the Second Amendment right to bear arms. See *PruneYard*, 447 U.S. at 81, 100 S.Ct. 2035 (noting that the state may exercise its police power to adopt individual liberties more expansive than those conferred by the Federal Constitution). Because we cannot say the Amendments have no reasonably conceivable rational basis, Plaintiffs' due process claim must fail.

For the foregoing reasons, we reverse the district court's grant of a permanent injunction.

15 Other Laws Impacting the Safety Function

CHAPTER OBJECTIVES

1. Acquire a working knowledge of other laws impacting the safety function.
2. Acquire an understanding of EEO Laws.
3. Acquire an understanding of the NDAA as related to FLMA.
4. Acquire an understanding of ADEA.

INTRODUCTION

In a safety and loss prevention professional's daily activities, a myriad of laws may impact each and every decision or action taken without the safety and loss prevention professional even possessing knowledge of the specific law or laws. Often, safety and loss prevention professionals find out that they potentially could have violated a law after the fact, when notified by the human resource department and/or a claim, complaint, or charge has already been filed by the employee or other party.

When was the last time you heard that Congress, a state legislature, or even a local government entity revoked or discontinued any law? New laws are continually being created and existing laws are being adjusted and modified by the various legislative, executive, and, arguably, the judicial branches of government. It is important that safety and loss prevention professionals, although not usually within their direct scope of work, be aware of the various new laws and modifications to existing laws in order to ensure that they do not inadvertently violate a law while performing their specific job functions. For this end, several laws and changes that may impact safety and loss prevention professionals in their job functions are provided for review.

LILLY LEDBETTER FAIR PAY ACT OF 2009

After a defeat in the U.S. Supreme Court in *Ledbetter v. Goodyear Tire & Rubber Co., Inc.,* Congress enacted and President Obama signed the Lilly Ledbetter Fair Pay Act of 2009, thus superseding the earlier U.S. Supreme Court defeat, restoring the pre-Ledbetter position of the Equal Employment Opportunity Commission and making this new law retroactive to May 28, 2007. In summary, the Lilly Ledbetter Fair Pay Act of 2009 requires that a compensation discrimination charge be filed within 180 days of the discriminatory pay-setting decision (or 300 days in jurisdictions that have a local or state law prohibiting the same form of compensation discrimination). Furthermore, this new law restores the position that each paycheck that delivers discriminatory compensation is a wrong, regardless of when the discrimination began, and is actionable under existing federal Equal Employment Opportunity statutes.

Under the Lilly Ledbetter Fair Pay Act, an individual subject to compensation discrimination under Title VII of the Civil Rights Act of 1964, the Age Discrimination in Employment Act of 1967, or the Americans with Disabilities Act of 1990 may file a charge within 180 (or 300) days of any of the following:

- when a discriminatory compensation decision or other discriminatory practice affecting compensation is adopted;
- when the individual becomes subject to a discriminatory compensation decision or other discriminatory practice affecting compensation; or
- when the individual's compensation is affected by the application of a discriminatory compensation decision or other discriminatory practice, including each time the individual receives compensation that is based in whole or part on such compensation decision or other practice.

Given this federal law, prudent safety and loss prevention professionals should work closely with their human resource department or legal counsel when hiring, promoting, demoting, or implementing any other modification directly or indirectly affecting compensation to ensure compliance with this law. Further details regarding this new law can be found on the Equal Employment Opportunity Commission website located at EEOC.gov.

NATIONAL DEFENSE AUTHORIZATION ACT AMENDS THE FMLA

For many safety and loss prevention professionals, the Family and Medical Leave Act (FMLA) does impact various aspects of the safety and loss prevention function. In summary, the Family and Medical Leave Act provides up to 12 weeks of job-protected leave within the applicable 12-month period for specific situations and circumstances. The FMLA applies to all public agencies; federal, state and local government employers; local education agencies; and private-sector employers who employ 50 or more employees (for 20 or more workweeks in the current or preceding calendar year). To be eligible for benefits under the FMLA, the employee must be working for a covered employer for a total of 12 months for at least 1,250 hours. Safety and loss prevention professionals should be aware that there is also a poster requirement for the FMLA. Additionally, the employee must request the leave. The covered employers must grant an eligible employee up to a total of 12 workweeks of unpaid leave during any 12-month period for one or more of the following reasons:

- for the birth and care of the newborn child of the employee;
- for placement with the employee of a son or daughter for adoption or foster care;

- to care for an immediate family member (spouse, child, or parent) with a serious health condition; or
- to take medical leave when the employee is unable to work because of a serious health condition.

In 2008, the FMLA was amended with the National Defense Authorization Act (NDAA) providing eligible employees up to 12 weeks of job-protected leave for any "qualifying exigency" arising out of the active duty or call to active duty status in the U.S. military of a spouse, son, daughter, or parent. The NDAA also allowed eligible employees to take up to 26 weeks of job-protected leave within a 12-month period to care for a covered service member with a serious injury or illness.

The NDAA, also known as "military family leave entitlements," applied to all public and private sector employers with 50 or more employees and the employee's eligibility mirrors that of the FMLA. A "covered service member" is defined as a current member of the armed forces, including any member of the National Guard or Reserves, "who is undergoing medical treatment, recuperation, or therapy, is otherwise in outpatient status, or is otherwise on the temporary disability retired list, for a serious injury or illness." Safety and loss prevention professionals should be aware that eligible employees are limited to a combined total of 26 workweeks of leave for any FMLA-qualifying reason within a 12-month period.

Safety and loss prevention professionals should be aware of the "qualifying exigencies" identified in the NDAA, which include:

- issues arising from a covered military member's short notice deployment (usually seven days or less);
- military events and related activities;
- certain childcare and related activities arising from the active duty or call to active duty status;
- financial or legal arrangements;
- up to five days of leave for short-term temporary rest and recuperation leave during deployment;
- attending certain post-deployment activities;
- "any other event that the employee and the employer agree is a qualifying exigency."

Additionally, safety and loss prevention professionals should be aware that the FMLA leave can be intermittent when medically necessary or for qualifying exigencies under the NDAA. The FMLA encourages qualified employees to schedule the leave with their employers when possible. FMLA permits either the employer or the qualified employee to substitute accrued paid leave, such as sick leave or vacation, to cover some of the FMLA leave. Safety and loss prevention professionals should be aware that the ability to substitute depends on your company's normal leave policies and many companies have specific policies in place to address requests for FMLA or NDAA leaves.

Safety and loss prevention professionals can acquire additional information regarding the FMLA or NDAA on the U.S. Department of Labor website located at

DOL.gov. The FMLA or NDAA is enforced by the U.S. Department of Labor and the Wage and Hour Division is the location for the filing and investigation of complaints for violation of these laws.

EEO LAWS

Safety and loss prevention professionals should be aware of the various federal laws as well as state laws prohibiting job discrimination. In addition to the ADA, there are five other laws at the federal level, including Title VII of the Civil Rights Act of 1964, the Civil Rights Act of 1991, the Equal Pay Act of 1963, the Age Discrimination in Employment Act of 1967, and the Rehabilitation Act of 1973 (specifically Sections 501 and 505). Additionally, safety and loss prevention professionals should be aware that most states possess state statutes or laws prohibiting discrimination in the workplace.

More specifically, Title VII of the Civil Rights Act of 1964 (often referred to as Title VII) prohibits employment discrimination based on race, color, religion, sex, and national origin. The subsequent Civil Rights Act of 1991 provided for jury trials and monetary damages for intentional employment discrimination, among other modifications. Safety and loss prevention professionals should be ever vigilant in all aspects of the job, especially in areas such as training, personal interaction, and disciplinary actions, to ensure that there are no discriminatory actions or inactions against a protected employee.

Harassment "is unwelcome conduct that is based on race, color, sex, religion, national origin, disability, and/or age. Harassment becomes unlawful where 1) enduring the offensive conduct becomes a condition of continued employment, or 2) the conduct is severe or pervasive enough to create a work environment that a reasonable person would consider intimidating, hostile or abusive." Safety and loss prevention professionals should be aware that same-sex sexual harassment can be actionable. "Offensive conduct can include, but is not limited to, offensive jokes, slurs, epithets or name calling, physical assaults or threats, intimidation, ridicule or mockery, insults or put-downs, offensive objects or pictures, and interference with work performance." Safety and loss prevention professionals should also be aware that harassment is prohibited against employees in retaliation for filing a discrimination charge or being involved in the action.

Sexual harassment can occur in a variety of circumstances, including but not limited to the following:

- The victim as well as the harasser may be a woman or man. The victim does not have to be of the opposite sex.
- The harasser can be the victim's supervisor, an agent of the employer, a supervisor in another area, a co-worker, or a non-employee.
- The victim does not have to be the person harassed but could be anyone affected by the offensive conduct.
- Unlawful sexual harassment may occur without economic injury to or discharge of the victim.
- The harasser's conduct must be unwelcome.

Safety and loss prevention professionals should be aware that the governing agency for Title VII is the Equal Employment Opportunity Commission (EEOC). When investigating a charge of discrimination, the EEOC will look at the whole record, including the circumstances and context, and evaluates the charges on a case-by-case basis. Like safety, prevention is the best method of eliminating any type of discrimination from the workplace. Many human resource departments have developed policies and procedures, posting, and training in this area.

Safety and loss prevention professionals should exercise caution to avoid any type of retaliation actionable under Title VII. More specifically, "an employer may not fire, demote, harass or otherwise retaliate against an individual for filing a charge of discrimination, participating in a discrimination proceeding, or otherwise opposing discrimination." Retaliation can occur when a safety or loss prevention professional, as the agent for the employer, takes adverse action against the covered employee (or individual) because he or she engaged in a protected activity. An adverse action can include the following activities:

- employment actions such as termination, refusal to hire, and denial of promotion;
- other actions affecting employment such as threats, unjustified negative evaluations, unjustified negative references, or increased surveillance; and
- any other actions such as an assault or unfounded civil or criminal charges that are likely to deter reasonable people from pursuing their rights.

Safety and loss prevention professionals should be aware that "protected activities" can include, but are not limited to "complaining to anyone about alleged discrimination against oneself or others, threatening to file a charge of discrimination, picketing in opposition to discrimination, or refusing to obey an order reasonably believed to be discriminatory." "Participation" means taking part in an employment discrimination proceeding, which can be filing a charge, cooperating in an investigation, or serving as a witness.

Safety and loss prevention professionals should become knowledgeable in the various aspects of Title VII and the other related laws prohibiting discrimination in the workplace. Additional information regarding Title VII and other laws can be found at the Equal Employment Opportunity Commission website located at EEOC. gov. If your company or organization does not already possess policies, procedures, and training in this area, safety and loss prevention professionals are encouraged to bring this to the attention of the management team.

NO FEAR ACT

For safety and loss prevention professionals working for a federal agency, the Notification and Federal Employee Antidiscrimination and Retaliation Act of 2002 (known as the No Fear Act) was enacted to reestablish the need to create and maintain a working environment that is free of discrimination and retaliation. Safety and loss prevention professionals should be aware of the following requirements under the No Fear Act:

- A Federal agency must reimburse the Judgment Fund for payments made to employees, former employees, or applicants for federal employment because of actual or alleged violations of federal employment discrimination laws, federal whistle-blower protection laws, and retaliation claims arising from the assertion of rights under those laws.
- An agency must provide annual notice to its employees, former employees, and applicants for federal employment concerning the rights and remedies applicable to them under the employment discrimination and whistle-blower protection laws.
- At least every two years, an agency must provide training to its employees, including managers, regarding the rights and remedies available under the employment discrimination and whistle-blower protection laws.
- An agency must submit to Congress, EEOC, the Department of Justice, and OPM, an annual report setting forth information about the agency's efforts to improve compliance with the employment discrimination and whistle-blower protection laws and detailing the status of complaints brought against the agency under these laws.
- An agency must post quarterly on its public website summary statistical data pertaining to EEO complaints filed with the agency.

Although this new law is narrow in focus and only applicable to safety and loss prevention professionals working for a federal agency, all safety and loss prevention professionals should be cognizant of the laws and regulations regarding discrimination and retaliation in the workplace. Although achieving and maintaining compliance with the antidiscrimination and retaliation laws is usually a human resource function, safety and loss prevention professionals are often the "eyes and ears" within the operation who can identify and react to situations in which discriminatory or retaliatory actions or inactions are taking place to eliminate these prohibited events.

AGE DISCRIMINATION IN EMPLOYMENT ACT

Safety and loss prevention professionals should be aware that the Age Discrimination in Employment Act of 1967 (ADEA) provides protection against discrimination and retaliation for individuals and employees who are 40 years of age or older. Of particular importance to safety and loss prevention professionals is the fact that the ADEA protects not only current employees but also applicants for employment. Specifically, under the ADEA, "it is unlawful to discriminate against a person because of his/her age with respect to any term, condition, or privilege of employment, including hiring, firing, promotion, layoff, compensation, benefits, job assignments, and training." Additionally, safety and loss prevention professionals should be aware that the ADEA makes it unlawful to retaliate against employees for filing a claim or participating in any way in an ADEA investigation, proceeding, or any type of ADEA litigation. However, for safety and loss prevention professionals working for smaller employers, the ADEA applies only to employers with 20 or more employees. Of particular note for safety and loss prevention professionals is the fact that the Older Workers Benefit Protection Act of 1990 (OWBPA) amended the ADEA and provides a prohibition against denying benefits to older protected employees.

Safety and loss prevention professionals should be aware that the ADEA generally prohibits age preferences or limitations in job notices and advertisements unless the age restriction is shown to be a bona fide occupational qualification (BFOQ) and these circumstances are very limited. Additionally, safety and loss prevention professionals should be aware that the ADEA does not specifically prohibit the asking of an applicant's age or date of birth. However, any request should be reviewed by your human resource or legal departments before asking because this inquiry may reflect an intent to discriminate against the applicant on the basis of age and possibly generate a charge.

When settling workers' compensation and other claims, safety and loss prevention professionals should be aware that there are very specific requirements for an employee to waive their ADEA rights. The following requirements apply:

[A] valid ADEA waiver must:

- be in writing and be understandable;
- specifically refer to ADEA rights or claims;
- not waive rights or claims that may arise in the future;
- be in exchange for valuable consideration;
- advise the individual in writing to consult an attorney before signing the waiver; and
- provide the individual at least 21 days to consider the agreement and at least seven days to revoke the agreement after signing it.

Prudent safety and loss prevention professionals should consult with their human resource department and/or legal counsel before attempting to include a waiver of ADEA rights within any settlement agreement. The ADEA, although part of the web of EEO laws and regulations, possesses different requirements than many of the other related laws.

FAIR LABOR STANDARDS ACT

The Fair Labor Standard Act (FLSA) addresses the requirements for minimum wage, overtime pay, record-keeping, and child labor. Certain employees, such as most safety and loss prevention professionals, are exempt from both the minimum wage and overtime pay requirements because of their administrative or professional status. It is important that safety and loss prevention professionals be aware of the requirements of the FLSA, especially in the area of overtime pay and child labor, depending on the type of their operations.

In the area of overtime, the FLSA does not limit the number of hours of overtime that may be required. However, the FLSA requires employers to pay covered employees not less than one and one-half times their regular rate of pay for all hours worked in excess *of 40 in a workweek. Although overtime pay is usually the responsibility of the human resource department*, safety and loss prevention professionals should be aware of this requirement when authorizing overtime in emergency situations and for budgetary concerns.

In the area of child labor governed under the FLSA, safety and loss prevention professionals should be aware that there are numerous provisions, depending on the applicant/employee's age and the type of work to be performed. Safety and loss prevention professionals are cautioned to review any applicant/employee younger than the age of 18 years with your human resource and/or legal counsel to ensure compliance.

SELECTED CASE STUDY

Case modified for the purposes of this text.
United States Court of Appeals, Eleventh Circuit.

JANICE MORGAN, BARBARA RICHARDSON, ON BEHALF OF THEMSELVES AND ALL OTHERS SIMILARLY SITUATED, ET AL., PLAINTIFFS-APPELLEES, v. FAMILY DOLLAR STORES, INC., DEFENDANT-APPELLANT.

No. 07-12398
Dec. 16, 2008
As Amended Dec. 22, 2008

Background

Managers of employer's stores brought collective action under the Fair Labor Standards Act (FLSA) for employer's alleged failure to pay overtime pay allegedly required by the FLSA, and employer moved to decertify and asserted, by way of affirmative defense, that its store managers were executive employees, exempt from the FLSA's overtime pay requirement. The United States District Court for the Northern District of Alabama, No. 01-00303-CV-UWC-W, U.W. Clemon, J., denied motion to decertify, entered judgment as matter of law against employer as to portion of store managers as to which it asserted its executive employee defense, entered judgment on jury verdict in plaintiffs' favor as to remaining employees, and denied employer's motion for judgment notwithstanding the verdict. Employer appealed.

Holdings

The Court of Appeals, Hull, Circuit Judge, held that:

1. district court's denial of employer's motion to decertify plaintiff class was not abuse of discretion;
2. whether employer satisfied its burden of proving that its store managers had duties that were primarily managerial in nature, as required for them to qualify as executive employees under the FLSA, was question for jury;

3. finding that 163 of the 1,424 store managers opting into action did not customarily and regularly direct work of two or more employees, as required for them to qualify as executive employees, did not rise to level of reversible error;

4. finding that employer had willfully violated overtime pay requirement of the FLSA, so as to extend, to three years, the ordinary two-year statute of limitations on claim for unpaid overtime wages, was sufficiently supported by evidence; and

5. any error by district court in its instructions did not rise to level of reversible error.

Affirmed.

Appeal from the United States District Court for the Northern District of Alabama.

Before DUBINA, HULL and FAY, Circuit Judges.

HULL, Circuit Judge:

The Court sua sponte issues this corrected opinion.

An opt-in class of 1,424 store managers, in a collective action certified by the district court, sued Family Dollar Stores, Inc. ("Family Dollar") for unpaid overtime wages under the Fair Labor Standards Act ("FLSA"), 29 U.S.C. §§ 201-219. During an eight-day trial, the Plaintiffs used Family Dollar's payroll records to establish that 1,424 store managers routinely worked 60 to 70 hours a week and to quantify the overtime wages owed to each Plaintiff. Family Dollar focused on its affirmative defense that the store managers were executives within the meaning of the FLSA and exempt from its overtime pay requirements.

The jury found that the Plaintiff store managers were not exempt executives and that Family Dollar had willfully denied them overtime pay. The jury awarded $19,092,003.39 in overtime wages. The court entered a final judgment of $35,576,059.48 against Family Dollar consisting of $17,788,029.74 in overtime wages and an equal amount in liquidated damages.

Because of the complex procedural history from 2001 to 2005 that led to the case being certified as a collective action, the subsequent eight-day trial in 2006, and Family Dollar's myriad challenges on appeal, we preface the opinion with a table of contents:

I. Procedural History from 2001–2005

A. Complaint

Family Dollar is a nationwide retailer that operates over 6,000 discount stores that sell a wide assortment of products, including groceries, clothing, household items, automotive supplies, general merchandise, and seasonal goods. In January 2001, Janice Morgan and Barbara Richardson, two store managers, filed a Complaint on behalf of themselves "and all other similarly situated persons," alleging that Family Dollar willfully violated the FLSA by refusing to pay its store managers overtime compensation.

The first Complaint asserted that Family Dollar paid store managers a salary, required them to work 60 to 90 hours a week, and refused to compensate them for overtime. According to Plaintiffs, store managers are managers only in name and actually spend the vast majority of their time performing manual labor, such as stocking shelves, running the cash registers, unloading trucks, and cleaning the parking lots, floors, and bathrooms. Store managers spend only five to 10 hours of their time managing anything. Plaintiffs sought unpaid benefits, overtime compensation, and liquidated damages as a result of Family Dollar's willful FLSA violations.

The Complaint urged the district court to issue notice of the action to all similarly situated Family Dollar employees nationwide, and to inform them of their right to opt into the suit as a collective action. Plaintiffs relied on 29 U.S.C. § 216(b), which authorizes courts to maintain a case as one collective action so long as the employee-plaintiffs are similarly situated.

Family Dollar's Answer raised a number of affirmative defenses. It asserted that its store managers were exempt executives and denied any violations were willful. Family Dollar also argued that a collective action, under § 216(b), was impermissible because (1) the store managers were not similarly situated, (2) Plaintiffs' claims were not representative of others in the group, and (3) Plaintiffs could not satisfy § 216(b)'s requirements for maintaining a collective action.

In May 2001, Plaintiffs filed their Third Amended Complaint on behalf of Morgan and Richardson, and added Cora Cannon and Laurie Trout-Wilson as Plaintiffs. The Third Amended Complaint raised the same claims for overtime pay and, again, urged the district court to notify other similarly situated store managers of the action.

B. April 2001 Motion to Facilitate Nationwide Notice

In April 2001, Plaintiffs moved the district court to (1) certify the case as a collective action, (2) authorize notice "by first class mail to all similarly situated management employees employed by Family Dollar Stores, Inc. at any time during the three years prior to the filing of this action to inform them of the nature of the action and their right to opt-into this lawsuit," and (3) order Family Dollar to "produce a computer-readable data file containing the names, addresses, Social Security number and telephone numbers of such potential opt-ins so that notice may be implemented." In May 2001, the court denied the motion for immediate notice, but indicated the motion was "overruled without prejudice."

In September 2001, the district court issued a scheduling order pursuant to Rule 16(b). The order indicated that the parties mutually agreed to "an initial period of discovery limited to identification of claims and their factual basis," and that, despite Family Dollar's opposition, Plaintiffs would request the court to facilitate notice on or before February 2002. Discovery was to expire on October 1, 2002.

C. October 2001—Second Motion to Facilitate Nationwide Notice

In October 2001, Plaintiffs renewed their motion to facilitate notice. Family Dollar twice opposed Plaintiffs' motion and urged the district court to delay ruling to allow more discovery. At oral argument in April 2002, the court withheld ruling pending additional discovery by Family Dollar and ordered Plaintiffs' counsel to make all named Plaintiffs available for deposition. In April 2002, before the court ruled on the renewed motion, the parties jointly agreed to send limited notice of the suit to current and former store managers that worked in the regions where the named Plaintiffs worked from July 1, 1999 to the present. As a result, the court denied Plaintiffs' October 2001 motion as moot.

D. July 2002 Notice

In July 2002, the parties notified 784 potential class members in Region 4 (which contains 15 Family Dollar districts in Alabama, Mississippi, Louisiana, Georgia, and Tennessee), district 39 (in Georgia), and district 118 (in New York). The jointly sent notices required the recipients to mail their consent forms by October 22, 2002. In August 2002, the court extended the discovery deadline by 120 days.

By October 2002, 142 store managers from different states had filed consent forms. Plaintiffs' counsel subsequently sent each of those store managers an 11-page questionnaire with 17 questions and 75 total subparts. The questionnaire asked about employment dates, weekly work hours, day-to-day duties, amount of hours spent on manual labor, what independent authority store managers had, whether district managers made all important managerial decisions, whether hourly assistant managers performed the same duties, and a host of other questions relating to Family Dollar store operations.

E. October 2002—Third Motion to Facilitate Nationwide Notice

In October 2002, Plaintiffs renewed their motion to facilitate nationwide notice. Plaintiffs' motion estimated that approximately 11,164 current or former store managers had no notice of the action and that, on the basis of the approximately 20% response rate, a sizeable number of potential class members would opt into the suit. Plaintiffs' motion and counsel's affidavit summarized the responses to the 11-page questionnaire. Plaintiffs argued that the opt-ins' responses showed that the store managers were similarly situated and that there were enough initial responses to warrant nationwide notice.

Plaintiffs also offered the Rule 30(b)(6) deposition testimony of Bruce Barkus, the Executive Vice President of Store Operations at Family Dollar. Barkus admitted that the store manager job description is the "current and only job description[ ] for Family Dollar Store Managers," and acknowledged that Family Dollar made no inquiry into how many hours a week store managers actually work or whether store managers' actual day-to-day duties mirror the ones outlined in the job description. Plaintiffs argued that Barkus's testimony bolstered the questionnaire responses that showed store managers spent 90% of their time on manual tasks.

F. November 2002 Order and Fact Findings

In November 2002, the district court, acting pursuant to § 216(b), granted Plaintiffs' motion to facilitate nationwide class notice to "all former and current Store Managers who work and/or worked for the Defendant over the last three years." The court found Family Dollar's store managers were similarly situated within the meaning of § 216(b) because they: (1) worked 60 to 80 hours a week; (2) received a fixed salary and no overtime pay; (3) spent 75 to 90% of their time on non-managerial tasks such as stocking shelves, running the cash registers, unloading trucks, and performing janitorial duties; (4) did not consistently supervise two or more employees; (5) lacked the authority to hire, discipline, or terminate employees without first obtaining permission from their district managers; (6) could not select outside vendors without their district managers' permission; (7) worked no less than 48 hours a week under the threat of pay cuts or loss of leave time; and (8) arrived at work before the store opened and stayed until after closing.

Although the district court acknowledged that there existed "some differences between the named-Plaintiffs and the opt-ins in terms of pay scale and job duties," it concluded that "these differences do not preclude the facilitation of nationwide service." The court stressed that Plaintiffs must only be "similarly situated"—not "identically situated." The court considered Family Dollar's contention that its stores have "different locations, are of various sizes, and sell different volumes of merchandise." But the court found that those differences did not undermine the factual basis for concluding that Family Dollar's store managers were similarly situated. The court emphasized that it had the benefit of making its decision after twenty months of litigation, considering Plaintiffs' motion to facilitate nationwide notice on two previous occasions, and giving Family Dollar an opportunity to depose the named Plaintiffs. The court found that a sufficient number of similarly situated employees probably were interested in joining the suit and that the case could be managed and resolved in a single litigation.

G. December 2002 Notice to Potential Opt-Ins

In December 2002, Plaintiffs' counsel mailed 12,145 notices nationwide to current and former Family Dollar store managers employed on or after July 1, 1999. Each had until February 25, 2003 to return the enclosed consent forms. Each mailing included the 11-page questionnaire. By March 2003, nearly 2,500 current and former Family Dollar store managers had joined the litigation.

H. Discovery Disputes

Throughout the litigation, the district court resolved scores of discovery-related motions. For example, Plaintiffs refused to turn over certain questionnaire responses on the basis of attorney–client privilege. This triggered various motions to compel. In another instance, Family Dollar refused to provide Plaintiffs with information related to the identity of its district managers. Plaintiffs responded with their own motions to compel. Meanwhile, a fight over depositions was brewing. In March 2003, Family Dollar notified Plaintiffs of its intent to depose, either in person or by written questions, all opt-in Plaintiffs.

The court extended discovery until August 29, 2003. And in June 2003, in a comprehensive order, the court (1) required Plaintiffs' counsel to produce the questionnaire responses used to support Plaintiffs' motion to facilitate nationwide notice; (2) ordered Family Dollar to produce the names, addresses, and telephone numbers of all former Family Dollar district managers since June 1999; (3) prohibited Plaintiffs' counsel from engaging in ex parte communication with former Family Dollar district managers; and (4) clarified that "this Court shall … treat each opt-in Plaintiff as a separate party for purposes of enforcement of the Scheduling Order."

Discovery issues continued to surface. The court again extended deadlines for discovery to December 12, 2003 and for dispositive motions to January 12, 2004.

In October 2003, Family Dollar informed Plaintiffs that it intended to depose all remaining opt-in Plaintiffs, using written questions, pursuant to Rule 31. In November 2003, Family Dollar's counsel sent a letter stating it planned to take 2,100 depositions in seven days, using 338 written questions per deponent, from December 6 to 12, 2003, in Birmingham.

Plaintiffs moved for protective orders to limit the number of depositions to two a day until the end of discovery and to prevent Family Dollar from deposing the nearly 2,100 opt-in Plaintiffs. Plaintiffs' motion noted that discovery had been ongoing for two and a half years, and that prior to October 2003, Family Dollar failed to depose any of the opt-in Plaintiffs. Plaintiffs objected to Family Dollar's attempt to depose 2,100 opt-in Plaintiffs during the last 29 days of discovery as burdensome, unreasonable, and expensive. In November 2003, Family Dollar moved for a protective order, prohibiting disclosure of the written deposition questions to the opt-in Plaintiffs.

In January 2004, the district court issued a discovery management order resolving many issues. The court limited Family Dollar to "not more than" 250 depositions of the opt-in Plaintiffs, "including those who [had] already been deposed." The court did not restrict Family Dollar to written questions, but limited depositions to five per day (each three hours long). The order authorized Plaintiffs to select 250 opt-ins for Family Dollar to depose in-person. The court pushed the discovery deadline back to April 12, 2004, with dispositive motions due May 12, 2004. It denied both parties' motions for protective orders as moot.

In late January 2004, Family Dollar moved to clarify or alter the discovery management order. In early February 2004, the district court denied Family Dollar's request to depose the rest of the opt-in Plaintiffs under Rule 31 (written deposition questions), but granted Family Dollar leave to use Rule 33 (interrogatories) to obtain discovery from the remaining opt-in Plaintiffs. In March 2004, the court issued an order clarifying that Family Dollar was entitled to serve 25 interrogatories on every opt-in Plaintiff.

By mid-February 2004, 152 opt-in Plaintiffs (of the 250) had not been deposed in-person. The court gave Plaintiffs' counsel seven days to provide Family Dollar with a list of the remaining 152 opt-in Plaintiffs and the dates that each would be available for deposition in Birmingham. To ensure Family Dollar had an opportunity to depose the remaining 152 opt-in Plaintiffs in Birmingham, the district court, in late February 2004, also ordered Plaintiffs' counsel to provide Family Dollar with three days' notice of any change in scheduled depositions and threatened to dismiss opt-in Plaintiffs who failed to attend. For the next several months, the court dismissed with prejudice various opt-in Plaintiffs for their failure to appear at depositions in Birmingham. By the end of discovery, Family Dollar deposed, in person, 250 opt-in Plaintiffs and the named Plaintiffs.

In addition to the 250 depositions of the opt-in Plaintiffs, the parties deposed Family Dollar's executives, district managers, various experts, and other witnesses. Family Dollar produced voluminous payroll records, store manuals, emails, and other communications. Plaintiffs produced the individual responses to the questionnaire. The record was fully developed before the next critical step in this case.

I. May 2004 Motion to Decertify the Collective Action

In May 2004, Family Dollar moved to decertify the collective action under § 216(b). Relying on affidavits and a wealth of information revealed during discovery, the parties briefed whether the case should proceed as 1,424 individual actions or as a § 216(b) collective action. Family Dollar argued (1) the opt-in Plaintiffs were not similarly situated under the FLSA because their day-to-day job duties were too different; (2) the executive exemption defense is inherently individualized; and (3) discrepancies in the store managers' duties made a collective trial impossible and unfair.

In response, Plaintiffs pointed out that discovery established that all store managers were similarly situated because they (1) have the same job description, (2) spend 75 to 90% of their time on the same non-management duties, and (3) spend little time on the management duties in their job description. In addition to Barkus's testimony, Plaintiffs emphasized the deposition testimony of two other Family Dollar executives, Bill Broome and Dennis Heskett, indicating that Family Dollar applied the overtime exemption across the board without any consideration of store-by-store variables, and that store size and location did not affect Family Dollar's decision to exempt all store managers from overtime pay requirements.

J. January 2005 Order and Fact Findings

In a January 2005 order, the district court denied Family Dollar's motion to decertify, determined that none of the factual findings in its November 2002 order had been called into question, and made additional fact-findings. The court's order also expressly incorporated those 2002 findings by reference.

In addition, the court found that the "evidence confirm[ed] that substantial similarities exist in the job duties of the named and opt-in Plaintiffs." The court found that 90% of the named and opt-in Plaintiffs (1) interview and train employees, (2) direct work of employees, and (3) maintain production and sales records.

The court also found that the named and opt-in Plaintiffs had similar restrictions on the scope of their responsibilities. Although classified as store managers, they lacked independent authority to hire, promote, discipline, or terminate assistant managers; award employees pay raises; or change weekly schedules of hourly employees. And 90% lacked the power to close the store in an emergency without the district manager's permission. The court concluded that none of the named and opt-in Plaintiffs were responsible for the "total operation of their stores" and that, in reality, district managers performed the relevant managerial duties. The court found:

> [m]uch of the management discretion which would ordinarily be exercised by store managers is exercised by the district management pursuant to corporate policies and practices set at headquarters. The store managers are very closely supervised by the district managers. In terms of managerial duties, the district manager is more directly responsible for the operation of a Family Dollar store than the store manager.

The court also determined that Plaintiffs similarly spent their time between managerial and non-managerial duties. It found that most (90%) of the named and optin Plaintiffs (1) "spend only a small fraction of their time performing managerial duties"; (2) "spend the vast majority of their time on essentially non-managerial duties such as unloading trucks, stocking shelves.

References

A.E. Staley Mfg. Co. v. Sec'y of Labor, 295 F.3d 1341, 353 U.S. App. D.C. 74, 19 O.S.H. Cas. (BNA) 1937, 2002 O.S.H.D. (CCH) P 32,606: C.A.D.C., 2002.

Baker v. Heartland Food Corp, — N.E. 2d —, 2009 WL 2705780 (Ind. App.), 2009.

Chao v. Occupational Safety and Health Review Com'n, 540 F.3d 519, 22 O.S.H. Cas. (BNA) 1313, 2008 O.S.H.D. (CCH) P 32,974, C.A. 6, 2008.

Dillon v. Mountain Coal Co., LLC, 569 F.3d 1215 (10th Cir. 6/23/09).

Marshall v. Barlow's Inc., 436 U.S. 307, 98 S. Ct. 1816, 56 L. Ed. 2d 305, 8 Envtl. L. Rep. 20,434, 6 O.S.H. Cas. (BNA) 1571, 1978 O.S.H.D. (CCH) P 22,735, U.S. Idaho, 1978.

Morgan v. Family Dollar Stores, Inc., 551 F.3d 1233, 157 Lab. Cas. P 35,515, 14 Wage & Hour Cas. 2d (BNA) 587, 21 Fla. L. Weekly Fed. C 1304 (2008).

New York State Elec. and Gas Corp. v. System Council U-7 of Intern. Broth. of Elec. Workers, 328 F. Supp. 2d 313, 175 L.R.R.M. (BNA) 2508, N.D.N.Y., 2004.

Perez v. Mountaire Farms, Inc., 610 F. Supp. 2d 499 (2009).

Ramsey Winch Inc. v. Henry, 555 F.3d 1199, 2009 O.S.H.D. (CCH) P 32,986, C.A. 10 (Okla) 2009.

Sec'y of Labor v. Summit Contractors, Inc., OSHRC Docket No. 03-1622 (2009).

Sec'y of Labor v. The Barbosa Group, Inc., d/b/a/Executive Security, OSHRC 02-0865 (2008).

Sec'y of Labor v. Dierzen-Kewanee Heavy Industries, Ltd., OSHRC Docket No. 07-0675 and 07-0676 (2009).

Toyota Motor Mfg., Kentucky, Inc. v. Williams, 534 U.S. 184, 122 S. Ct. 681, 200 A.L.R. Fed. 667, 151 L. Ed. 2d 615, 70 USLW 4050, 67 Cal. Comp. Cases 60, 12 A.D. Cases 993, 22 NDLR P 97, 02 Cal. Daily Op. Serv. 149, 2002 Daily Journal D.A.R. 197, 15 Fla. L. Weekly Fed. S 39, U.S., January 08, 2002 (NO. 00-1089).

U.S. v. L.E. Myers Co., 562 F.3d 845, 22 O.S.H. Cas. (BNA) 1621, 2009 O.S.H.D. (CCH) P 32,996, C.A. 7 (Ill.), 2009.

U.S. v. MYR Group, Inc., 361 F.3d 364, 20 O.S.H. Cas. (BNA) 1614, 2002 O.S.H.D. (CCH) P 32,706: C.A. 7 (Ill.), 2004.

APPENDIX A: OCCUPATIONAL SAFETY AND HEALTH ACT of 1970

Public Law 91-596
84 STAT. 1590
91st Congress, S.2193
December 29, 1970,
as amended through January 1, 2004. (1)

AN ACT

To assure safe and healthful working conditions for working men and women; by authorizing enforcement of the standards developed under the Act; by assisting and encouraging the States in their efforts to assure safe and healthful working conditions; by providing for research, information, education, and training in the field of occupational safety and health; and for other purposes.

Be it enacted by the Senate and House of Representatives of the United States of America in Congress assembled, **That this Act may be cited as the "Occupational Safety and Health Act of 1970."**

Footnote (1) See Historical notes at the end of this document for changes and amendments affecting the OSH Act since its passage in 1970 through January 1, 2004.

SEC. 2. Congressional Findings and Purpose

(a)   29 USC 651   The Congress finds that personal injuries and illnesses arising out of work situations impose a substantial burden upon, and are a hindrance to, interstate commerce in terms of lost production, wage loss, medical expenses, and disability compensation payments.

(b)   The Congress declares it to be its purpose and policy, through the exercise of its powers to regulate commerce among the several States and with foreign nations and to provide for the general welfare, to assure so far as possible every working man and woman in the Nation safe and healthful working conditions and to preserve our human resources –

(1) by encouraging employers and employees in their efforts to reduce the number of occupational safety and health hazards at their places of employment,

and to stimulate employers and employees to institute new and to perfect existing programs for providing safe and healthful working conditions;

(2) by providing that employers and employees have separate but dependent responsibilities and rights with respect to achieving safe and healthful working conditions;

(3) by authorizing the Secretary of Labor to set mandatory occupational safety and health standards applicable to businesses affecting interstate commerce, and by creating an Occupational Safety and Health Review Commission for carrying out adjudicatory functions under the Act;

(4) by building upon advances already made through employer and employee initiative for providing safe and healthful working conditions;

(5) by providing for research in the field of occupational safety and health, including the psychological factors involved, and by developing innovative methods, techniques, and approaches for dealing with occupational safety and health problems;

(6) by exploring ways to discover latent diseases, establishing causal connections between diseases and work in environmental conditions, and conducting other research relating to health problems, in recognition of the fact that occupational health standards present problems often different from those involved in occupational safety;

(7) by providing medical criteria which will assure insofar as practicable that no employee will suffer diminished health, functional capacity, or life expectancy as a result of his work experience;

(8) by providing for training programs to increase the number and competence of personnel engaged in the field of occupational safety and health; affecting the OSH Act since its passage in 1970 through January 1, 2004.

(9) by providing for the development and promulgation of occupational safety and health standards;

(10) by providing an effective enforcement program which shall include a prohibition against giving advance notice of any inspection and sanctions for any individual violating this prohibition;

(11) by encouraging the States to assume the fullest responsibility for the administration and enforcement of their occupational safety and health laws by providing grants to the States to assist in identifying their needs and responsibilities in the area of occupational safety and health, to develop plans in accordance with the provisions of this Act, to improve the administration and enforcement of State occupational safety and health laws, and to conduct experimental and demonstration projects in connection therewith;

(12) by providing for appropriate reporting procedures with respect to occupational safety and health which procedures will help achieve the objectives

of this Act and accurately describe the nature of the occupational safety and health problem;

(13) by encouraging joint labor-management efforts to reduce injuries and disease arising out of employment.

SEC. 3. Definitions

29 USC 652   For the purposes of this Act –

(1) The term "Secretary" means the Secretary of Labor.

(2) The term "Commission" means the Occupational Safety and Health Review Commission established under this Act.

(3) For Trust Territory coverage, including the Northern Mariana Islands, see Historical notes

The term "commerce" means trade, traffic, commerce, transportation, or communication among the several States, or between a State and any place outside thereof, or within the District of Columbia, or a possession of the United States (other than the Trust Territory of the Pacific Islands), or between points in the same State but through a point outside thereof.

(4) The term "person" means one or more individuals, partnerships, associations, corporations, business trusts, legal representatives, or any organized group of persons.

(5) Pub. L. 105-241 United States Postal Service is an employer subject to the Act. See Historical notes.

The term "employer" means a person engaged in a business affecting commerce who has employees, but does not include the United States (not including the United States Postal Service) or any State or political subdivision of a State.

(6) The term "employee" means an employee of an employer who is employed in a business of his employer which affects commerce.

(7) The term "State" includes a State of the United States, the District of Columbia, Puerto Rico, the Virgin Islands, American Samoa, Guam, and the Trust Territory of the Pacific Islands.

(8) The term "occupational safety and health standard" means a standard which requires conditions, or the adoption or use of one or more practices, means, methods, operations, or processes, reasonably necessary or appropriate to provide safe or healthful employment and places of employment.

(9) The term "national consensus standard" means any occupational safety and health standard or modification thereof which (1), has been adopted and promulgated by a nationally recognized standards-producing organization under procedures whereby it can be determined by the Secretary that persons interested and affected by the scope or provisions of the standard have reached

substantial agreement on its adoption, (2) was formulated in a manner which afforded an opportunity for diverse views to be considered and (3) has been designated as such a standard by the Secretary, after consultation with other appropriate Federal agencies.

(10) The term "established Federal standard" means any operative occupational safety and health standard established by any agency of the United States and presently in effect, or contained in any Act of Congress in force on the date of enactment of this Act.

(11) The term "Committee" means the National Advisory Committee on Occupational Safety and Health established under this Act.

(12) The term "Director" means the Director of the National Institute for Occupational Safety and Health.

(13) The term "Institute" means the National Institute for Occupational Safety and Health established under this Act.

(14) The term "Workmen's Compensation Commission" means the National Commission on State Workmen's Compensation Laws established under this Act.

SEC. 4. Applicability of This Act

(a)   29 USC 653   For Canal Zone and Trust Territory coverage, including the Northern Mariana Islands, see Historical notes.

This Act shall apply with respect to employment performed in a workplace in a State, the District of Columbia, the Commonwealth of Puerto Rico, the Virgin Islands, American Samoa, Guam, the Trust Territory of the Pacific Islands, Wake Island, Outer Continental Shelf Lands defined in the Outer Continental Shelf Lands Act, Johnston Island, and the Canal Zone. The Secretary of the Interior shall, by regulation, provide for judicial enforcement of this Act by the courts established for areas in which there are no United States district courts having jurisdiction.

(b)

(1) Nothing in this Act shall apply to working conditions of employees with respect to which other Federal agencies, and State agencies acting under section 274 of the Atomic Energy Act of 1954, as amended (42 U.S.C. 2021), exercise statutory authority to prescribe or enforce standards or regulations affecting occupational safety or health.

(2) The safety and health standards promulgated under the Act of June 30, 1936, commonly known as the Walsh-Healey Act (41 U.S.C. 35 et seq.), the Service Contract Act of 1965 (41 U.S.C. 351 et seq.), Public Law 91-54, Act of August 9, 1969 (40 U.S.C. 333), Public Law 85-742, Act of August 23, 1958 (33 U.S.C. 941), and the National Foundation on Arts and Humanities Act (20 U.S.C. 951 et seq.) are superseded on the effective date of corresponding standards, promulgated under this Act, which are determined by the Secretary to be more effective. Standards issued under the laws listed in this paragraph and in effect on

or after the effective date of this Act shall be deemed to be occupational safety and health standards issued under this Act, as well as under such other Acts.

(3) The Secretary shall, within three years after the effective date of this Act, report to the Congress his recommendations for legislation to avoid unnecessary duplication and to achieve coordination between this Act and other Federal laws.

(4) Nothing in this Act shall be construed to supersede or in any manner affect any workmen's compensation law or to enlarge or diminish or affect in any other manner the common law or statutory rights, duties, or liabilities of employers and employees under any law with respect to injuries, diseases, or death of employees arising out of, or in the course of, employment.

SEC. 5. Duties

(a)   Each employer –

(1)   29 USC 654   shall furnish to each of his employees employment and a place of employment which are free from recognized hazards that are causing or are likely to cause death or serious physical harm to his employees;

(2) shall comply with occupational safety and health standards promulgated under this Act.

(b)   Each employee shall comply with occupational safety and health standards and all rules, regulations, and orders issued pursuant to this Act which are applicable to his own actions and conduct.

SEC. 6. Occupational Safety and Health Standards

(a)   29 USC 655   Without regard to chapter 5 of title 5, United States Code, or to the other subsections of this section, the Secretary shall, as soon as practicable during the period beginning with the effective date of this Act and ending two years after such date, by rule promulgate as an occupational safety or health standard any national consensus standard, and any established Federal standard, unless he determines that the promulgation of such a standard would not result in improved safety or health for specifically designated employees. In the event of conflict among any such standards, the Secretary shall promulgate the standard which assures the greatest protection of the safety or health of the affected employees.

(b)   The Secretary may by rule promulgate, modify, or revoke any occupational safety or health standard in the following manner:

(1) Whenever the Secretary, upon the basis of information submitted to him in writing by an interested person, a representative of any organization of employers or employees, a nationally recognized standards-producing organization, the Secretary of Health and Human Services, the National Institute for Occupational Safety and Health, or a State or political subdivision, or on the basis of information developed by the Secretary or otherwise available to him, determines

that a rule should be promulgated in order to serve the objectives of this Act, the Secretary may request the recommendations of an advisory committee appointed under section 7 of this Act. The Secretary shall provide such an advisory committee with any proposals of his own or of the Secretary of Health and Human Services, together with all pertinent factual information developed by the Secretary or the Secretary of Health and Human Services, or otherwise available, including the results of research, demonstrations, and experiments. An advisory committee shall submit to the Secretary its recommendations regarding the rule to be promulgated within ninety days from the date of its appointment or within such longer or shorter period as may be prescribed by the Secretary, but in no event for a period which is longer than two hundred and seventy days.

(2) The Secretary shall publish a proposed rule promulgating, modifying, or revoking an occupational safety or health standard in the Federal Register and shall afford interested persons a period of thirty days after publication to submit written data or comments. Where an advisory committee is appointed and the Secretary determines that a rule should be issued, he shall publish the proposed rule within sixty days after the submission of the advisory committee's recommendations or the expiration of the period prescribed by the Secretary for such submission.

(3) On or before the last day of the period provided for the submission of written data or comments under paragraph (2), any interested person may file with the Secretary written objections to the proposed rule, stating the grounds therefor and requesting a public hearing on such objections. Within thirty days after the last day for filing such objections, the Secretary shall publish in the Federal Register a notice specifying the occupational safety or health standard to which objections have been filed and a hearing requested, and specifying a time and place for such hearing.

(4) Within sixty days after the expiration of the period provided for the submission of written data or comments under paragraph (2), or within sixty days after the completion of any hearing held under paragraph (3), the Secretary shall issue a rule promulgating, modifying, or revoking an occupational safety or health standard or make a determination that a rule should not be issued. Such a rule may contain a provision delaying its effective date for such period (not in excess of ninety days) as the Secretary determines may be necessary to insure that affected employers and employees will be informed of the existence of the standard and of its terms and that employers affected are given an opportunity to familiarize themselves and their employees with the existence of the requirements of the standard.

(5) The Secretary, in promulgating standards dealing with toxic materials or harmful physical agents under this subsection, shall set the standard which most adequately assures, to the extent feasible, on the basis of the best available evidence, that no employee will suffer material impairment of health or functional capacity even if such employee has regular exposure to the hazard dealt with by such standard for the period of his working life. Development of

standards under this subsection shall be based upon research, demonstrations, experiments, and such other information as may be appropriate. In addition to the attainment of the highest degree of health and safety protection for the employee, other considerations shall be the latest available scientific data in the field, the feasibility of the standards, and experience gained under this and other health and safety laws. Whenever practicable, the standard promulgated shall be expressed in terms of objective criteria and of the performance desired.

(6)

(A) Any employer may apply to the Secretary for a temporary order granting a variance from a standard or any provision thereof promulgated under this section. Such temporary order shall be granted only if the employer files an application which meets the requirements of clause (B) and establishes that –

(i) he is unable to comply with a standard by its effective date because of unavailability of professional or technical personnel or of materials and equipment needed to come into compliance with the standard or because necessary construction or alteration of facilities cannot be completed by the effective date,

(ii) he is taking all available steps to safeguard his employees against the hazards covered by the standard, and

(iii) he has an effective program for coming into compliance with the standard as quickly as practicable.

Any temporary order issued under this paragraph shall prescribe the practices, means, methods, operations, and processes which the employer must adopt and use while the order is in effect and state in detail his program for coming into compliance with the standard. Such a temporary order may be granted only after notice to employees and an opportunity for a hearing: *Provided,* That the Secretary may issue one interim order to be effective until a decision is made on the basis of the hearing. No temporary order may be in effect for longer than the period needed by the employer to achieve compliance with the standard or one year, whichever is shorter, except that such an order may be renewed not more that twice (I) so long as the requirements of this paragraph are met and (II) if an application for renewal is filed at least 90 days prior to the expiration date of the order. No interim renewal of an order may remain in effect for longer than 180 days.

(B) An application for temporary order under this paragraph (6) shall contain:

(i) a specification of the standard or portion thereof from which the employer seeks a variance,

(ii) a representation by the employer, supported by representations from qualified persons having firsthand knowledge of the facts represented, that he is unable to comply with the standard or portion thereof and a detailed statement of the reasons therefor,

(iii) a statement of the steps he has taken and will take (with specific dates) to protect employees against the hazard covered by the standard,

(iv) a statement of when he expects to be able to comply with the standard and what steps he has taken and what steps he will take (with dates specified) to come into compliance with the standard, and

(v) a certification that he has informed his employees of the application by giving a copy thereof to their authorized representative, posting a statement giving a summary of the application and specifying where a copy may be examined at the place or places where notices to employees are normally posted, and by other appropriate means.

A description of how employees have been informed shall be contained in the certification. The information to employees shall also inform them of their right to petition the Secretary for a hearing.

(C) The Secretary is authorized to grant a variance from any standard or portion thereof whenever he determines, or the Secretary of Health and Human Services certifies, that such variance is necessary to permit an employer to participate in an experiment approved by him or the Secretary of Health and Human Services designed to demonstrate or validate new and improved techniques to safeguard the health or safety of workers.

(7) Any standard promulgated under this subsection shall prescribe the use of labels or other appropriate forms of warning as are necessary to insure that employees are apprised of all hazards to which they are exposed, relevant symptoms and appropriate emergency treatment, and proper conditions and precautions of safe use or exposure. Where appropriate, such standard shall also prescribe suitable protective equipment and control or technological procedures to be used in connection with such hazards and shall provide for monitoring or measuring employee exposure at such locations and intervals, and in such manner as may be necessary for the protection of employees. In addition, where appropriate, any such standard shall prescribe the type and frequency of medical examinations or other tests which shall be made available, by the employer or at his cost, to employees exposed to such hazards in order to most effectively determine whether the health of such employees is adversely affected by such exposure. In the event such medical examinations are in the nature of research, as determined by the Secretary of Health and Human Services, such examinations may be furnished at the expense of the Secretary of Health and Human Services. The results of such examinations or tests shall be furnished only to the Secretary or the Secretary of Health and Human Services, and, at the request of the employee, to his physician. The Secretary, in consultation with the Secretary of Health and Human Services, may by rule promulgated pursuant to section 553 of title 5, United States Code, make appropriate

modifications in the foregoing requirements relating to the use of labels or other forms of warning, monitoring or measuring, and medical examinations, as may be warranted by experience, information, or medical or technological developments acquired subsequent to the promulgation of the relevant standard.

(8) Whenever a rule promulgated by the Secretary differs substantially from an existing national consensus standard, the Secretary shall, at the same time, publish in the Federal Register a statement of the reasons why the rule as adopted will better effectuate the purposes of this Act than the national consensus standard.

(c)

(1) The Secretary shall provide, without regard to the requirements of chapter 5, title 5, Unites States Code, for an emergency temporary standard to take immediate effect upon publication in the Federal Register if he determines –

(A) that employees are exposed to grave danger from exposure to substances or agents determined to be toxic or physically harmful or from new hazards, and

(B) that such emergency standard is necessary to protect employees from such danger.

(2) Such standard shall be effective until superseded by a standard promulgated in accordance with the procedures prescribed in paragraph (3) of this subsection.

(3) Upon publication of such standard in the Federal Register the Secretary shall commence a proceeding in accordance with section 6 (b) of this Act, and the standard as published shall also serve as a proposed rule for the proceeding. The Secretary shall promulgate a standard under this paragraph no later than six months after publication of the emergency standard as provided in paragraph (2) of this subsection.

(d) Any affected employer may apply to the Secretary for a rule or order for a variance from a standard promulgated under this section. Affected employees shall be given notice of each such application and an opportunity to participate in a hearing. The Secretary shall issue such rule or order if he determines on the record, after opportunity for an inspection where appropriate and a hearing, that the proponent of the variance has demonstrated by a preponderance of the evidence that the conditions, practices, means, methods, operations, or processes used or proposed to be used by an employer will provide employment and places of employment to his employees which are as safe and healthful as those which would prevail if he complied with the standard. The rule or order so issued shall prescribe the conditions the employer must maintain, and the practices, means, methods, operations, and processes which he must adopt and utilize to the extent they differ from the standard in question. Such a rule or order may be modified or revoked upon application by an employer, employees, or by the Secretary on his own motion, in the manner prescribed for its issuance under this subsection at any time after six months from its issuance.

(e) Whenever the Secretary promulgates any standard, makes any rule, order, or decision, grants any exemption or extension of time, or compromises, mitigates, or settles any penalty assessed under this Act, he shall include a statement of the reasons for such action, which shall be published in the Federal Register.

(f) Any person who may be adversely affected by a standard issued under this section may at any time prior to the sixtieth day after such standard is promulgated file a petition challenging the validity of such standard with the United States court of appeals for the circuit wherein such person resides or has his principal place of business, for a judicial review of such standard. A copy of the petition shall be forthwith transmitted by the clerk of the court to the Secretary. The filing of such petition shall not, unless otherwise ordered by the court, operate as a stay of the standard. The determinations of the Secretary shall be conclusive if supported by substantial evidence in the record considered as a whole.

(g) In determining the priority for establishing standards under this section, the Secretary shall give due regard to the urgency of the need for mandatory safety and health standards for particular industries, trades, crafts, occupations, businesses, workplaces or work environments. The Secretary shall also give due regard to the recommendations of the Secretary of Health and Human Services regarding the need for mandatory standards in determining the priority for establishing such standards.

SEC. 7. Advisory Committees; Administration

(a)   29 USC 656

(1) There is hereby established a National Advisory Committee on Occupational Safety and Health consisting of twelve members appointed by the Secretary, four of whom are to be designated by the Secretary of Health and Human Services, without regard to the provisions of title 5, United States Code, governing appointments in the competitive service, and composed of representatives of management, labor, occupational safety and occupational health professions, and of the public. The Secretary shall designate one of the public members as Chairman. The members shall be selected upon the basis of their experience and competence in the field of occupational safety and health.

(2) The Committee shall advise, consult with, and make recommendations to the Secretary and the Secretary of Health and Human Services on matters relating to the administration of the Act. The Committee shall hold no fewer than two meetings during each calendar year. All meetings of the Committee shall be open to the public and a transcript shall be kept and made available for public inspection.

(3) The members of the Committee shall be compensated in accordance with the provisions of section 3109 of title 5, United States Code.

(4) The Secretary shall furnish to the Committee an executive secretary and such secretarial, clerical, and other services as are deemed necessary to the conduct of its business.

(b)   An advisory committee may be appointed by the Secretary to assist him in his standard-setting functions under section 6 of this Act. Each such committee shall consist of not more than fifteen members and shall include as a member one or more designees of the Secretary of Health and Human Services, and shall include among its members an equal number of persons qualified by experience and affiliation to present the viewpoint of the employers involved, and of persons similarly qualified to present the viewpoint of the workers involved, as well as one or more representatives of health and safety agencies of the States. An advisory committee may also include such other persons as the Secretary may appoint who are qualified by knowledge and experience to make a useful contribution to the work of such committee, including one or more representatives of professional organizations of technicians or professionals specializing in occupational safety or health, and one or more representatives of nationally recognized standards producing organizations, but the number of persons so appointed to any such advisory committee shall not exceed the number appointed to such committee as representatives of Federal and State agencies. Persons appointed to advisory committees from private life shall be compensated in the same manner as consultants or experts under section 3109 of title 5, United States Code. The Secretary shall pay to any State which is the employer of a member of such a committee who is a representative of the health or safety agency of that State, reimbursement sufficient to cover the actual cost to the State resulting from such representative's membership on such committee. Any meeting of such committee shall be open to the public and an accurate record shall be kept and made available to the public. No member of such committee (other than representatives of employers and employees) shall have an economic interest in any proposed rule.

(c)   In carrying out his responsibilities under this Act, the Secretary is authorized to –

(1) use, with the consent of any Federal agency, the services, facilities, and personnel of such agency, with or without reimbursement, and with the consent of any State or political subdivision thereof, accept and use the services, facilities, and personnel of any agency of such State or subdivision with reimbursement; and

(2) employ experts and consultants or organizations thereof as authorized by section 3109 of title 5, United States Code, except that contracts for such employment may be renewed annually; compensate individuals so employed at rates not in excess of the rate specified at the time of service for grade GS-18 under section 5332 of title 5, United States Code, including travel time, and allow them while away from their homes or regular places of business, travel expenses (including per diem in lieu of subsistence) as authorized by section 5703 of title 5, United States Code, for persons in the Government service employed intermittently, while so employed.

SEC. 8. Inspections, Investigations, and Recordkeeping

(a)   29 USC 657   In order to carry out the purposes of this Act, the Secretary, upon presenting appropriate credentials to the owner, operator, or agent in charge, is authorized –

(1) to enter without delay and at reasonable times any factory, plant, establishment, construction site, or other area, workplace or environment where work is performed by an employee of an employer; and

(2) to inspect and investigate during regular working hours and at other reasonable times, and within reasonable limits and in a reasonable manner, any such place of employment and all pertinent conditions, structures, machines, apparatus, devices, equipment, and materials therein, and to question privately any such employer, owner, operator, agent or employee.

(b)   In making his inspections and investigations under this Act the Secretary may require the attendance and testimony of witnesses and the production of evidence under oath. Witnesses shall be paid the same fees and mileage that are paid witnesses in the courts of the United States. In case of a contumacy, failure, or refusal of any person to obey such an order, any district court of the United States or the United States courts of any territory or possession, within the jurisdiction of which such person is found, or resides or transacts business, upon the application by the Secretary, shall have jurisdiction to issue to such person an order requiring such person to appear to produce evidence if, as, and when so ordered, and to give testimony relating to the matter under investigation or in question, and any failure to obey such order of the court may be punished by said court as a contempt thereof.

(c)

(1) Each employer shall make, keep and preserve, and make available to the Secretary or the Secretary of Health and Human Services, such records regarding his activities relating to this Act as the Secretary, in cooperation with the Secretary of Health and Human Services, may prescribe by regulation as necessary or appropriate for the enforcement of this Act or for developing information regarding the causes and prevention of occupational accidents and illnesses. In order to carry out the provisions of this paragraph such regulations may include provisions requiring employers to conduct periodic inspections. The Secretary shall also issue regulations requiring that employers, through posting of notices or other appropriate means, keep their employees informed of their protections and obligations under this Act, including the provisions of applicable standards.

(2) The Secretary, in cooperation with the Secretary of Health and Human Services, shall prescribe regulations requiring employers to maintain accurate records of, and to make periodic reports on, work-related deaths, injuries and illnesses other than minor injuries requiring only first aid treatment and which do not involve medical treatment, loss of consciousness, restriction of work or motion, or transfer to another job.

(3) The Secretary, in cooperation with the Secretary of Health and Human Services, shall issue regulations requiring employers to maintain accurate records of employee exposures to potentially toxic materials or harmful physical agents which are required to be monitored or measured under section 6. Such regulations shall provide employees or their representatives with an opportunity to observe such monitoring or measuring, and to have access to the records thereof. Such regulations shall also make appropriate provision for each employee or former employee to have access to such records as will indicate his own exposure to toxic materials or harmful physical agents. Each employer shall promptly notify any employee who has been or is being exposed to toxic materials or harmful physical agents in concentrations or at levels which exceed those prescribed by an applicable occupational safety and health standard promulgated under section 6, and shall inform any employee who is being thus exposed of the corrective action being taken.

(d)   Any information obtained by the Secretary, the Secretary of Health and Human Services, or a State agency under this Act shall be obtained with a minimum burden upon employers, especially those operating small businesses. Unnecessary duplication of efforts in obtaining information shall be reduced to the maximum extent feasible.

(e)   Subject to regulations issued by the Secretary, a representative of the employer and a representative authorized by his employees shall be given an opportunity to accompany the Secretary or his authorized representative during the physical inspection of any workplace under subsection (a) for the purpose of aiding such inspection. Where there is no authorized employee representative, the Secretary or his authorized representative shall consult with a reasonable number of employees concerning matters of health and safety in the workplace.

(f)

(1) Any employees or representative of employees who believe that a violation of a safety or health standard exists that threatens physical harm, or that an imminent danger exists, may request an inspection by giving notice to the Secretary or his authorized representative of such violation or danger. Any such notice shall be reduced to writing, shall set forth with reasonable particularity the grounds for the notice, and shall be signed by the employees or representative of employees, and a copy shall be provided the employer or his agent no later than at the time of inspection, except that, upon the request of the person giving such notice, his name and the names of individual employees referred to therein shall not appear in such copy or on any record published, released, or made available pursuant to subsection (g) of this section. If upon receipt of such notification the Secretary determines there are reasonable grounds to believe that such violation or danger exists, he shall make a special inspection in accordance with the provisions of this section as soon as practicable, to determine if such violation or danger exists. If the Secretary determines there are no reasonable grounds to believe that a violation or danger exists he

shall notify the employees or representative of the employees in writing of such determination.

(2) Prior to or during any inspection of a workplace, any employees or representative of employees employed in such workplace may notify the Secretary or any representative of the Secretary responsible for conducting the inspection, in writing, of any violation of this Act which they have reason to believe exists in such workplace. The Secretary shall, by regulation, establish procedures for informal review of any refusal by a representative of the Secretary to issue a citation with respect to any such alleged violation and shall furnish the employees or representative of employees requesting such review a written statement of the reasons for the Secretary's final disposition of the case.

(g)

(1) The Secretary and Secretary of Health and Human Services are authorized to compile, analyze, and publish, either in summary or detailed form, all reports or information obtained under this section.

(2) The Secretary and the Secretary of Health and Human Services shall each prescribe such rules and regulations as he may deem necessary to carry out their responsibilities under this Act, including rules and regulations dealing with the inspection of an employer's establishment.

(h)   Pub. L. 105-198 added subsection (h).   The Secretary shall not use the results of enforcement activities, such as the number of citations issued or penalties assessed, to evaluate employees directly involved in enforcement activities under this Act or to impose quotas or goals with regard to the results of such activities.

SEC. 9. Citations

(a)   29 USC 658   If, upon inspection or investigation, the Secretary or his authorized representative believes that an employer has violated a requirement of section 5 of this Act, of any standard, rule or order promulgated pursuant to section 6 of this Act, or of any regulations prescribed pursuant to this Act, he shall with reasonable promptness issue a citation to the employer. Each citation shall be in writing and shall describe with particularity the nature of the violation, including a reference to the provision of the Act, standard, rule, regulation, or order alleged to have been violated. In addition, the citation shall fix a reasonable time for the abatement of the violation. The Secretary may prescribe procedures for the issuance of a notice in lieu of a citation with respect to de minimis violations which have no direct or immediate relationship to safety or health.

(b)   Each citation issued under this section, or a copy or copies thereof, shall be prominently posted, as prescribed in regulations issued by the Secretary, at or near each place a violation referred to in the citation occurred.

(c)   No citation may be issued under this section after the expiration of six months following the occurrence of any violation.

SEC. 10. Procedure for Enforcement

(a)   29 USC 659   If, after an inspection or investigation, the Secretary issues a citation under section 9(a), he shall, within a reasonable time after the termination of such inspection or investigation, notify the employer by certified mail of the penalty, if any, proposed to be assessed under section 17 and that the employer has fifteen working days within which to notify the Secretary that he wishes to contest the citation or proposed assessment of penalty. If, within fifteen working days from the receipt of the notice issued by the Secretary the employer fails to notify the Secretary that he intends to contest the citation or proposed assessment of penalty, and no notice is filed by any employee or representative of employees under subsection (c) within such time, the citation and the assessment, as proposed, shall be deemed a final order of the Commission and not subject to review by any court or agency.

(b)   If the Secretary has reason to believe that an employer has failed to correct a violation for which a citation has been issued within the period permitted for its correction (which period shall not begin to run until the entry of a final order by the Commission in the case of any review proceedings under this section initiated by the employer in good faith and not solely for delay or avoidance of penalties), the Secretary shall notify the employer by certified mail of such failure and of the penalty proposed to be assessed under section 17 by reason of such failure, and that the employer has fifteen working days within which to notify the Secretary that he wishes to contest the Secretary's notification or the proposed assessment of penalty. If, within fifteen working days from the receipt of notification issued by the Secretary, the employer fails to notify the Secretary that he intends to contest the notification or proposed assessment of penalty, the notification and assessment, as proposed, shall be deemed a final order of the Commission and not subject to review by any court or agency.

(c)   If an employer notifies the Secretary that he intends to contest a citation issued under section 9(a) or notification issued under subsection (a) or (b) of this section, or if, within fifteen working days of the issuance of a citation under section 9(a), any employee or representative of employees files a notice with the Secretary alleging that the period of time fixed in the citation for the abatement of the violation is unreasonable, the Secretary shall immediately advise the Commission of such notification, and the Commission shall afford an opportunity for a hearing (in accordance with section 554 of title 5, United States Code, but without regard to subsection (a)(3) of such section). The Commission shall thereafter issue an order, based on findings of fact, affirming, modifying, or vacating the Secretary's citation or proposed penalty, or directing other appropriate relief, and such order shall become final thirty days after its issuance. Upon a showing by an employer of a good faith effort to comply with the abatement requirements of a citation, and that abatement has not been completed because of factors beyond his reasonable control, the Secretary, after an opportunity for a hearing as provided in this subsection, shall issue an order affirming or modifying the abatement requirements in such citation. The rules of procedure prescribed by the Commission shall provide affected employees or representatives of affected employees an opportunity to participate as parties to hearings under this subsection.

SEC. 11. Judicial Review

(a)   29 USC 660   Any person adversely affected or aggrieved by an order of the Commission issued under subsection (c) of section 10 may obtain a review of such order in any United States court of appeals for the circuit in which the violation is alleged to have occurred or where the employer has its principal office, or in the Court of Appeals for the District of Columbia Circuit, by filing in such court within sixty days following the issuance of such order a written petition praying that the order be modified or set aside. A copy of such petition shall be forthwith transmitted by the clerk of the court to the Commission and to the other parties, and thereupon the Commission shall file in the court the record in the proceeding as provided in section 2112 of title 28, United States Code. Upon such filing, the court shall have jurisdiction of the proceeding and of the question determined therein, and shall have power to grant such temporary relief or restraining order as it deems just and proper, and to make and enter upon the pleadings, testimony, and proceedings set forth in such record a decree affirming, modifying, or setting aside in whole or in part, the order of the Commission and enforcing the same to the extent that such order is affirmed or modified. The commencement of proceedings under this subsection shall not, unless ordered by the court, operate as a stay of the order of the Commission. No objection that has not been urged before the Commission shall be considered by the court, unless the failure or neglect to urge such objection shall be excused because of extraordinary circumstances. The findings of the Commission with respect to questions of fact, if supported by substantial evidence on the record considered as a whole, shall be conclusive. If any party shall apply to the court for leave to adduce additional evidence and shall show to the satisfaction of the court that such additional evidence is material and that there were reasonable grounds for the failure to adduce such evidence in the hearing before the Commission, the court may order such additional evidence to be taken before the Commission and to be made a part of the record. The Commission may modify its findings as to the facts, or make new findings, by reason of additional evidence so taken and filed, and it shall file such modified or new findings, which findings with respect to questions of fact, if supported by substantial evidence on the record considered as a whole, shall be conclusive, and its recommendations, if any, for the modification or setting aside of its original order. Upon the filing of the record with it, the jurisdiction of the court shall be exclusive and its judgment and decree shall be final, except that the same shall be subject to review by the Supreme Court of the United States, as provided in section 1254 of title 28, United States Code.

(b)   Pub. L. 98-620   The Secretary may also obtain review or enforcement of any final order of the Commission by filing a petition for such relief in the United States court of appeals for the circuit in which the alleged violation occurred or in which the employer has its principal office, and the provisions of subsection (a) shall govern such proceedings to the extent applicable. If no petition for review, as provided in subsection (a), is filed within sixty days after service of the Commission's order, the Commission's findings of fact and order shall be conclusive in connection with any petition for enforcement which is filed by the Secretary after the expiration of such sixty-day period. In any such case, as well as in the case of a noncontested citation or notification by the Secretary which has become a final order of the Commission under subsection (a) or (b) of section 10, the clerk of the court, unless otherwise ordered by the court, shall forthwith enter a decree enforcing the order and shall transmit a copy of such decree to the Secretary and the employer named in the petition. In any contempt proceeding brought to enforce a decree of a court of appeals entered pursuant to this subsection or subsection (a), the court of appeals may assess the penalties provided in section 17, in addition to invoking any other available remedies.

(c)

(1) No person shall discharge or in any manner discriminate against any employee because such employee has filed any complaint or instituted or caused to be instituted any proceeding under or related to this Act or has testified or is about to testify in any such proceeding or because of the exercise by such employee on behalf of himself or others of any right afforded by this Act.

(2) Any employee who believes that he has been discharged or otherwise discriminated against by any person in violation of this subsection may, within thirty days after such violation occurs, file a complaint with the Secretary alleging such discrimination. Upon receipt of such complaint, the Secretary shall cause such investigation to be made as he deems appropriate. If upon such investigation, the Secretary determines that the provisions of this subsection have been violated, he shall bring an action in any appropriate United States district court against such person. In any such action the United States district courts shall have jurisdiction, for cause shown to restrain violations of paragraph (1) of this subsection and order all appropriate relief including rehiring or reinstatement of the employee to his former position with back pay.

(3) Within 90 days of the receipt of a complaint filed under this subsection the Secretary shall notify the complainant of his determination under paragraph 2 of this subsection.

SEC. 12. The Occupational Safety and Health Review Commission

(a)   29 USC 661   The Occupational Safety and Health Review Commission is hereby established. The Commission shall be composed of three members who shall be appointed by the President, by and with the advice and consent of the Senate, from among persons who by reason of training, education, or experience are qualified to carry out the functions of the Commission under this Act. The President shall designate one of the members of the Commission to serve as Chairman.

(b)   The terms of members of the Commission shall be six years except that

(1) the members of the Commission first taking office shall serve, as designated by the President at the time of appointment, one for a term of two years, one for a term of four years, and one for a term of six years, and

(2) a vacancy caused by the death, resignation, or removal of a member prior to the expiration of the term for which he was appointed shall be filled only for the remainder of such unexpired term.

A member of the Commission may be removed by the President for inefficiency, neglect of duty, or malfeasance in office.

(c)   See notes on omitted text.   (Text omitted.)

(d)   The principal office of the Commission shall be in the District of Columbia. Whenever the Commission deems that the convenience of the public or of the parties may be promoted, or delay or expense may be minimized, it may hold hearings or conduct other proceedings at any other place.

(e)   Pub. L. 95-251   The Chairman shall be responsible on behalf of the Commission for the administrative operations of the Commission and shall appoint such administrative law judges and other employees as he deems necessary to assist in the performance of the Commission's functions and to fix their compensation in accordance with the provisions of chapter 51 and subchapter III of chapter 53 of title 5, United States Code, relating to classification and General Schedule pay rates: *Provided,* That assignment, removal and compensation of administrative law judges shall be in accordance with sections 3105, 3344, 5372, and 7521 of title 5, United States Code.

(f)   For the purpose of carrying out its functions under this Act, two members of the Commission shall constitute a quorum and official action can be taken only on the affirmative vote of at least two members.

(g)   Every official act of the Commission shall be entered of record, and its hearings and records shall be open to the public. The Commission is authorized to make such rules as are necessary for the orderly transaction of its proceedings. Unless the Commission has adopted a different rule, its proceedings shall be in accordance with the Federal Rules of Civil Procedure.

(h)   The Commission may order testimony to be taken by deposition in any proceedings pending before it at any state of such proceeding. Any person may be compelled to appear and depose, and to produce books, papers, or documents, in the same manner as witnesses may be compelled to appear and testify and produce like documentary evidence before the Commission. Witnesses whose depositions are taken under this subsection, and the persons taking such depositions, shall be entitled to the same fees as are paid for like services in the courts of the United States.

(i)   For the purpose of any proceeding before the Commission, the provisions of section 11 of the National Labor Relations Act (29 U.S.C. 161) are hereby made applicable to the jurisdiction and powers of the Commission.

(j)   An administrative law judge appointed by the Commission shall hear, and make a determination upon, any proceeding instituted before the Commission and any motion in connection therewith, assigned to such administrative law judge by the Chairman of the Commission, and shall make a report of any such determination which constitutes his final disposition of the proceedings. The report of the administrative law judge shall become the final order of the Commission within thirty days after such report by the administrative law judge, unless within such period any Commission member has directed that such report shall be reviewed by the Commission.

(k)   Except as otherwise provided in this Act, the administrative law judges shall be subject to the laws governing employees in the classified civil service, except that appointments shall be made without regard to section 5108 of title 5, United States Code. Each administrative law judge shall receive compensation at a rate not less than that prescribed for GS-16 under section 5332 of title 5, United States Code.

SEC. 13. Procedures to Counteract Imminent Dangers

(a)   29 USC 662   The United States district courts shall have jurisdiction, upon petition of the Secretary, to restrain any conditions or practices in any place of employment which are such that a danger exists which could reasonably be expected to cause death or serious physical harm immediately or before the imminence of such danger can be eliminated through the enforcement procedures otherwise provided by this Act. Any order issued under this section may require such steps to be taken as may be necessary to avoid, correct, or remove such imminent danger and prohibit the employment or presence of any individual in locations or under conditions where such imminent danger exists, except individuals whose presence is necessary to avoid, correct, or remove such imminent danger or to maintain the capacity of a continuous process operation to resume normal operations without a complete cessation of operations, or where a cessation of operations is necessary, to permit such to be accomplished in a safe and orderly manner.

(b)   Upon the filing of any such petition the district court shall have jurisdiction to grant such injunctive relief or temporary restraining order pending the outcome of an enforcement proceeding pursuant to this Act. The proceeding shall be as provided by Rule 65 of the Federal Rules, Civil Procedure, except that no temporary restraining order issued without notice shall be effective for a period longer than five days.

(c)   Whenever and as soon as an inspector concludes that conditions or practices described in subsection (a) exist in any place of employment, he shall inform the affected employees and employers of the danger and that he is recommending to the Secretary that relief be sought.

(d)   If the Secretary arbitrarily or capriciously fails to seek relief under this section, any employee who may be injured by reason of such failure, or the representative of such employees, might bring an action against the Secretary in the United States district court for the district in which the imminent danger is alleged to exist or the employer has its principal office, or for the District of Columbia, for a writ of mandamus to compel the Secretary to seek such an order and for such further relief as may be appropriate.

SEC. 14. Representation in Civil Litigation

29 USC 663   Except as provided in section 518(a) of title 28, United States Code, relating to litigation before the Supreme Court, the Solicitor of Labor may appear for and represent the Secretary in any civil litigation brought under this Act but all such litigation shall be subject to the direction and control of the Attorney General.

SEC. 15. Confidentiality of Trade Secrets

29 USC 664   All information reported to or otherwise obtained by the Secretary or his representative in connection with any inspection or proceeding under this Act which contains or which might reveal a trade secret referred to in section 1905 of title 18 of the United States Code shall be considered confidential for the purpose of that section, except that such information may be disclosed to other officers or employees concerned with carrying out this Act or when relevant in any proceeding under this Act. In any such proceeding the Secretary, the Commission, or the court shall issue such orders as may be appropriate to protect the confidentiality of trade secrets.

SEC. 16. Variations, Tolerances, and Exemptions

29 USC 665   The Secretary, on the record, after notice and opportunity for a hearing may provide such reasonable limitations and may make such rules and regulations allowing reasonable variations, tolerances, and exemptions to and from any or all provisions of this Act as he may find necessary and proper to avoid serious impairment of the national defense. Such action shall not be in effect for more than six months without notification to affected employees and an opportunity being afforded for a hearing.

SEC. 17. Penalties

(a)   29 USC 666   Pub. L. 101-508 increased the civil penalties in subsections (a)-(d) & (i). See Historical notes.

Any employer who willfully or repeatedly violates the requirements of section 5 of this Act, any standard, rule, or order promulgated pursuant to section 6 of this Act, or regulations prescribed pursuant to this Act, may be assessed a civil penalty of not more than $70,000 for each violation, but not less than $5,000 for each willful violation.

(b)   Any employer who has received a citation for a serious violation of the requirements of section 5 of this Act, of any standard, rule, or order promulgated pursuant to section 6 of this Act, or of any regulations prescribed pursuant to this Act, shall be assessed a civil penalty of up to $7,000 for each such violation.

(c)   Any employer who has received a citation for a violation of the requirements of section 5 of this Act, of any standard, rule, or order promulgated pursuant to section 6 of this Act, or of regulations prescribed pursuant to this Act, and such violation is specifically determined not to be of a serious nature, may be assessed a civil penalty of up to $7,000 for each violation.

(d)   Any employer who fails to correct a violation for which a citation has been issued under section 9(a) within the period permitted for its correction (which period shall not begin to run until the date of the final order of the Commission in the case of any review proceeding under section 10 initiated by the employer in good faith and not solely for delay or avoidance of penalties), may be assessed a civil penalty of not more than $7,000 for each day during which such failure or violation continues.

(e)   Pub. L. 98-473 Maximum criminal fines are increased by the Sentencing Reform Act of 1984, 18 USC § 3551 et seq. See Historical notes.

Any employer who willfully violates any standard, rule, or order promulgated pursuant to section 6 of this Act, or of any regulations prescribed pursuant to this Act, and that violation caused death to any employee, shall, upon conviction, be punished by a fine of not more than $10,000 or by imprisonment for not more than six months, or by both; except that if the conviction is for a violation committed after a first conviction of such person, punishment shall be by a fine of not more than $20,000 or by imprisonment for not more than one year, or by both.

(f)   See historical notes.

Any person who gives advance notice of any inspection to be conducted under this Act, without authority from the Secretary or his designees, shall, upon conviction, be punished by a fine of not more than $1,000 or by imprisonment for not more than six months, or by both.

(g)   Whoever knowingly makes any false statement, representation, or certification in any application, record, report, plan, or other document filed or required to be maintained pursuant to this Act shall, upon conviction, be punished by a fine of not more than $10,000, or by imprisonment for not more than six months, or by both.

(h)

(1) Section 1114 of title 18, United States Code, is hereby amended by striking out "designated by the Secretary of Health and Human Services to conduct investigations, or inspections under the Federal Food, Drug, and Cosmetic Act" and inserting in lieu thereof "or of the Department of Labor assigned to perform investigative, inspection, or law enforcement functions".

(2) Notwithstanding the provisions of sections 1111 and 1114 of title 18, United States Code, whoever, in violation of the provisions of section 1114 of such title, kills a person while engaged in or on account of the performance of investigative, inspection, or law enforcement functions added to such section 1114 by paragraph (1) of this subsection, and who would otherwise be subject to the penalty provisions of such section 1111, shall be punished by imprisonment for any term of years or for life.

(i)   Any employer who violates any of the posting requirements, as prescribed under the provisions of this Act, shall be assessed a civil penalty of up to $7,000 for each violation.

(j)   The Commission shall have authority to assess all civil penalties provided in this section, giving due consideration to the appropriateness of the penalty with respect to the size of the business of the employer being charged, the gravity of the violation, the good faith of the employer, and the history of previous violations.

(k)   For purposes of this section, a serious violation shall be deemed to exist in a place of employment if there is a substantial probability that death or serious physical harm could result from a condition which exists, or from one or more practices, means, methods, operations, or processes which have been adopted or are in use, in such place of employment unless the employer did not, and could not with the exercise of reasonable diligence, know of the presence of the violation.

(l)   Civil penalties owed under this Act shall be paid to the Secretary for deposit into the Treasury of the United States and shall accrue to the United States and may be recovered in a civil action in the name of the United States brought in the United States district court for the district where the violation is alleged to have occurred or where the employer has its principal office.

SEC. 18. State Jurisdiction and State Plans

(a)   29 USC 667   Nothing in this Act shall prevent any State agency or court from asserting jurisdiction under State law over any occupational safety or health issue with respect to which no standard is in effect under section 6.

(b)   Any State which, at any time, desires to assume responsibility for development and enforcement therein of occupational safety and health standards relating to any occupational safety or health issue with respect to which a Federal standard has been promulgated under section 6 shall submit a State plan for the development of such standards and their enforcement.

(c)   The Secretary shall approve the plan submitted by a State under subsection (b), or any modification thereof, if such plan in his judgement –

(1) designates a State agency or agencies as the agency or agencies responsible for administering the plan throughout the State,

(2) provides for the development and enforcement of safety and health standards relating to one or more safety or health issues, which standards (and the enforcement of which standards) are or will be at least as effective in providing safe and healthful employment and places of employment as the standards promulgated under section 6 which relate to the same issues, and which standards, when applicable to products which are distributed or used in interstate commerce, are required by compelling local conditions and do not unduly burden interstate commerce,

(3) provides for a right of entry and inspection of all workplaces subject to the Act which is at least as effective as that provided in section 8, and includes a prohibition on advance notice of inspections,

(4) contains satisfactory assurances that such agency or agencies have or will have the legal authority and qualified personnel necessary for the enforcement of such standards,

(5) gives satisfactory assurances that such State will devote adequate funds to the administration and enforcement of such standards,

(6) contains satisfactory assurances that such State will, to the extent permitted by its law, establish and maintain an effective and comprehensive occupational safety and health program applicable to all employees of public agencies of the State and its political subdivisions, which program is as effective as the standards contained in an approved plan,

(7) requires employers in the State to make reports to the Secretary in the same manner and to the same extent as if the plan were not in effect, and

(8) provides that the State agency will make such reports to the Secretary in such form and containing such information, as the Secretary shall from time to time require.

(d)   If the Secretary rejects a plan submitted under subsection (b), he shall afford the State submitting the plan due notice and opportunity for a hearing before so doing.

(e)   After the Secretary approves a State plan submitted under subsection (b), he may, but shall not be required to, exercise his authority under sections 8, 9, 10, 13, and 17 with respect to comparable standards promulgated under section 6, for the period specified in the next sentence. The Secretary may exercise the authority referred to above until he determines, on the basis of actual operations under the State plan, that the criteria set forth in subsection (c) are being applied, but he shall not make such determination for at least three years after the plan's approval under subsection (c). Upon making the determination referred to in the preceding sentence, the provisions of sections 5(a)(2), 8 (except for the purpose of carrying out subsection (f) of this section), 9, 10, 13, and 17, and standards promulgated under section 6 of this Act, shall not apply with respect to any occupational safety or health issues covered under the plan, but the Secretary may retain jurisdiction under the above provisions in any proceeding commenced under section 9 or 10 before the date of determination.

(f)   The Secretary shall, on the basis of reports submitted by the State agency and his own inspections make a continuing evaluation of the manner in which each State having a plan approved under this section is carrying out such plan. Whenever the Secretary finds, after affording due notice and opportunity for a hearing, that in the administration of the State plan there is a failure to comply substantially with any provision of the State plan (or any assurance contained therein), he shall notify the State agency of his withdrawal of approval of such plan and upon receipt of such notice such plan shall cease to be in effect, but the State may retain jurisdiction in any case commenced before the withdrawal of the plan in order to enforce standards under the plan whenever the issues involved do not relate to the reasons for the withdrawal of the plan.

(g)   The State may obtain a review of a decision of the Secretary withdrawing approval of or rejecting its plan by the United States court of appeals for the circuit in which the State is located by filing in such court within thirty days following receipt of notice of such decision a petition to modify or set aside in whole or in part the action of the Secretary. A copy of such petition shall forthwith be served upon the Secretary, and thereupon the Secretary shall certify and file in the court the record upon which the decision complained of was issued as provided in section 2112 of title 28, United States Code. Unless the court finds that the Secretary's decision in rejecting a proposed State plan or withdrawing his approval of such a plan is not supported by substantial evidence the court shall affirm the Secretary's decision. The judgment of the court shall be subject to review by the Supreme Court of the United States upon certiorari or certification as provided in section 1254 of title 28, United States Code.

(h)   The Secretary may enter into an agreement with a State under which the State will be permitted to continue to enforce one or more occupational health and safety standards in effect in such State until final action is taken by the Secretary with respect to a plan submitted by a State under subsection (b) of this section, or two years from the date of enactment of this Act, whichever is earlier.

SEC. 19. Federal Agency Safety Programs and Responsibilities

(a)   29 USC 668   It shall be the responsibility of the head of each Federal agency (not including the United States Postal Service) to establish and maintain an effective and comprehensive occupational safety and health program which is consistent with the standards promulgated under section 6. The head of each agency shall (after consultation with representatives of the employees thereof) –

(1) Pub. L. 50-241   provide safe and healthful places and conditions of employment, consistent with the standards set under section 6;

(2) acquire, maintain, and require the use of safety equipment, personal protective equipment, and devices reasonably necessary to protect employees;

(3) keep adequate records of all occupational accidents and illnesses for proper evaluation and necessary corrective action;

(4) consult with the Secretary with regard to the adequacy as to form and content of records kept pursuant to subsection (a)(3) of this section; and

(5) make an annual report to the Secretary with respect to occupational accidents and injuries and the agency's program under this section. Such report shall include any report submitted under section 7902(e)(2) of title 5, United States Code.

(b)   Pub. L. 97-375   The Secretary shall report to the President a summary or digest of reports submitted to him under subsection (a)(5) of this section, together with his evaluations of and recommendations derived from such reports.

(c)   Section 7902(c)(1) of title 5, United States Code, is amended by inserting after "agencies" the following: "and of labor organizations representing employees".

(d)   The Secretary shall have access to records and reports kept and filed by Federal agencies pursuant to subsections (a)(3) and (5) of this section unless those records and reports are specifically required by Executive order to be kept secret in the interest of the national defense or foreign policy, in which case the Secretary shall have access to such information as will not jeopardize national defense or foreign policy.

SEC. 20. Research and Related Activities

(a)

(1) 29 USC 669   The Secretary of Health and Human Services, after consultation with the Secretary and with other appropriate Federal departments or agencies, shall conduct (directly or by grants or contracts) research, experiments, and demonstrations relating to occupational safety and health, including studies of psychological factors involved, and relating to innovative methods, techniques, and approaches for dealing with occupational safety and health problems.

(2) The Secretary of Health and Human Services shall from time to time consult with the Secretary in order to develop specific plans for such research, demonstrations, and experiments as are necessary to produce criteria, including criteria identifying toxic substances, enabling the Secretary to meet his responsibility for the formulation of safety and health standards under this Act; and the Secretary of Health and Human Services, on the basis of such research, demonstrations, and experiments and any other information available to him, shall develop and publish at least annually such criteria as will effectuate the purposes of this Act.

(3) The Secretary of Health and Human Services, on the basis of such research, demonstrations, and experiments, and any other information available to him, shall develop criteria dealing with toxic materials and harmful physical agents and substances which will describe exposure levels that are safe for various periods of employment, including but not limited to the exposure levels at which no employee will suffer impaired health or functional capacities or diminished life expectancy as a result of his work experience.

(4) The Secretary of Health and Human Services shall also conduct special research, experiments, and demonstrations relating to occupational safety and health as are necessary to explore new problems, including those created by new technology in occupational safety and health, which may require ameliorative action beyond that which is otherwise provided for in the operating provisions of this Act. The Secretary of Health and Human Services shall also conduct research into the motivational and behavioral factors relating to the field of occupational safety and health.

(5) The Secretary of Health and Human Services, in order to comply with his responsibilities under paragraph (2), and in order to develop needed information regarding potentially toxic substances or harmful physical agents, may prescribe regulations requiring employers to measure, record, and make reports on

the exposure of employees to substances or physical agents which the Secretary of Health and Human Services reasonably believes may endanger the health or safety of employees. The Secretary of Health and Human Services also is authorized to establish such programs of medical examinations and tests as may be necessary for determining the incidence of occupational illnesses and the susceptibility of employees to such illnesses. Nothing in this or any other provision of this Act shall be deemed to authorize or require medical examination, immunization, or treatment for those who object thereto on religious grounds, except where such is necessary for the protection of the health or safety of others. Upon the request of any employer who is required to measure and record exposure of employees to substances or physical agents as provided under this subsection, the Secretary of Health and Human Services shall furnish full financial or other assistance to such employer for the purpose of defraying any additional expense incurred by him in carrying out the measuring and recording as provided in this subsection.

(6) The Secretary of Health and Human Services shall publish within six months of enactment of this Act and thereafter as needed but at least annually a list of all known toxic substances by generic family or other useful grouping, and the concentrations at which such toxicity is known to occur. He shall determine following a written request by any employer or authorized representative of employees, specifying with reasonable particularity the grounds on which the request is made, whether any substance normally found in the place of employment has potentially toxic effects in such concentrations as used or found; and shall submit such determination both to employers and affected employees as soon as possible. If the Secretary of Health and Human Services determines that any substance is potentially toxic at the concentrations in which it is used or found in a place of employment, and such substance is not covered by an occupational safety or health standard promulgated under section 6, the Secretary of Health and Human Services shall immediately submit such determination to the Secretary, together with all pertinent criteria.

(7) Within two years of enactment of the Act, and annually thereafter the Secretary of Health and Human Services shall conduct and publish industry wide studies of the effect of chronic or low-level exposure to industrial materials, processes, and stresses on the potential for illness, disease, or loss of functional capacity in aging adults.

(b)   The Secretary of Health and Human Services is authorized to make inspections and question employers and employees as provided in section 8 of this Act in order to carry out his functions and responsibilities under this section.

(c)   The Secretary is authorized to enter into contracts, agreements, or other arrangements with appropriate public agencies or private organizations for the purpose of conducting studies relating to his responsibilities under this Act. In carrying out his responsibilities under this subsection, the Secretary shall cooperate with the Secretary of Health and Human Services in order to avoid any duplication of efforts under this section.

(d)   Information obtained by the Secretary and the Secretary of Health and Human Services under this section shall be disseminated by the Secretary to employers and employees and organizations thereof.

(e)   The functions of the Secretary of Health and Human Services under this Act shall, to the extent feasible, be delegated to the Director of the National Institute for Occupational Safety and Health established by section 22 of this Act.

EXPANDED RESEARCH ON WORKER SAFETY AND HEALTH

29 USC 669a   Pub. L. 107-188, Title I, § 153 added this text.

The Secretary of Health and Human Services (referred to in this section as the "Secretary"), acting through the Director of the National Institute of Occupational Safety and Health, shall enhance and expand research as deemed appropriate on the health and safety of workers who are at risk for bioterrorist threats or attacks in the workplace, including research on the health effects of measures taken to treat or protect such workers for diseases or disorders resulting from a bioterrorist threat or attack. Nothing in this section may be construed as establishing new regulatory authority for the Secretary or the Director to issue or modify any occupational safety and health rule or regulation.

SEC. 21. Training and Employee Education

(a)   29 USC 670   The Secretary of Health and Human Services, after consultation with the Secretary and with other appropriate Federal departments and agencies, shall conduct, directly or by grants or contracts –

(1) education programs to provide an adequate supply of qualified personnel to carry out the purposes of this Act, and

(2) informational programs on the importance of and proper use of adequate safety and health equipment.

(b)   The Secretary is also authorized to conduct, directly or by grants or contracts, short-term training of personnel engaged in work related to his responsibilities under this Act.

(c)   The Secretary, in consultation with the Secretary of Health and Human

Services, shall –

(1) provide for the establishment and supervision of programs for the education and training of employers and employees in the recognition, avoidance, and prevention of unsafe or unhealthful working conditions in employments covered by this Act, and

(2) Pub. L. 105-97, §2 added subsection (d). See Historical notes.

consult with and advise employers and employees, and organizations representing employers and employees as to effective means of preventing occupational injuries and illnesses.

(d)

(1) The Secretary shall establish and support cooperative agreements with the States under which employers subject to this Act may consult with State personnel with respect to –

(A) the application of occupational safety and health requirements under this Act or under State plans approved under section 18; and

(B) voluntary efforts that employers may undertake to establish and maintain safe and healthful employment and places of employment. Such agreements may provide, as a condition of receiving funds under such agreements, for contributions by States towards meeting the costs of such agreements.

(2) Pursuant to such agreements the State shall provide on-site consultation at the employer's worksite to employers who request such assistance. The State may also provide other education and training programs for employers and employees in the State. The State shall ensure that on-site consultations conducted pursuant to such agreements include provision for the participation by employees.

(3) Activities under this subsection shall be conducted independently of any enforcement activity. If an employer fails to take immediate action to eliminate employee exposure to an imminent danger identified in a consultation or fails to correct a serious hazard so identified within a reasonable time, a report shall be made to the appropriate enforcement authority for such action as is appropriate.

(4) The Secretary shall, by regulation after notice and opportunity for comment, establish rules under which an employer –

(A) which requests and undergoes an on-site consultative visit provided under this subsection;

(B) which corrects the hazards that have been identified during the visit within the time frames established by the State and agrees to request a subsequent consultative visit if major changes in working conditions or work processes occur which introduce new hazards in the workplace; and

(C) which is implementing procedures for regularly identifying and preventing hazards regulated under this Act and maintains appropriate involvement of, and training for, management and non-management employees in achieving safe and healthful working conditions, may be exempt from an inspection (except an inspection requested under section 8(f) or an inspection to determine the cause of a workplace accident which resulted in the death of one or more employees or hospitalization for three or more employees) for a period of 1 year from the closing of the consultative visit.

(5) A State shall provide worksite consultations under paragraph (2) at the request of an employer. Priority in scheduling such consultations shall be assigned to requests from small businesses which are in higher hazard industries or have the most hazardous conditions at issue in the request.

SEC. 22. National Institute for Occupational Safety and Health

(a)   29 USC 671   It is the purpose of this section to establish a National Institute for Occupational Safety and Health in the Department of Health and Human Services in order to carry out the policy set forth in section 2 of this Act and to perform the functions of the Secretary of Health and Human Services under sections 20 and 21 of this Act.

(b)   There is hereby established in the Department of Health and Human Services a National Institute for Occupational Safety and Health. The Institute shall be headed by a Director who shall be appointed by the Secretary of Health and Human Services, and who shall serve for a term of six years unless previously removed by the Secretary of Health and Human Services.

(c)   The Institute is authorized to –

(1) develop and establish recommended occupational safety and health standards; and

(2) perform all functions of the Secretary of Health and Human Services under sections 20 and 21 of this Act.

(d)   Upon his own initiative, or upon the request of the Secretary of Health and Human Services, the Director is authorized (1) to conduct such research and experimental programs as he determines are necessary for the development of criteria for new and improved occupational safety and health standards, and (2) after consideration of the results of such research and experimental programs make recommendations concerning new or improved occupational safety and health standards. Any occupational safety and health standard recommended pursuant to this section shall immediately be forwarded to the Secretary of Labor, and to the Secretary of Health and Human Services.

(e)   In addition to any authority vested in the Institute by other provisions of this section, the Director, in carrying out the functions of the Institute, is authorized to –

(1) prescribe such regulations as he deems necessary governing the manner in which its functions shall be carried out;

(2) receive money and other property donated, bequeathed, or devised, without condition or restriction other than that it be used for the purposes of the Institute and to use, sell, or otherwise dispose of such property for the purpose of carrying out its functions;

(3) receive (and use, sell, or otherwise dispose of, in accordance with paragraph (2)), money and other property donated, bequeathed, or devised to the Institute with a condition or restriction, including a condition that the Institute use other funds of the Institute for the purposes of the gift;

(4) in accordance with the civil service laws, appoint and fix the compensation of such personnel as may be necessary to carry out the provisions of this section;

(5) obtain the services of experts and consultants in accordance with the provisions of section 3109 of title 5, United States Code;

(6) accept and utilize the services of voluntary and noncompensated personnel and reimburse them for travel expenses, including per diem, as authorized by section 5703 of title 5, United States Code;

(7) enter into contracts, grants or other arrangements, or modifications thereof to carry out the provisions of this section, and such contracts or modifications thereof may be entered into without performance or other bonds, and without regard to section 3709 of the Revised Statutes, as amended (41 U.S.C. 5), or any other provision of law relating to competitive bidding;

(8) make advance, progress, and other payments which the Director deems necessary under this title without regard to the provisions of section 3324 (a) and (b) of Title 31; and

(9)   Pub. L. 97-258   make other necessary expenditures.

(f)   The Director shall submit to the Secretary of Health and Human Services, to the President, and to the Congress an annual report of the operations of the Institute under this Act, which shall include a detailed statement of all private and public funds received and expended by it, and such recommendations as he deems appropriate.

(g)   Pub. L. 102-550 added subsection (g).   Lead-Based Paint Activities.

(1) Training Grant Program.

(A) The Institute, in conjunction with the Administrator of the Environmental Protection Agency, may make grants for the training and education of workers and supervisors who are or may be directly engaged in lead-based paint activities.

(B) Grants referred to in subparagraph (A) shall be awarded to nonprofit organizations (including colleges and universities, joint labor-management trust funds, States, and nonprofit government employee organizations) –

(i) which are engaged in the training and education of workers and supervisors who are or who may be directly engaged in lead-based paint activities (as defined in Title IV of the Toxic Substances Control Act),

(ii) which have demonstrated experience in implementing and operating health and safety training and education programs, and

(iii) with a demonstrated ability to reach, and involve in lead-based paint training programs, target populations of individuals who are or will be engaged in lead-based paint activities. Grants under this subsection shall be awarded only to those organizations that fund at least 30 percent of their lead-based paint activities training programs from non-Federal sources, excluding in-kind contributions. Grants may also be made to local governments to carry out such training and education for their employees.

(C) There are authorized to be appropriated, a minimum, $10,000,000 to the Institute for each of the fiscal years 1994 through 1997 to make grants under this paragraph.

(2) Evaluation of Programs. The Institute shall conduct periodic and comprehensive assessments of the efficacy of the worker and supervisor training programs developed and offered by those receiving grants under this section. The Director shall prepare reports on the results of these assessments addressed to the Administrator of the Environmental Protection Agency to include recommendations as may be appropriate for the revision of these programs. The sum of $500,000 is authorized to be appropriated to the Institute for each of the fiscal years 1994 through 1997 to carry out this paragraph.

WORKERS' FAMILY PROTECTION

(a) 29 USC 671a Short title

This section may be cited as the "Workers' Family Protection Act".

(b) Findings and purpose

(1) Pub. L. 102-522, Title II, §209 added this text.

Findings
Congress finds that–

(A) hazardous chemicals and substances that can threaten the health and safety of workers are being transported out of industries on workers' clothing and persons;

(B) these chemicals and substances have the potential to pose an additional threat to the health and welfare of workers and their families;

(C) additional information is needed concerning issues related to employee transported contaminant releases; and

(D) additional regulations may be needed to prevent future releases of this type.

(2) Purpose
It is the purpose of this section to–

(A) increase understanding and awareness concerning the extent and possible health impacts of the problems and incidents described in paragraph (1);

(B) prevent or mitigate future incidents of home contamination that could adversely affect the health and safety of workers and their families;

(C) clarify regulatory authority for preventing and responding to such incidents; and

(D) assist workers in redressing and responding to such incidents when they occur.

(c)   Evaluation of employee transported contaminant releases

(1) Study

(A) In general
Not later than 18 months after October 26, 1992, the Director of the National Institute for Occupational Safety and Health (hereafter in this section referred to as the "Director"), in cooperation with the Secretary of Labor, the Administrator of the Environmental Protection Agency, the Administrator of the Agency for Toxic Substances and Disease Registry, and the heads of other Federal Government agencies as determined to be appropriate by the Director, shall conduct a study to evaluate the potential for, the prevalence of, and the issues related to the contamination of workers' homes with hazardous chemicals and substances, including infectious agents, transported from the workplaces of such workers.

(B) Matters to be evaluated
In conducting the study and evaluation under subparagraph (A), the Director shall–

(i) conduct a review of past incidents of home contamination through the utilization of literature and of records concerning past investigations and enforcement actions undertaken by–

(I) the National Institute for Occupational Safety and Health;

(II) the Secretary of Labor to enforce the Occupational Safety and Health Act of 1970 (29 U.S.C. 651 et seq.);

(III) States to enforce occupational safety and health standards in accordance with section 18 of such Act (29 U.S.C. 667); and

(IV) other government agencies (including the Department of Energy and the Environmental Protection Agency), as the Director may determine to be appropriate;

(ii) evaluate current statutory, regulatory, and voluntary industrial hygiene or other measures used by small, medium and large employers to prevent or remediate home contamination;

(iii) compile a summary of the existing research and case histories conducted on incidents of employee transported contaminant releases, including–

(I) the effectiveness of workplace housekeeping practices and personal protective equipment in preventing such incidents;

(II) the health effects, if any, of the resulting exposure on workers and their families;

(III) the effectiveness of normal house cleaning and laundry procedures for removing hazardous materials and agents from workers' homes and personal clothing;

(IV) indoor air quality, as the research concerning such pertains to the fate of chemicals transported from a workplace into the home environment; and

(V) methods for differentiating exposure health effects and relative risks associated with specific agents from other sources of exposure inside and outside the home;

(iv) identify the role of Federal and State agencies in responding to incidents of home contamination;

(v) prepare and submit to the Task Force established under paragraph (2) and to the appropriate committees of Congress, a report concerning the results of the matters studied or evaluated under clauses (i) through (iv); and

(vi) study home contamination incidents and issues and worker and family protection policies and practices related to the special circumstances of firefighters and prepare and submit to the appropriate committees of Congress a report concerning the findings with respect to such study.

(2) Development of investigative strategy

(A) Task Force
Not later than 12 months after October 26, 1992, the Director shall establish a working group, to be known as the "Workers' Family Protection Task Force". The Task Force shall–

(i) be composed of not more than 15 individuals to be appointed by the Director from among individuals who are representative of workers, industry, scientists, industrial hygienists, the National Research Council, and government agencies, except that not more than one such individual shall be from each appropriate government agency and the number of individuals appointed to represent industry and workers shall be equal in number;

(ii) review the report submitted under paragraph (1)(B)(v);

(iii) determine, with respect to such report, the additional data needs, if any, and the need for additional evaluation of the scientific issues related to and the feasibility of developing such additional data; and

(iv) if additional data are determined by the Task Force to be needed, develop a recommended investigative strategy for use in obtaining such information.

(B) Investigative strategy

(i) Content
The investigative strategy developed under subparagraph (A)(iv) shall identify data gaps that can and cannot be filled, assumptions and uncertainties associated with various components of such strategy, a timetable

for the implementation of such strategy, and methodologies used to gather any required data.

(ii) Peer review

The Director shall publish the proposed investigative strategy under subparagraph (A)(iv) for public comment and utilize other methods, including technical conferences or seminars, for the purpose of obtaining comments concerning the proposed strategy.

(iii) Final strategy

After the peer review and public comment is conducted under clause (ii), the Director, in consultation with the heads of other government agencies, shall propose a final strategy for investigating issues related to home contamination that shall be implemented by the National Institute for Occupational Safety and Health and other Federal agencies for the period of time necessary to enable such agencies to obtain the information identified under subparagraph (A)(iii).

(C) Construction

Nothing in this section shall be construed as precluding any government agency from investigating issues related to home contamination using existing procedures until such time as a final strategy is developed or from taking actions in addition to those proposed in the strategy after its completion.

(3) Implementation of investigative strategy

Upon completion of the investigative strategy under subparagraph (B)(iii), each Federal agency or department shall fulfill the role assigned to it by the strategy.

(d) Regulations

(1) In general

Not later than 4 years after October 26, 1992, and periodically thereafter, the Secretary of Labor, based on the information developed under subsection (c) of this section and on other information available to the Secretary, shall–

(A) determine if additional education about, emphasis on, or enforcement of existing regulations or standards is needed and will be sufficient, or if additional regulations or standards are needed with regard to employee transported releases of hazardous materials; and

(B) prepare and submit to the appropriate committees of Congress a report concerning the result of such determination.

(2) Additional regulations or standards If the Secretary of Labor determines that additional regulations or standards are needed under paragraph (1), the Secretary shall promulgate, pursuant to the Secretary's authority under the Occupational Safety and Health Act of 1970 (29 U.S.C. 651 et seq.), such regulations or standards as determined to be appropriate not later than 3 years after such determination.

(e) Authorization of appropriations There are authorized to be appropriated from sums otherwise authorized to be appropriated, for each fiscal year such sums as may be necessary to carry out this section.

SEC. 23. Grants to the States

(a) 29 USC 672 The Secretary is authorized, during the fiscal year ending June 30, 1971, and the two succeeding fiscal years, to make grants to the States which have designated a State agency under section 18 to assist them –

(1) in identifying their needs and responsibilities in the area of occupational safety and health,

(2) in developing State plans under section 18, or

(3) in developing plans for –

(A) establishing systems for the collection of information concerning the nature and frequency of occupational injuries and diseases;

(B) increasing the expertise and enforcement capabilities of their personnel engaged in occupational safety and health programs; or

(C) otherwise improving the administration and enforcement of State occupational safety and health laws, including standards thereunder, consistent with the objectives of this Act.

(b) The Secretary is authorized, during the fiscal year ending June 30, 1971, and the two succeeding fiscal years, to make grants to the States for experimental and demonstration projects consistent with the objectives set forth in subsection (a) of this section.

(c) The Governor of the State shall designate the appropriate State agency for receipt of any grant made by the Secretary under this section.

(d) Any State agency designated by the Governor of the State desiring a grant under this section shall submit an application therefor to the Secretary.

(e) The Secretary shall review the application, and shall, after consultation with the Secretary of Health and Human Services, approve or reject such application.

(f) The Federal share for each State grant under subsection (a) or (b) of this section may not exceed 90 per centum of the total cost of the application. In the event the Federal share for all States under either such subsection is not the same, the differences among the States shall be established on the basis of objective criteria.

(g) The Secretary is authorized to make grants to the States to assist them in administering and enforcing programs for occupational safety and health contained in State plans approved by the Secretary pursuant to section 18 of this Act. The Federal share for each State grant under this subsection may not exceed 50 per centum of the total cost to the State of such a program. The last sentence of subsection (f) shall be applicable in determining the Federal share under this subsection.

(h) Prior to June 30, 1973, the Secretary shall, after consultation with the Secretary of Health and Human Services, transmit a report to the President and to the Congress, describing the experience under the grant programs authorized by this section and making any recommendations he may deem appropriate.

SEC. 24. Statistics

(a)   In order to further the purposes of this Act, the Secretary, in consultation with the Secretary of Health and Human Services, shall develop and maintain an effective program of collection, compilation, and analysis of occupational safety and health statistics. Such program may cover all employments whether or not subject to any other provisions of this Act but shall not cover employments excluded by section 4 of the Act. The Secretary shall compile accurate statistics on work injuries and illnesses which shall include all disabling, serious, or significant injuries and illnesses, whether or not involving loss of time from work, other than minor injuries requiring only first aid treatment and which do not involve medical treatment, loss of consciousness, restriction of work or motion, or transfer to another job.

(b)   To carry out his duties under subsection (a) of this section, the Secretary may –

(1) promote, encourage, or directly engage in programs of studies, information and communication concerning occupational safety and health statistics;

(2) make grants to States or political subdivisions thereof in order to assist them in developing and administering programs dealing with occupational safety and health statistics; and

(3) arrange, through grants or contracts, for the conduct of such research and investigations as give promise of furthering the objectives of this section.

(c)   The Federal share for each grant under subsection (b) of this section may be up to 50 per centum of the State's total cost.

(d)   The Secretary may, with the consent of any State or political subdivision thereof, accept and use the services, facilities, and employees of the agencies of such State or political subdivision, with or without reimbursement, in order to assist him in carrying out his functions under this section.

(e)   On the basis of the records made and kept pursuant to section 8(c) of this Act, employers shall file such reports with the Secretary as he shall prescribe by regulation, as necessary to carry out his functions under this Act.

(f)   Agreements between the Department of Labor and States pertaining to the collection of occupational safety and health statistics already in effect on the effective date of this Act shall remain in effect until superseded by grants or contracts made under this Act.

SEC. 25. Audits

(a)   29 USC 674   Each recipient of a grant under this Act shall keep such records as the Secretary or the Secretary of Health and Human Services shall prescribe, including records which fully disclose the amount and disposition by such recipient of the proceeds of such grant, the total cost of the project or undertaking in connection with which such grant is made or used, and the amount of that portion of the cost of the project or undertaking supplied by other sources, and such other records as will facilitate an effective audit.

(b)   The Secretary or the Secretary of Health and Human Services, and the Comptroller General of the United States, or any of their duly authorized representatives, shall have access for the purpose of audit and examination to any books, documents, papers, and records of the recipients of any grant under this Act that are pertinent to any such grant.

SEC. 26. Annual Report

29 USC 675 Pub. L. 104-66 §3003 terminated provision relating to transmittal of report to Congress.

Within one hundred and twenty days following the convening of each regular session of each Congress, the Secretary and the Secretary of Health and Human Services shall each prepare and submit to the President for transmittal to the Congress a report upon the subject matter of this Act, the progress toward achievement of the purpose of this Act, the needs and requirements in the field of occupational safety and health, and any other relevant information. Such reports shall include information regarding occupational safety and health standards, and criteria for such standards, developed during the preceding year; evaluation of standards and criteria previously developed under this Act, defining areas of emphasis for new criteria and standards; an evaluation of the degree of observance of applicable occupational safety and health standards, and a summary of inspection and enforcement activity undertaken; analysis and evaluation of research activities for which results have been obtained under governmental and nongovernmental sponsorship; an analysis of major occupational diseases; evaluation of available control and measurement technology for hazards for which standards or criteria have been developed during the preceding year; description of cooperative efforts undertaken between Government agencies and other interested parties in the implementation of this Act during the preceding year; a progress report on the development of an adequate supply of trained manpower in the field of occupational safety and health, including estimates of future needs and the efforts being made by Government and others to meet those needs; listing of all toxic substances in industrial usage for which labeling requirements, criteria, or standards have not yet been established; and such recommendations for additional legislation as are deemed necessary to protect the safety and health of the worker and improve the administration of this Act.

SEC. 27. National Commission on State Workmen's Compensation Laws

29 USC 676   (Text omitted.)

SEC. 28. Economic Assistance to Small Businesses

See notes on omitted text.   (Text omitted.)

SEC. 29. Additional Assistant Secretary of Labor

See notes on omitted text.   (Text omitted.)

SEC. 30. Additional Positions

See notes on omitted text.　(Text omitted.)

SEC. 31. Emergency Locator Beacons

See notes on omitted text.　(Text omitted.)

SEC. 32. Separability

29 USC 677　If any provision of this Act, or the application of such provision to any person or circumstance, shall be held invalid, the remainder of this Act, or the application of such provision to persons or circumstances other than those as to which it is held invalid, shall not be affected thereby.

SEC. 33. Appropriations

29 USC 678　There are authorized to be appropriated to carry out this Act for each fiscal year such sums as the Congress shall deem necessary.

SEC. 34. Effective Date

This Act shall take effect one hundred and twenty days after the date of its enactment.

Approved December 29, 1970.

As amended through January 1, 2004.

HISTORICAL NOTES

This reprint generally retains the section numbers originally created by Congress in the Occupational Safety and Health (OSH) Act of 1970, Pub. L. 91-596, 84 Stat 1590. This document includes some editorial changes, such as changing the format to make it easier to read, correcting typographical errors, and updating some of the margin notes. Because Congress enacted amendments to the Act since 1970, this version differs from the original version of the OSH Act. It also differs slightly from the version published in the United States Code at 29 U.S.C. 661 *et seq.* For example, this reprint refers to the statute as the "Act" rather than the "chapter."

This reprint reflects the provisions of the OSH Act that are in effect as of January 1, 2004. Citations to Public Laws which made important amendments to the OSH Act since 1970 are set forth in the margins and explanatory notes are included below.

NOTE: Some provisions of the OSH Act may be affected by the enactment of, or amendments to, other statutes. Section 17(h)(1), 29 U.S.C. 666, is an example. The original provision amended section 1114 of title 18 of the United States Code to include employees of "the Department of Labor assigned to perform investigative, inspection, or law enforcement functions" within the list of persons protected by the provisions to allow prosecution of persons who have killed or attempted to kill an officer or employee of the U.S. government while performing official duties. This reprint sets forth the text of section 17(h) as enacted in 1970. However, since

1970, Congress has enacted multiple amendments to 18 U.S.C. 1114. The current version does not specifically include the Department of Labor in a list; rather it states that "Whoever kills or attempts to kill any officer or employee of the United States or of any agency in any branch of the United States Government (including any member of the uniformed services) while such officer or employee is engaged in or on account of the performance of official duties, or any person assisting such an officer or employee in the performance of such duties or on account of that assistance shall be punished ..." as provided by the statute. Readers are reminded that the official version of statutes can be found in the current volumes of the United States Code, and more extensive historical notes can be found in the current volumes of the United States Code Annotated.

AMENDMENTS

On January 2, 1974, section 2(c) of Pub. L. 93-237 replaced the phrase "7(b)(6)" in section 28(d) of the OSH Act with "7(b)(5)". 87 Stat. 1023. Note: The text of Section 28 (Economic Assistance to Small Business) amended Sections 7(b) and Section 4(c)(1) of the Small Business Act. Because these amendments are no longer current, the text of section 28 is omitted in this reprint. For the current version, see 15 U.S.C. 636.

In 1977, the U.S. entered into the Panama Canal Treaty of 1977, Sept. 7, 1977, U.S.-Panama, T.I.A.S. 10030, 33 U.S.T. 39. In 1979, Congress enacted implementing legislation. Panama Canal Act of 1979, Pub. L. 96-70, 93 Stat. 452 (1979). Although no corresponding amendment to the OSH Act was enacted, the Canal Zone ceased to exist in 1979. The U.S. continued to manage, operate and facilitate the transit of ships through the Canal under the authority of the Panama Canal Treaty until December 31, 1999, at which time authority over the Canal was transferred to the Republic of Panama.

On March 27, 1978, Pub. L. 95-251, 92 Stat. 183, replaced the term "hearing examiner(s)" with "administrative law judge(s)" in all federal laws, including sections 12(e), 12(j), and 12(k) of the OSH Act, 29 U.S.C. 661.

On October 13, 1978, Pub. L. 95-454, 92 Stat. 1111, 1221, which redesignated section numbers concerning personnel matters and compensation, resulted in the substitution of section 5372 of Title 5 for section 5362 in section 12(e) of the OSH Act, 29 U.S.C. 661.

On October 17, 1979, Pub. L. 96-88, Title V, section 509(b), 93 Stat. 668, 695, redesignated references to the Department of Health, Education, and Welfare to the Department of Health and Human Services and redesignated references to the Secretary of Health, Education, and Welfare to the Secretary of Health and Human Services.

On September 13, 1982, Pub. L. 97-258, §4(b), 96 Stat. 877, 1067, effectively substituted "Section 3324(a) and (b) of Title 31" for "Section 3648 of the Revised Statutes, as amended (31 U.S.C. 529)" in section 22 (e)(8), 29 U.S.C. 671, relating to NIOSH procurement authority.

On December 21, 1982, Pub. L. 97-375, 96 Stat. 1819, deleted the sentence in section 19(b) of the Act, 29 U.S.C. 668, that directed the President of the United States to transmit annual reports of the activities of federal agencies to the House of Representatives and the Senate.

On October 12, 1984, Pub. L. 98-473, Chapter II, 98 Stat. 1837, 1987, (commonly referred to as the "Sentencing Reform Act of 1984") instituted a classification system for criminal offenses punishable under the United States Code. Under this system, an offense with imprisonment terms of "six months or less but more than thirty days," such as that found in 29 U.S.C. 666(e) for a willful violation of the OSH Act, is classified as a criminal "Class B misdemeanor." 18 U.S.C. 3559(a)(7).

The criminal code increases the monetary penalties for criminal misdemeanors beyond what is provided for in the OSH Act: a fine for a Class B misdemeanor resulting in death, for example, is not more than $250,000 for an individual, and is not more than $500,000 for an organization. 18 U.S.C. 3571(b)(4), (c)(4). The criminal code also provides for authorized terms of probation for both individuals and organizations. 18 U.S.C. 3551, 3561. The term of imprisonment for individuals is the same as that authorized by the OSH Act. 18 U.S.C. 3581(b)(7).

On November 8, 1984, Pub. L. 98-620, 98 Stat. 3335, deleted the last sentence in section 11(a) of the Act, 29 U.S.C. 660, that required petitions filed under the subsection to be heard expeditiously.

On November 5, 1990, Pub. L. 101-508, 104 Stat. 1388, amended section 17 of the Act, 29 U.S.C. 666, by increasing the penalties in section 17(a) from $10,000 for each violation to "$70,000 for each violation, but not less than $5,000 for each willful violation," and increased the limitation on penalties in sections (b), (c), (d), and (i) from $1,000 to $7,000.

On October 26, 1992, Pub. L. 102-522, 106 Stat. 3410, 3420, added to Title 29, section 671a "Workers' Family Protection" to grant authority to the Director of NIOSH to evaluate, investigate and if necessary, for the Secretary of Labor to regulate employee transported releases of hazardous material that result from contamination on the employee's clothing or person and may adversely affect the health and safety of workers and their families. Note: section 671a was enacted as section 209 of the Fire Administration Authorization Act of 1992, but it is reprinted here because it is codified within the chapter that comprises the OSH Act.

On October 28, 1992, the Housing and Community Development Act of 1992, Pub. L. 102-550, 106 Stat. 3672, 3924, amended section 22 of the Act, 29 U.S.C. 671, by adding subsection (g), which requires NIOSH to institute a training grant program for lead-based paint activities.

On July 5, 1994, section 7(b) of Pub. L. 103-272, 108 Stat. 745, repealed section 31 of the OSH Act, "Emergency Locator Beacons." Section 1(e) of the same Public Law, however, enacted a modified version of section 31 of the OSH Act. This provision, titled "Emergency Locator Transmitters," is codified at 49 U.S.C. 44712.

On December 21, 1995, Section 3003 of Pub. L. 104-66, 109 Stat. 707, as amended, effective May 15, 2000, terminated the provisions relating to the transmittal to Congress of reports under section 26 of the OSH Act. 29 U.S.C. 675.

On July 16, 1998, Pub. L. 105-197, 112 Stat. 638, amended section 21 of the Act, 29 U.S.C. 670, by adding subsection (d), which required the Secretary to establish a compliance assistance program by which employers can consult with state personnel regarding the application of and compliance with OSHA standards.

On July 16, 1998, Pub. L. 105-198, 112 Stat. 640, amended section 8 of the Act, 29 U.S.C. 657, by adding subsection (h), which forbids the Secretary to use the results of enforcement activities to evaluate the employees involved in such enforcement or to impose quotas or goals.

On September 29, 1998, Pub. L. 105-241, 112 Stat. 1572, amended sections 3(5) and 19(a) of the Act, 29 U.S.C. 652 and 668, to include the United States Postal Service as an "employer" subject to OSHA enforcement.

On June 12, 2002, Pub. L. 107-188, Title I, Section 153, 116 Stat. 631, Congress enacted 29 U.S.C. 669a, to expand research on the "health and safety of workers who are at risk for bioterrorist threats or attacks in the workplace."

JURISDICTIONAL NOTE

Although no corresponding amendments to the OSH Act have been made, OSHA no longer exercises jurisdiction over the entity formerly known as the Trust Territory of the Pacific Islands. The Trust Territory, which consisted of the Former Japanese Mandated Islands, was established in 1947 by the Security Council of the United Nations, and administered by the United States. *Trusteeship Agreement for the Former Japanese Mandated Islands,* Apr. 2-July 18, 1947, 61 Stat. 3301, T.I.A.S. 1665, 8 U.N.T.S. 189.

From 1947 to 1994, the people of these islands exercised the right of self-determination conveyed by the Trusteeship four times, resulting in the division of the Trust Territory into four separate entities. Three entities: the Republic of Palau, the Federated States of Micronesia, and the Republic of the Marshall Islands, became "Freely Associated States," to which U.S. Federal Law does not apply. Since the OSH Act is a generally applicable law that applies to Guam, it applies to the Commonwealth of Northern Mariana Islands, which elected to become a "Flag Territory" of the United States. *See Covenant to Establish a Commonwealth of the Northern Mariana Islands in Political Union with the United States of America,* Article V, section 502(a) as contained in Pub. L. 94-24, 90 Stat. 263 (Mar. 24, 1976) [citations to amendments omitted]; 48 U.S.C. 1801 and note (1976); *see also Saipan Stevedore Co., Inc. v. Director, Office of Workers' Compensation Programs,* 133 F.3d 717, 722 (9th Cir. 1998) (Longshore and Harbor Workers' Compensation Act applies to the Commonwealth of Northern Mariana Islands pursuant to section 502(a) of the Covenant because the Act has general application to the states and to Guam). For up-to-date information on the legal status of these freely associated states and territories, contact the Office of Insular Affairs of the Department of the Interior. (Web address: http://www.doi.gov/oia/)

Omitted Text. Reasons for textual deletions vary. Some deletions may result from amendments to the OSH Act; others to subsequent amendments to other statutes which the original provisions of the OSH Act may have amended in 1970. In some instances, the original provision of the OSH Act was date-limited and is no longer operative.

The text of section 12(c), 29 U.S.C. 661, is omitted. Subsection (c) amended sections 5314 and 5315 of Title 5, United States Code, to add the positions of Chairman and members of the Occupational Safety and Health Review Commission.

The text of section 27, 29 U.S.C. 676, is omitted. Section 27 listed Congressional findings on workers' compensation and established the National Commission on

State Workmen's Compensation Laws, which ceased to exist ninety days after the submission of its final report, which was due no later than July 31, 1972.

The text of section 28 (Economic Assistance to Small Business) amended sections 7(b) and section 4(c)(1) of the Small Business Act to allow for small business loans in order to comply with applicable standards. Because these amendments are no longer current, the text is omitted here. For the current version see 15 U.S.C. 636.

The text of section 29, (Additional Assistant Secretary of Labor), created an Assistant Secretary for Occupational Safety and Health, and section 30 (Additional Positions) created additional positions within the Department of Labor and the Occupational Safety and Health Review Commission in order to carry out the provisions of the OSH Act. The text of these sections is omitted here because it no longer reflects the current statutory provisions for staffing and pay. For current provisions, see 29 U.S.C. 553 and 5 U.S.C. 5108 (c).

Section 31 of the original OSH Act amended 49 U.S.C. 1421 by inserting a section entitled "Emergency Locator Beacons." The text of that section is omitted in this reprint because Pub. L. 103-272, 108 Stat.745, (July 5, 1994), repealed the text of section 31 and enacted a modified version of the provision, entitled "Emergency Locator Transmitters," which is codified at 49 U.S.C. 44712.

Notes on other legislation affecting the administration of the Occupational Safety and Health Act. Sometimes legislation does not directly amend the OSH Act, but does place requirements on the Secretary of Labor either to act or to refrain from acting under the authority of the OSH Act. Included below are some examples of such legislation. Please note that this is not intended to be a comprehensive list.

STANDARDS PROMULGATION

For example, legislation may require the Secretary to promulgate specific standards pursuant to authority under section 6 of the OSH Act, 29 U.S.C. 655. Some examples include the following:

Hazardous Waste Operations. Pub. L. 99-499, Title I, section 126(a)-(f), 100 Stat. 1613 (1986), as amended by Pub. L. 100-202, section 101(f), Title II, section 201, 101 Stat. 1329 (1987), required the Secretary of Labor to promulgate standards concerning hazardous waste operations.

Chemical Process Safety Management. Pub. L. 101-549, Title III, section 304, 104 Stat. 2399 (1990), required the Secretary of Labor, in coordination with the Administrator of the Environmental Protection Agency, to promulgate a chemical process safety standard.

Hazardous Materials. Pub. L. 101-615, section 29, 104 Stat. 3244 (1990), required the Secretary of Labor, in consultation with the Secretaries of Transportation and Treasury, to issue specific standards concerning the handling of hazardous materials.

Bloodborne Pathogens Standard. Pub. L. 102-170, Title I, section 100, 105 Stat. 1107 (1991), required the Secretary of Labor to promulgate a final Bloodborne Pathogens standard.

Lead Standard. The Housing and Community Development Act of 1992, Pub. L. 102-550, Title X, sections 1031 and 1032, 106 Stat. 3672 (1992), required the Secretary of Labor to issue an interim final lead standard.

EXTENSION OF COVERAGE

Sometimes a statute may make some OSH Act provisions applicable to certain entities that are not subject to those provisions by the terms of the OSH Act. For example, the Congressional Accountability Act of 1995, Pub. L. 104-1, 109 Stat. 3, (1995), extended certain OSH Act coverage, such as the duty to comply with Section 5 of the OSH Act, to the Legislative Branch. Among other provisions, this legislation authorizes the General Counsel of the Office of Compliance within the Legislative Branch to exercise the authority granted to the Secretary of Labor in the OSH Act to inspect places of employment and issue a citation or notice to correct the violation found. This statute does not make all the provisions of the OSH Act applicable to the Legislative Branch. Another example is the Medicare Prescription Drug, Improvement, and Modernization Act of 2003, Title IX, Section 947, Pub. L. 108-173, 117 Stat. 2066 (2003), which requires public hospitals not otherwise subject to the OSH Act to comply with OSHA's Bloodborne Pathogens standard, 29 CFR 1910.1030. This statute provides for the imposition and collection of civil money penalties by the Department of Health and Human Services in the event that a hospital fails to comply with OSHA's Bloodborne Pathogens standard.

PROGRAM CHANGES ENACTED THROUGH
APPROPRIATIONS LEGISLATION

Sometimes an appropriations statute may allow or restrict certain substantive actions by OSHA or the Secretary of Labor. For example, sometimes an appropriations statute may restrict the use of money appropriated to run the Occupational Safety and Health Administration or the Department of Labor. One example of such a restriction, that has been included in OSHA's appropriation for many years, limits the applicability of OSHA requirements with respect to farming operations that employ ten or fewer workers and do not maintain a temporary labor camp. Another example is a restriction that limits OSHA's authority to conduct certain enforcement activity with respect to employers of ten or fewer employees in low hazard industries. See Consolidated Appropriations Act, 2004, Pub. L. 108-199, Div. E - Labor, Health and Human Services, and Education, and Related Agencies Appropriations, 2004, Title I - Department of Labor, 118 Stat. 3 (2004). Sometimes an appropriations statute may allow OSHA to retain some money collected to use for occupational safety and health training or grants. For example, the Consolidated Appropriations Act, 2004, Div. E, Title I, cited above, allows OSHA to retain up to $750,000 of training institute course tuition fees per fiscal year for such uses. For the statutory text of currently applicable appropriations provisions, consult the OSHA appropriations statute for the fiscal year in question.

APPENDIX B: OSHA SEVERE VIOLATOR ENFORCEMENT PROGRAM (SVEP)

(Modified for the purposes of this text)

- **Record Type:** OSHA Instruction
- **Current Directive Number:** CPL 02-00-149
- **Old Directive Number:** CPL 02-00-149
- **Title:** Severe Violator Enforcement Program (SVEP)
- **Information Date:** 06/18/2010

EXECUTIVE SUMMARY

This Instruction establishes enforcement policies and procedures for OSHA's Severe Violator Enforcement Program (SVEP), which concentrates resources on inspecting employers who have demonstrated indifference to their OSH Act obligations by committing willful, repeated, or failure-to-abate violations. Enforcement actions for severe violator cases include mandatory follow-up inspections, increased company/corporate awareness of OSHA enforcement, corporate-wide agreements, where appropriate, enhanced settlement provisions, and federal court enforcement under Section 11(b) of the OSH Act. In addition, this Instruction provides for nationwide referral procedures, which includes OSHA's State Plan States. This Instruction replaces OSHA's Enhanced Enforcement Program (EEP).

SIGNIFICANT CHANGES FROM THE ENHANCED ENFORCEMENT PROGRAM (EEP)

- High-Emphasis Hazards are targeted, which include fall hazards and specific hazards identified from selected National Emphasis Programs.
- The Assistant Secretary has determined that Nationwide Inspections of Related Workplaces/Worksites are critical inspections for the purpose of 29 CFR §1908.7(b)(2)(iv).
- Creates a nationwide referral procedure for Regions and State Plan States.

TABLE OF CONTENTS

    I.   **Purpose**. This Instruction establishes enforcement policies and pro-
cedures for OSHA's Severe Violator Enforcement Program (SVEP),
which concentrates resources on inspecting employers who have demonstrated indifference to their OSH Act obligations by committing willful,
repeated, or failure-to-abate violations. This Instruction replaces OSHA's
Enhanced Enforcement Program (EEP).

EXECUTIVE SUMMARY

This Instruction establishes enforcement policies and procedures for OSHA's
Severe Violator Enforcement Program (SVEP), which concentrates resources
on inspecting employers who have demonstrated indifference to their OSH
Act obligations by committing willful, repeated, or failure-to-abate violations.
Enforcement actions for severe violator cases include mandatory follow-up
inspections, increased company/corporate awareness of OSHA enforcement,
corporate-wide agreements, where appropriate, enhanced settlement provisions, and federal court enforcement under Section 11(b) of the OSH Act. In
addition, this Instruction provides for nationwide referral procedures, which
includes OSHA's State Plan States. This Instruction replaces OSHA's Enhanced
Enforcement Program (EEP).

SIGNIFICANT CHANGES FROM THE ENHANCED
ENFORCEMENT PROGRAM (EEP)

- High-Emphasis Hazards are targeted, which include fall hazards and spe-
cific hazards identified from selected National Emphasis Programs.
- The Assistant Secretary has determined that Nationwide Inspections of
Related Workplaces/Worksites are critical inspections for the purpose of
29 CFR §1908.7(b)(2)(iv).
- Creates a nationwide referral procedure for Regions and State Plan
States.

I. **Purpose**. This Instruction establishes enforcement policies and procedures for OSHA's Severe Violator Enforcement Program (SVEP), which concentrates resources on inspecting employers who have demonstrated indifference to their OSH Act obligations by committing willful, repeated, or failure-to-abate violations. This Instruction replaces OSHA's Enhanced Enforcement Program (EEP).

II. **Scope**.

III. This Instruction applies OSHA-wide.

IV. **Background**.

V. The SVEP is intended to focus enforcement efforts on significant hazards and violations by concentrating inspection resources on employers who have demonstrated recalcitrance or indifference to their OSH Act obligations by committing willful, repeated, or failure-to-abate violations in one or more of the following circumstances: (1) a fatality or catastrophe situation; (2) in industry operations or processes that expose employees to the most severe occupational hazards and those identified as "High-Emphasis Hazards," as defined in Section XII. of this Instruction; (3) exposing employees to hazards related to the potential release of a highly hazardous chemical; or (4) all egregious enforcement actions.

VI. Cases meeting the severe violator enforcement criteria are those in which the employer is found to be recalcitrant or indifferent to its obligations under the OSH Act, thereby endangering employees. The SVEP procedures in Section XV are intended to increase attention on the correction of hazards found in these workplaces and, where appropriate, in other worksites of the same employer where similar hazards and deficiencies may be present. This program applies to all employers regardless of size.

VII. **Transition between the EEP and the SVEP**.

VIII. On the effective date of this Instruction, both the original EEP (referred to as EEP or EEP1) and the revised EEP (referred to as EEP2) will terminate (including any follow-up inspections that have not yet been performed). However, Area Directors have the discretion to conduct any follow-up inspections related to the EEP in accordance with the policies and procedures in the FOM (CPL 02-00-148, Chapter 2, *Program Planning*).

IX. **Significant Changes**.

    A. High-Emphasis Hazards are targeted and include fall hazards and hazards identified from the following National Emphasis Programs (NEPs): amputations, combustible dust, crystalline silica, excavation/trenching, lead, and shipbreaking. See section XII.

    B. The Assistant Secretary by this Instruction determines that the specific inspections under the Nationwide Inspections of Related Workplaces/Worksites process are critical inspections for the purpose of 29 CFR §1908.7(b)(2)(iv) and delegates to the Director of the Directorate of Enforcement Programs the authority to determine site selections of those related workplaces/worksites. See paragraph XV.B.7.b.

C. A nationwide referral procedure is being initiated in which OSHA may inspect related worksites/workplaces of a SVEP employer. See section XV.B.

X. **Handling Severe Violator Enforcement Cases**.

A. Compliance Officers (CSHOs) must become familiar with Appendix B to effectively evaluate employers during all inspections likely to result in a severe violator enforcement case.

B. The Area Director will identify severe violator enforcement cases no later than at the time the citations are issued, in accordance with criteria set forth in this Instruction.

C. Federal Agency cases that meet the SVEP case criteria will also be classified as severe violator enforcement cases, and where the term "employer-wide" or "company-wide" is used, it will apply agency-wide or department-wide, as appropriate. Appropriate SVEP actions for such cases will be determined by the Area Director in consultation with the Regional Administrator.

D. When a case meets the severe violator enforcement case criteria, the Area Director will notify the Regional Administrator, who in turn will notify the Directorate of Enforcement Programs (DEP).

E. Regional Administrator notification to DEP must be by e-mail using the SVEP-group e-mail address on OSHA's Global Address list: "zzOS-HA-SVEP." The notification must be at least monthly (by the 20th of the month) and include the information requested in Appendix A. Regions must use the Excel spreadsheet format that will be sent to the SVEP Regional Coordinators shortly after this Instruction becomes effective.

XI. **Criteria for a Severe Violator Enforcement Case**.

XII. Any inspection that meets one or more of criteria A. through D., at the time that the citations are issued, will be considered a severe violator enforcement case.

XIII. **Willful** and **repeated** citations and **failure-to-abate** notices must be based on serious violations, except for recordkeeping, which must be egregious (e.g., per-instance citations). See *FOM*, CPL 02-00-148, Chapter 6, paragraphs V.A.1. and VI.A.1.

A. **Fatality/Catastrophe Criterion**.

B. A fatality/catastrophe inspection in which OSHA finds **one** or more **willful** or **repeated** violations or **failure-to-abate** notices based on a serious violation related to a death of an employee or three or more hospitalizations.

C. NOTE: The violations under this criterion **do not have to be High-Emphasis Hazards** as defined in Section XII.

D. **Non-Fatality/Catastrophe Criterion Related to High-Emphasis Hazards**.

E. An inspection in which OSHA finds **two** or more **willful** or **repeated** violations or **failure-to-abate** notices (or any combination of these violations/notices), based on **high gravity serious** violations related to a **High-Emphasis Hazard** as defined in Section XII.

F. **Non-Fatality/Catastrophe Criterion for Hazards Due to the Potential Release of a Highly Hazardous Chemical (Process Safety Management)**.

G. An inspection in which OSHA finds **three** or more **willful** or **repeated** violations or **failure-to-abate** notices (or any combination of these violations/notices), based on **high gravity serious** violations related to hazards due to the potential release of a highly hazardous chemical, as defined in the PSM standard.

H. **Egregious Criterion**.

I. All **egregious** (e.g., per-instance citations) enforcement actions will be considered SVEP cases.

XIV. **Definition of High-Emphasis Hazards.**

XV. High-Emphasis Hazards as used in this Instruction means only **high gravity serious** violations of the following specific standards covered under falls or the National Emphasis Programs (NEPs) listed in paragraphs F. through K. below, **regardless of the type of inspection** being conducted (e.g., complaint, SST, Local Emphasis Programs, National Emphasis Programs). Low and moderate gravity violations **will not** be considered for a SVEP case. See Chapter 4, section II, *Serious Violations*, and Chapter 6, section III.A *Gravity of Violation* of the FOM (CPL 02-00-148) for determining what constitutes a **high gravity serious** violation.

XVI. Example 1: A CSHO conducts an **SST inspection** and cites the employer for one high gravity willful violation of 29 CFR §1910.23 and one low gravity willful violation of 29 CFR §1910.28. The inspection has not met the Non-Fatality/Catastrophe Criterion Related to High-Emphasis Hazards and is not subject to the SVEP.

XVII. Example 2: A CSHO conducts a **Local Emphasis Program** inspection for Residential Construction. While on-site, the CSHO observes employees working in an unsupported trench and cites the employer for two high gravity willful violations of 29 CFR §1926.651. The inspection has met the Non-Fatality/Catastrophe Criterion Related to High-Emphasis Hazards and the case is subject to the SVEP.

XVIII. Example 3: A CSHO conducts a **National Emphasis Program** inspection for Shipbreaking. While on-site, the CSHO observes a piece of scrap metal from the dismantled vessel being lifted with a crane over top of an employee with a kinked wire rope. The employer is cited for one high gravity repeat violation of 29 CFR §1915.112 and one high gravity willful violation of 29 CFR §1915.116. The inspection has met the Non-Fatality/Catastrophe Criterion Related to High-Emphasis Hazards and the case is subject to the SVEP.

A. **Fall hazards covered under the following general industry standards:**

1. 29 CFR §1910.23 - Guarding floor and wall openings and holes [Walking-Working Surfaces]

2. 29 CFR §1910.28 - Safety requirements for scaffolding [Walking-Working Surfaces]

3. 29 CFR §1910.29 - Manually propelled mobile ladder stands and scaffolds (towers) [Walking-Working Surfaces]

4. 29 CFR §1910.66 - Powered platforms for building maintenance [Powered Platforms, Manlifts, and Vehicle-Mounted Work Platforms]

5. 29 CFR §1910.67 - Vehicle-mounted elevating and rotating work platforms [Powered Platforms, Manlifts, and Vehicle-Mounted Work Platforms]

6. 29 CFR §1910.68 - Manlifts [Powered Platforms, Manlifts, and Vehicle-Mounted Work Platforms]

B. **Fall hazards covered under the following construction industry standards:**

1. 29 CFR §1926.451 - General requirements [Scaffolds]

2. 29 CFR §1926.452 - Additional requirements applicable to specific types of scaffolds

3. 29 CFR §1926.453 - Aerial lifts [Scaffolds]

4. 29 CFR §1926.501 - Duty to have fall protection

5. 29 CFR §1926.502 - Fall protection systems criteria and practices

6. 29 CFR §1926.760 - Fall protection [Steel Erection]

7. 29 CFR §1926.1052 - Stairways [Ladders]

C. **Fall hazards covered under the following shipyard standards:**

1. 29 CFR §1915.71 - Scaffolds or staging [Scaffolds, ladders and Other Working Surfaces]

2. 29 CFR §1915.73 - Guarding of deck openings and edges [Scaffolds, ladders and Other Working Surfaces]

3. 29 CFR §1915.74 - Access to vessels [Scaffolds, ladders and Other Working Surfaces]

4. 29 CFR §1915.75 - Access to and guarding of dry docks and marine railways [Scaffolds, ladders and Other Working Surfaces]

5. 29 CFR §1915.159 - Personal fall arrest systems (PFAS) [Personal Protective Equipment (PPE)]

D. **Fall hazards covered under the following marine terminal standards:**

1. 29 CFR §1917.45 - Cranes and derricks [Cargo Handling Gear and Equipment]

2. 29 CFR §1917.49 - Spouts, chutes, hoppers, bins, and associated equipment [Cargo Handling Gear and Equipment]

3. 29 CFR §1917.112 - Guarding of edges [Terminal Facilities]

E. **Fall hazards covered under the following longshoring standards:**

1. 29 CFR §1918.22 - Gangways [Gangways and Other Means of Access]

2. 29 CFR §1918.85 - Containerized cargo operations [Handling Cargo]

F. **Amputation hazards** specified below that are covered under the National Emphasis Program on Amputations. (See CPL03-00-003):

1. 29 CFR §1910.147 - The control of hazardous energy (lockout/tagout)

    2.   29 CFR §1910.212 - General requirements for all machines

    3.   29 CFR §1910.213 - Woodworking machinery requirements

    4.   29 CFR §1910.217 - Mechanical power presses

    5.   29 CFR §1910.219 - Mechanical power-transmission apparatus

G. **Combustible dust hazards** specified below that are covered by the Combustible Dust National Emphasis Program (Reissued), including the General Duty Clause (Sec. 5(a)(1) of the OSH Act). (See CPL 03-00-008):

    1.   29 CFR §1910.22 - General requirements [Walking-Working Surfaces]

    2.   29 CFR §1910.307 - Hazardous (classified) locations [Electrical]

    3.   Sec. 5(a)(1) of the OSH Act

    4.   Any General Duty Clause violation concerning hazards related to dust collectors inside buildings, deflagration isolation systems, and duct-work issues.

H. **Crystalline silica hazards** specified below that are covered by the National Emphasis Program — Crystalline Silica (See CPL 03-00-007):

    1.   **Overexposure.**

        a.   29 CFR Part §1910.1000 and 29 CFR Part §1915.1000 - Air Contaminants

        b.   29 CFR §1926.55 - Gases, vapors, fumes, dusts, and mists

    2.   **Failure to Implement Engineering Controls.**

        a.   29 CFR §1910.1000(e) - Air Contaminants

        b.   29 CFR §1926.55(b) - Gases, vapors, fumes, dusts, and mists

    3.   **When Overexposure Occurs**.

    4.   29 CFR §1910.134; 29 CFR §1926.103; and 29 CFR §1915.154 - Respiratory protection

NOTE: The Silica NEP requires a mandatory follow-up inspection when overexposures to crystalline silica are found. If a follow-up inspection finds the same violations as previously cited, the follow-up inspection will most likely qualify as a SVEP case. See paragraph XV.A.4.

I. **Lead hazards** specified below that are covered by the National Emphasis Program — Lead (only violations based on sampling). See CPL 03-00-009.

    1.   29 CFR §1910.1025 - Lead

    2.   29 CFR §1926.62 - Lead

    3.   29 CFR §1915.1025 - Lead and 29 CFR §1915 Subpart D Welding, Cutting, and Heating

J. **Excavation/trenching hazards** specified below that are covered by the Special Emphasis Program - Trenching and Excavation (See CPL 02-00-069)

    1.   29 CFR §1926.651 - Specific excavation requirements

    2.   29 CFR §1926.652 - Requirements for protective systems [Excavations]

K. **Shipbreaking hazards** specified below that are covered by the National Emphasis Program — Shipbreaking. See CPL 02-00-136.
   1. 29 CFR §1915.12 - Precautions and the order of testing before entering confined and enclosed spaces and other dangerous atmospheres [Confined and Enclosed Spaces and Other Dangerous Atmospheres in Shipyard Employment]
   2. 29 CFR §1915.112 - Ropes, chains, and slings [Gear and Equipment for Rigging and Materials Handling]
   3. 29 CFR §1915.116 - Use of Gear [Gear and Equipment for Rigging and Materials Handling]
   4. 29 CFR §1915.159 - Personal fall arrest systems (PFAS) [Personal Protective Equipment (PPE)]
   5. 29 CFR §1915.503 - Precautions for hot work [Fire Protection in Shipyard Employment]

XIX. **Hazards Due to the Potential Release of a Highly Hazardous Chemical (Process Safety Management)**.

XX. **Petroleum refinery hazards** are those hazards that are covered by the Petroleum Refinery Process Safety Management National Emphasis Program (See CPL 03-00-004) and hazards due to the potential release of a highly hazardous chemical as covered by the PSM Covered Chemical Facilities National Emphasis Program See Instruction 09-06 (CPL 02):

29 CFR §1910.119 - Process safety management of highly hazardous chemicals

XXI. **Enforcement Considerations — Two or More Inspections of the Same Employer**.

XXII. For classification under the SVEP, each individual inspection must be evaluated separately to determine if it meets one of the criteria in XI.A., B., C., or D. If any of the inspections meet one of the severe violator criteria, it will be considered a SVEP case and coded according to paragraph XVIII.A.

XXIII. **Procedures of the Severe Violator Enforcement Program (SVEP)**.

XXIV. When the Area Director determines that a case meets one of the SVEP criteria, it will be treated in accordance with paragraphs XV.A. through E. below. Only those SVEP actions that are appropriate for the particular employer should be taken.
   A. **Enhanced Follow-up Inspections**.
      1. **General**.
      2. For any SVEP inspection issued on or after the effective date of this Instruction, a follow-up inspection must be conducted after the citations become final orders even if abatement verification of the cited violations has been received. The purpose of the follow-up inspection is to assess **not only** whether the cited violation(s) were abated **but also** whether the employer is committing similar violations.
      3. **Compelling Reason Not to Conduct**.
      4. If there is a compelling reason not to conduct a follow-up inspection, the reason must be documented in the file. **The Region shall**

also report these cases monthly to the Director of Enforcement Programs, along with the reason a follow-up was not initiated.

5. If a follow-up cannot be initiated, the follow-up column of the SVEP Log must be completed by giving the reason. Examples of compelling reasons not to conduct a follow-up inspection may include: (1) worksite/workplace closed, (2) employer is out of business, (3) operation cited has been discontinued at the worksite/workplace, or (4) case no longer meets any of the SVEP criteria because citation has been withdrawn/vacated.

6. NOTE: A Corrected During Inspection (CDI) situation does not take the place of a needed follow-up inspection.

7. If the Area Director learns that a cited operation has been moved from the cited location to a different location, the new location shall be inspected. If the new location is outside the area office jurisdiction, a referral shall be made.

8. **Construction Worksites**.

9. When the Area Office has reason to believe that a construction worksite is no longer active (or is nearing completion), thus making a follow-up inspection of the same worksite impossible or impractical, the provisions in paragraph XV.B.5 shall apply. When a construction follow-up is attempted but the employer is no longer at the site, it will not be added to the SVEP Log.

10. **Silica Overexposure Follow-ups**.

11. The Silica NEP (CPL 03-00-007) in paragraph XI.E.1. requires a mandatory follow-up inspection when citations are issued for overexposure to crystalline silica to determine whether the employer is eliminating silica exposures or reducing exposures below the PEL. If a follow-up inspection finds the same or similar violations as previously cited, the follow-up inspection will most likely qualify as a severe violator enforcement case under the criteria in section XI.

B. **Nationwide Inspections of Related Workplaces/Worksites**.

   1. **General**.

   2. OSHA has found that employer indifference to compliance responsibilities under the Act may be indicative of broader patterns of non-compliance at related employer worksites. When there are reasonable grounds to believe that compliance problems identified in the initial inspection may be indicative of a broader pattern of non-compliance, OSHA will inspect related worksites of the same employer. Appendix B of this Instruction provides guidance in evaluating whether compliance problems found during the initial SVEP inspection are localized or likely to exist at related facilities. This information should be gathered, to the extent possible, during the initial SVEP inspection. Such information may also be sought by letter, telephone, or if necessary, by subpoena.

3. The Regional Administrator shall be responsible for assuring that relevant information is gathered and for determining whether the information provides reasonable grounds to believe that a broader pattern of non-compliance may exist. The Regional Administrator shall consult with the Regional Solicitor as appropriate. When sufficient evidence is found, all related establishments of the employer that are in the same 3-digit NAICS code (or 2-digit SIC code) as the initial SVEP case will be identified; establishments will be selected for inspection in accordance with subsection 4 below. Establishments that are not in the same 3-digit NAICS code (or 2-digit SIC code) also may be inspected if there are reasonable grounds to believe hazards and violations may be present at the related sites.

4. The Directorate of Enforcement programs will serve as the National Office point of contact for all SVEP nationwide referrals. Any questions should be addressed to the Director or Deputy Director in DEP. All Regional Administrators will name a SVEP Coordinator.

5. **Office of Statistical Analysis (OSA).**

6. At the request of the Director of the Directorate of Enforcement Programs, the Regional Administrator, or the Regional Coordinator, OSA will assist in identifying similar and other related worksites nationwide (including in State Plan States) of the same employer.

7. Establishments are related when there is common ownership. Related establishments include establishments of corporations that are in the same corporate family, including subsidiary, affiliate, or parent corporations with substantial common ownership.

8. Similar related establishments are related establishments that are in the same 3-digit NAICS code (or 2-digit SIC code).

9. **State Plan State Referrals.**

10. OSHA will accept referrals, which include all relevant facts, from State Plan States regarding any inspections conducted pursuant to the State's SVEP. State Plan referrals to Federal OSHA are to be sent to the Regional Administrator, who will forward any referrals not in its Region to the appropriate OSHA Regional Administrator.

11. **General Industry Workplaces.**

    a. **Employer Has Three (3) or Fewer Similar Related Workplaces.**

    b. When the Regional Administrator determines that additional workplaces within that region should be inspected, and the employer has three or fewer similar related workplaces, **all such workplaces will be inspected** to determine whether those sites have hazardous conditions or violations similar to those in the severe violator enforcement case. The

Regional Administrator shall have overall responsibility for coordinating the inspections and planning investigative strategy. The Regional Administrator shall consult with the Regional Solicitor as appropriate.

c. When any of the three or fewer workplaces that the Regional Administrator believes should be inspected are in **one or more of the Region's State Plan States**, the information will be forwarded to the State Plan Designee for inspection. A copy of the referral will also be sent to the Director of DEP.

d. When any of the three or fewer workplaces that the Regional Administrator believes should be inspected are in **two or more Regions or a State Plan State in another Region**, the information will be forwarded to the appropriate Regional Administrator for inspection. A copy of the referral will also be sent to the Director of DEP.

e. **Employer Has Four (4) or More Similar Related Workplaces**.

f. When the Regional Administrator determines that additional workplaces should be inspected, and the employer has **four or more similar related establishments within the Region or in other Regions**, or the number of workplaces/worksites cannot be determined, the Regional Administrator will send the recommendation for inspection, including all relevant facts, to the Director of DEP. The Director shall consult with the Associate Solicitor as appropriate.

g. (1) When the Director of DEP determines that there are reasonable grounds for inspecting similar related establishments, he/she will issue a SVEP Nationwide inspection list. Normally, when the number of similar related establishments nationwide is 10 or less, all will be selected for inspection. When there are more than 10, the Office of Statistical Analysis will assign random numbers to the complete list of similar related establishments, sort those establishments in random number order, and select the first 10 for inspection.

h. **All establishments on the inspection list will be inspected to determine whether hazardous conditions or violations similar to those found in the initial SVEP inspection are present**. Based on the results of these inspections, the Director may determine whether inspections of additional establishments should be conducted. Any inspection conducted under a SVEP Nationwide inspection list is to be coded as an unprogrammed-referral, and is to be considered a referral from the National Office. An OSHA-90 is to be generated when a site is discovered where a SVEP Nationwide Referral employer is working.

i. (2) In addition to or in lieu of (1) above, when the Director has reasonable grounds to believe that hazards may exist at particular other related establishments, he/she may select those establishments for inspection.

j. (3) The Director shall be responsible for coordinating nation-wide inspections of related establishments under this paragraph. Where complex or systemic issue are present, the Director shall convene a team to advise on investigative strategies, such as the use of administrative depositions or experts, and share information among offices participating in the inspections. The team shall include representatives from the OSHA and SOL national offices and regional offices where inspections will be conducted. In the event the inspections result in multiple contested citations, the team will advise SOL on litigation strategies that take account of such matters as distribution of work among affected offices and budget.

k. **SVEP Nationwide Related Inspections that involve process safety management (PSM) hazards**.

l. A SVEP Nationwide inspection will be limited to investigations of the PSM standard for which the willful or repeated citations or failure-to-abate notices were issued, and will not include units that were inspected in the previous two years.

12. **Construction Worksites**.

a. **Regional Office**.

b. Whenever an employer in the construction industry has a SVEP case, the Regional Administrator must further investigate the employer's OSH Act compliance. If the initially inspected worksite is closed before a follow-up inspection can be conducted, at least one other worksite of the cited employer must be inspected to determine whether the employer is committing violations similar to those found in the initial severe violator enforcement inspection. Because the worksites of construction employers are often difficult to locate, the following means may be used to identify other worksites of the cited employer.

- If the severe violator enforcement case is resolved through a settlement, the agreement should require the employer to notify the Area Director of its other jobsites prior to when work starts at new construction sites during the following one-year period.

- An administrative subpoena may be issued to an employer prior to the issuance of the citation to identify the location of worksites where employees of that employer are presently working or are expected to be working within the next 12 months. See *FOM*, Chapter 15, section I, (*Administrative Subpoenas*).

- A subpoena may be issued at any time during an inspection if it appears that the inspection is likely to result in a SVEP case and the Area Director determines (after consultation with the RA) that the hazards disclosed by the inspection and the inadequacy of the employer's response to those hazards indicate that a broader response by OSHA is appropriate.
- Whenever a subpoena is to be issued pursuant to the SVEP, the Regional Administrator shall coordinate with the Regional Solicitor.

c. **National Office**.

- When a Regional Administrator determines that a SVEP construction employer is operating in a different region, the Regional Administrator will send a recommendation for inspection, including all relevant facts, to the Director of DEP. The Director shall consult with the Associate Solicitor as appropriate.
- When the Director of DEP deems it necessary to notify Regional Administrators and State Designees regarding activity of a particular construction employer with worksites in more than one Region and/or State Plan States, the Director will issue a SVEP Nationwide referral. The procedures outlined under XV.B.4.b. will be followed.
- Any inspection conducted under a SVEP Nationwide referral is to be coded as an unprogrammed-referral, and is to be considered a referral from the National Office. An OSHA-90 is to be generated when a site is discovered where a SVEP Nationwide Referral employer is operating.

13. **Scope of Related Inspections**.
14. The scope of inspection of a related establishment will depend upon the evidence gathered in the original SVEP inspection, and will mainly focus on the same or similar hazards to those found in the original case.
15. **Priority of the Inspection**.
   a. In accordance with inspection priorities of the FOM (CPL 02-00-148), in Chapter 2, in section IV.B., (*Inspection Priority Criteria*), the SVEP nationwide referral inspections will come after imminent danger, fatality, and complaints, but before other programmed inspections. But see section XVII. (*Relationship to Other Programs*) of this Instruction, regarding when other inspections may be conducted concurrently.
   b. The Assistant Secretary by this Instruction determines that the specific inspections under the Nationwide Inspections of Related Workplaces/Worksites process are critical inspections for the purpose of 29 CFR §1908.7(b)(2)(iv) and

delegates to the Director of the Directorate of Enforcement Programs the authority to determine site selections of those related workplaces/worksites.

C. **Increased Company Awareness of OSHA Enforcement**.

  1. **Sending Citations and Notifications of Penalty to Headquarters**.

      a. For all employers that are the subject of a SVEP case, the Area Director shall mail a copy of the Citations and Notifications of Penalty to the employer's national headquarters if the employer has more than one fixed establishment. See sample cover letter in Appendix C.

      b. Employee representatives (e.g., unions) shall receive a copy of the Citations and Notifications of Penalty that is mailed to the employer's national headquarters.

  2. **Issuing News Releases**.

      a. Regional News Releases.

      b. The Regional Offices may issue a News Release for all SVEP cases upon issuance of the citations. Regional Administrators have the discretion to determine which SVEP cases will receive a News Release.

      c. **Nationwide Referral Inspection News Releases**.

      d. In SVEP cases that were a result of a Nationwide Referral, the Regional Office is to issue a News Release at the time the citations are issued. In certain SVEP cases, the National Office may issue a News Release in coordination with the Regional Office.

  3. **Sending Letters to Corporate Officers or Coordinating Meetings with the Regional or National Office**.

  4. In cases where OSHA determines that an establishment's safety and health problems should be addressed at the corporate level, the following actions should be considered:

      a. A letter sent from the Regional Administrator, or the appropriate National Office official, to the company President expressing OSHA's concern with the company's violations. A copy of the citations shall be sent with the letter if the citations and cover letter have not been sent to the company President previously.

      b. A meeting may be held between OSHA, company officials, employees and unions representing affected employees to discuss how the company intends to address safety and health compliance. If the company operates in more than one region, this normally will require National Office coordination.

      c. Employee representatives shall be notified by letter when OSHA determines that the establishment's safety and health problems need to be addressed at the corporate level.

D. **Enhanced Settlement Provisions**.

E. The following settlement provisions shall be considered to ensure future compliance both at the cited facility and at other related facilities of the employer:

1. Employers shall hire a qualified safety and health consultant to develop and implement an effective and comprehensive safety and health program or, where appropriate, a program to ensure full compliance with the subpart under which the employer was cited under the SVEP;

2. NOTE: Employers cannot be required in a settlement agreement to use OSHA's state consultation services; such services are strictly voluntary.

3. Applying the agreement company-wide (See CPL 02-00-090 Guidelines for Administration of Corporate-wide Settlement Agreements. June 3, 1991);

4. NOTE: Company-wide Settlement Agreements are to be coordinated with the National Office of the Solicitor.

5. Requiring interim abatement controls if OSHA is convinced that final abatement cannot be accomplished in a short period of time;

6. In construction (and, where appropriate, in general industry), using settlement agreements to obtain from the employer a list of its current jobsites, or future jobsites within a specified time period. The employer should be required to indicate to OSHA the specific protective measure to be used for each current or future jobsite;

7. Requiring the employer for a specified time period to submit to the Area Director its Log of Work-related Injuries and Illnesses on a quarterly basis, and to consent to OSHA conducting an inspection based on the information;

8. Requiring the employer for a specified time period to notify the Area Office of any serious injury or illness requiring medical attention and to consent to an inspection; and

9. Obtaining employer consent to entry of a court enforcement order under Section 11(b) of the Act.

F. **Federal Court Enforcement under Section 11(b) of the OSH Act**.

G. SVEP cases should be strongly considered for section 11(b) orders when it appears that such orders may be needed to assure compliance. An employer's obligation to abate a cited violation arises when there is a final order of the Review Commission affirming the citation. For guidance on drafting citations and settlement agreements that can maximize the deterrent effect of a Section 11(b) order, see *FOM* (CPL 02-00-148) Chapter 15, section XIV.

XXV. **SVEP Log**.

A. **General**.

B. The National Office will maintain a SVEP Log in which inspections that meet the SVEP criteria, or are SVEP-related inspections (i.e., SVEP follow-ups, or inspections at other worksites of the same employer), are logged as they are reported to the National Office by the Regional SVEP Coordinators.

C. **Lining-out Establishments from the SVEP Log**.

D. If an establishment has entered into a settlement agreement (informal or formal) in which a citation that qualified the establishment for SVEP designation is deleted, or if there has been an Administrative Law Judge, Review Commission, or court decision that has vacated such a citation, then the entry on the SVEP Log will be lined-out and the IMIS "SVEP" code will be removed from that establishment's Internet Inspection Detail summary. The Area Director must notify the Regional SVEP Coordinator of these changes, who in turn must notify DEP to line-out the inspection from the SVEP Log.

XXVI. **Relationship to Other Programs**.

    A. **Unprogrammed Inspections**.

    B. If an unprogrammed inspection arises with respect to an establishment that is to receive an SVEP-related inspection, the two inspections may be conducted either separately or concurrently. This Instruction does not affect in any way OSHA's ability to conduct unprogrammed inspections.

    C. **Programmed Inspections**.

    D. Some establishments selected for inspection under the SVEP may also fall under one or more other OSHA initiatives such as Site-Specific Targeting (SST) or Local Emphasis Programs (LEP). Inspections under these programs may be conducted either separately or concurrently with inspections under this Instruction.

XXVII. **Recording and Tracking Inspections**.

    A. **SVEP Code**.

    B. This applies to all severe violator enforcement cases issued on or after the effective date of this Instruction. Once a case is identified as a severe violator enforcement case, enter the NEP code "SVEP" from the drop-down list in field 25d, for the inspection.

    C. NOTE: **Only inspections** that meet one of the four criteria for a severe violator enforcement case will be coded with the SVEP NEP code.

    D. **NEP Codes for High-Emphasis Hazards**.

    E. If the SVEP criterion used is that described in paragraph XII.B., the appropriate NEP codes must be entered in field 25d.

    F. **Significant Enforcement Actions and Enhanced Settlement Codes**.

    G. If any inspection in a significant enforcement action qualifies as a severe violator case, it is to be coded "SIGCASE" in item 42, for that inspection.

    H. EXAMPLE: N 08 SIGCASE

    I. If a severe violator case receives an enhanced settlement agreement, it is to be coded "ENHSA" in item 42.

    J. EXAMPLE: N 08 ENHSA

    K. **Other Program Codes**.

    L. Remember to enter all applicable SST, REP, NEP, and LEP program codes in Item

M. 25c and 25d when an inspection is conducted and the inspection also meets the protocol for other program(s). Also, enter all applicable Strategic Management Plan hazard/industry codes in Item 25f.

Appendix A
Information Needed on Each SVEP Inspection
for Monthly Report to the National Office

Employer Name	Inspection Number	Regional Office	Area Office
Opening Date	SIC & NAICS codes	# of Employees Controlled	

Indicate if inspection is a SVEP, a Follow-up (FU), a Construction-Related (C-R), or a General Industry-Related (GI-R). If inspection is done based on an SVEP Nationwide Expansion Memo the inspection will either be a C-R or a GI-R.

If the inspection is other than a SVEP, give the name and inspection number of the SVEP case to which it is a follow-up or related.

Remember: any FU, C-R, or GI-R inspections can also be a SVEP.

Indicate if construction or non-construction.

What SVEP criteria apply (more than one can apply):

1. Fatality/Catastrophe – One/more W/R/FTA based on a serious violation of **any gravity** related to death or three or more hospitalized
2. Non-Fatality/Catastrophe – Two/more W/R/FTA based on high gravity serious violations related to a High-Emphasis Hazard (excluding Process Safety Management)
3. Non-Fatality/Catastrophe for PSM hazards – Three/more W/R/FTA based on high gravity serious violations
4. Egregious Case

What SVEP actions have been taken (do not report any planned activities):

1. Follow-up inspection conducted; or compelling reason not to conduct
2. Additional construction worksite inspected
3. Additional general industry worksite inspected
4. News Releases issued by Regional Office
5. Letter and citation sent to company headquarters by Region or National Office official
6. Meeting with company officials (separate from informal conference)
7. Enhanced settlement provisions used in informal/formal settlements
8. Court enforcement under Sec. 11(b)
    - Case submitted to RSOL
    - Case submitted to N.O. SOL
    - Petition filed with court [State which court & date]
    - Petition granted by court [State which court & date]
    - Other actions [State & give date]

Appendix B
CSHO Guidance: Considerations in Determining
Company Structure and Safety and Health Organization

When determining whether to inspect other worksites of a company that has been designated a severe violator enforcement case, it must first be determined whether compliance problems and issues found during the initial SVEP inspection are localized or are likely to exist at other, similar facilities owned and operated by that employer. If violations at a local workplace appear to be symptomatic of a broader company neglect for employee safety and health, either generally or with respect to conditions cited under the SVEP inspection, the company structure must be investigated to help identify other establishments and conditions similar to those found in the initial inspection. At the request of the Director of Directorate of Enforcement Programs, a Regional Administrator, or a Regional Coordinator, the Office of Statistical Analysis will be contacted to assist in identifying similar or related worksites of the employer.

Extent of Compliance Problems. Are violative conditions a result of a company decision or interpretation concerning a standard or hazardous condition? Have corporate safety personnel addressed the standard or condition? Ask the following types of questions of the plant manager, safety and health personnel, and line employees.

- Who made the decision concerning the violative operation, local management or company headquarters? Was the decision meant to apply to other facilities of the employer as well? If the decision was from company headquarters, what is their explanation?
- Is there a written company-wide safety program? If so, does it address this issue? If so, how is the issue addressed?
- Is there a company-wide safety department? If so, who are they and where are they located? How does company headquarters communicate with facilities/worksites? Are establishment/worksite management and safety and health personnel trained by the company?
- Do personnel from company headquarters visit facilities/worksites? Are visits on a regular or irregular basis? What subjects are covered during visits? Are there audits of safety and health conditions? Were the types of violative conditions being cited discussed during corporate visits?
- Are there insurance company or contractor safety and health audit reports that have been ignored? Are headquarters safety and health personnel aware of the reports and the inaction?
- Does the company have facilities or worksites other than the one being inspected that do similar or substantially similar work, use similar processes or equipment, or produce like products? If so, where are they?
- What is the overall company attitude concerning safety and health? Does the establishment or worksite receive good support from company headquarters on safety and health matters?

- Does the company provide appropriate safety and health training to its employees?
- Ask whether the establishment's/worksite's overall condition is better or worse at present compared to past years? If it is worse, ask why? Has new management or ownership stressed production over safety and health? Is the equipment outdated or in very poor condition? Does management allege that stressed financial conditions keep it from addressing safety and health issues?
- Is there an active and adequately funded maintenance department? Have they identified these problems and tried to fix them?
- Has the management person being interviewed worked at or visited other similar facilities or worksites owned by the company? How was this issue being treated there?

Identifying Company Structure. Inquire where other facilities or worksites are located and how they may be linked to the one being inspected. Sometimes establishment/worksite management will not have a clear understanding of the company structure, just an awareness of facts concerning control and influence from the corporate office.

- Is the establishment/worksite, or the company that owns the establishment or uses the worksite, owned by another legal entity (parent company)? If so, what is the name and location? Try to find out whether the inspected establishment/worksite is a "division" or a "subsidiary" of the parent company. (NOTE: A "division" is a wholly-owned part of the same company that may be differently named, e.g., Chevrolet is a division of GM. A "subsidiary" is a company controlled or owned by another company which owns all or a majority of its shares. Try to determine if the parent company has divisions or subsidiaries other than the one that owns or uses the establishment or worksite being inspected. If so, try to get the names and the type of business they are involved in. Sometimes this type of information can be found on a website or in Dun and Bradstreet. Another good source of information is the office of the Secretary of State within the state government.
- Are there other facilities or worksites controlled by these entities that do the same type of work and might have the same kinds of safety and health concerns?
- Are the company entities publicly held (have publicly traded shares) or are they closely held (owned by one or more individuals)?
- What are the names, positions, and business addresses of relevant company personnel of whom interviewees are aware? For which entities do the company safety and health personnel work?
- On what kind of safety and health-related issues or subjects do personnel from company headquarters give instructions?
- Are there other companies owned by the same or related persons that do similar work (especially in construction).

Appendix C
Sample Letter to Company Headquarters

Area Office Header
Date
Name of Employer's National Headquarters
Address of Headquarters
Dear _____:

Enclosed you will find a copy of a Citation and Notification of Penalties for violations of the Occupational Safety and Health Act of 1970, which were issued to [establishment name, located in city, state]. This case has been identified as a severe violator enforcement case under the Occupational Safety and Health Administration's (OSHA) Severe Violator Enforcement Program (SVEP).

The violations referred to in this Citation must be abated by the dates listed and the penalties paid, unless they are contested. This Citation and Notification of Penalties is being provided to you for informational purposes so that you are aware of the violations; the original was mailed to [establishment name] on [date]. We encourage you to work with all of your sites to ensure that these violations are corrected.

OSHA is dedicated to saving lives, preventing injuries and illnesses and protecting America's workers. For more information about OSHA programs, please visit our website at www.osha.gov.

Sincerely,
Area Director

APPENDIX C: OSHA PENALTIES

Below are the penalty amounts adjusted for inflation as of Jan. 2, 2018. (See OSHA Memo, 1/3/2018)

Type of Violation	Penalty
Serious Other-Than-Serious Posting Requirements	$12,934 per violation
Failure to Abate	$12,934 per day beyond the abatement date
Willful or Repeated	$129,336 per violation

STATE PLAN STATES

States that operate their own Occupational Safety and Health Plans are required to adopt maximum penalty levels that are at least as effective as Federal OSHA's.

FOR MORE ASSISTANCE

OSHA offers a variety of options for employers looking for compliance assistance.

The On-site Consultation Program provides professional, high-quality, individualized assistance to small businesses at no cost.

OSHA also has compliance assistance specialists in most of our 85 Area Offices across the nation who provide robust outreach and education programs for employers and workers.

For more information, please contact the Regional or Area Office nearest you.

APPENDIX D:
FILING A CHARGE
OF DISCRIMINATION

Log into the EEOC Public Portal to:

- Submit an inquiry online
- Schedule an intake interview

WITH THE EEOC

If you believe that you have been discriminated against at work because of your race, color, religion, sex (including pregnancy, gender identity, and sexual orientation), national origin, age (40 or older), disability or genetic information, you can file a **Charge of Discrimination**. A charge of discrimination is a signed statement asserting that an employer, union or labor organization engaged in employment discrimination. It requests EEOC to take remedial action.

All of the laws enforced by EEOC, except for the Equal Pay Act, require you to file a Charge of Discrimination with us before you can file a job discrimination lawsuit against your employer. In addition, an individual, organization, or agency may file a charge on behalf of another person in order to protect the aggrieved person's identity. There are time limits for filing a charge. The laws enforced by the EEOC require the agency to notify the employer that a charge has been filed against it.

A Charge of Discrimination can be completed through our EEOC Public Portal after you submit an online inquiry and we interview you. Filing a formal charge of employment discrimination is a serious matter. In the EEOC's experience, having the opportunity to discuss your concerns with an EEOC staff member in an interview is the best way to assess how to address your concerns about employment discrimination and determine whether filing a charge of discrimination is the appropriate path for you. In any event, the final decision to file a charge is your own.

If you have 60 days or fewer in which to file a timely charge, the EEOC Public Portal will provide special directions for quickly providing necessary information to the EEOC and how to file your charge quickly. Or, go to https://www.eeoc.gov/field/index.cfm and enter your zip code for the contact information of the EEOC office closest to you.

The laws enforced by the EEOC require the agency to accept charges alleging employment discrimination. If the laws do not apply to your claims, if the charge was not filed within the law's time limits, or if the EEOC decides to limit its investigation, the EEOC will dismiss the charge without any further investigation and notify you of your legal rights.

WITH A STATE OR LOCAL AGENCY

Many states and local jurisdictions have their own anti-discrimination laws, and agencies responsible for enforcing those laws (Fair Employment Practices Agencies, or FEPAs). If you file a charge with a FEPA, it will automatically be "dual-filed" with EEOC if federal laws apply. You do not need to file with both agencies.

Note: Federal employees and job applicants have similar protections, but a different complaint process.

- How to File a Charge
- What You Can Expect After You File a Charge
- Confidentiality
- Mediation
- Remedies
- Existing Charges
- Filing a Lawsuit

FEDERAL GOVERNMENT EMPLOYEES AND APPLICANTS

The procedures for filing a complaint of discrimination against a federal government agency differ from those for filing a charge against a private or public employer. For discrimination complaints against a federal government agency, the procedures are different. Go to https://www.eeoc.gov/federal/fed_employees/complaint_overview. cfm for a description of those procedures. Federal employees and applicants can request a hearing or file a n appeal with EEOC through the EEOC Public Portal, which allows individuals to:

- Create an account
- Request a hearing
- File an appeal
- Identify a representative and provide their contact information
- Submit and receive documents supporting their hearing request or appeal

(From the Equal Employment Opportunity Commission website located at EEOC. gov).

APPENDIX E: OSHA ONLINE COMPLAINT FORM

Notice of Alleged Safety or Health Hazards

EMERGENCY NOTICE

Do Not Report an Emergency Using this Form or Email!

To report an emergency, fatality, or imminent life threatening situation please contact our toll free number immediately:

1-800-321-OSHA (6742)
TTY 1-877-889-5627

Please fill out sections 1 through 19, but READ THIS FIRST. Items noted with an asterisk (*) are required in order to accept your submission.

Top of Form

* 1. Establishment Name:

Note: In order for OSHA to fully process your complaint, complete and accurate information about the worksite is necessary.

* 2. Site Street:

* 3. Site City:

* 4. Site State:

* 5. Site ZIP Code:

6. Mailing Address (if different):

7. Management Official:

8. Telephone Number:

9. Type of Business:

* 10. Hazard Description.
Describe briefly the hazards(s) which you believe exist.Include the approximate number of employees exposed to or threatened by each hazard:

* 11. Hazard Location.
Specify the particular building or worksite where the alleged violation exists:

12. This condition has been brought to the attention of: (*Choose all that apply*)
☐ Employer ☐ Other Government Agency (*specify*)

* 13. I am a(n):
○ Former Employee ○ Current Employee ○ Federal Safety and Health Committee
○ Representative of Employees ○ Other: (*specify*)

The OSH Act gives complainants the right to request that their names not be revealed to their employer. Providing your name and address, will only allow OSHA staff to communicate with you regarding your complaint.

14. Please indicate your desire:
◉ Do **NOT** reveal my name to my Employer ○ My name may be revealed to my Employer

* 15. Complainant Name:
☐ *This constitutes my electronic signature.* (**If this box is checked, this submission shall be considered as an authorized written signature.**)

* 16. Complainant Telephone Number:

17. Complainant Mailing Address

Street: [_____]

City: [_____]

State: [_____▼]

ZIP Code: [_____]

* 18. Complainant E-Mail Address:

19. If you are an authorized representative of employees affected by this complaint, please state the name of the organization that you represent and your title:

Organization Name: [_____]

Your Title: [_____]

SEND Clear Form

_____Bottom of Form_____

PUNISHMENT FOR UNLAWFUL STATEMENTS

Potential complainants also should keep in mind that it is unlawful to make any false statement, representation, or certification in any complaint. Violations can be punished under Section 17(g) of the OSH Act by a fine of not more than $10,000, or by imprisonment of not more than 6 months, or by both.

Public reporting burden for this voluntary collection of information is estimated to vary from 15 to 25 minutes per response with an average of 17 minutes per response, including the time for reviewing instructions, searching existing data sources, gathering and maintaining the data needed, and completing and reviewing the collection of information. An Agency may not conduct or sponsor, and persons are not required to respond to the collection of information unless it displays a valid OMB Control Number. Send comment regarding this burden estimate or any other aspect of this collection of information, including suggestions for reducing this burden to the Directorate of Enforcement Programs, Department of Labor, Room N-3119, 200 Constitution Ave., NW, Washington, DC; 20210.

OMB Approval# 1218-0064; Expires: 11-30-2020

DO NOT SEND THE COMPLETED FORM TO THIS OFFICE.

Index